T0192092

Ernst Kunz

Introduction to Plane Algebraic Curves

Translated from the original German by Richard G. Belshoff

Birkhäuser
Boston • Basel • Berlin

Ernst Kunz
Universität Regensburg
NWF I – Mathematik
D-93040 Regensburg
Germany

Richard G. Belshoff (Translator)
Southwest Missouri State University
Department of Mathematics
Springfield, MO 65804
U.S.A.

Cover design by Mary Burgess.

Mathematics Subject Classicification (2000): 14-xx, 14Hxx, 14H20, 14H45 (primary);
14-01, 13-02, 13A02, 13A30 (secondary)

Library of Congress Cataloging-in-Publication Data
Kunz, Ernst, 1933-
[Ebene algebraische Kurven. English]
Introduction to plane algebraic curves / Ernst Kunz; translated by Richard G. Belshoff.
p. cm.
Includes bibliographical references and index.
ISBN 0-8176-4381-8 (alk. paper)
1. Curves, Plane. 2. Curves, Algebraic. 3. Singularities (Mathematics) I. Title.

QA567.K8613 2005
516.3'52–dc22 2005048053

ISBN-10 0-8176-4381-8
ISBN-13 978-0-8176-4381-2
eISBN 0-8176-4443-1
Printed on acid-free paper.

Birkhäuser

Based on the original German edition, *Ebene algebraische Kurven*,
Der Regensburg Trichter, 23, Universität Regensburg, ISBN 3-88246-167-5, ©1991 Ernst Kunz

Printed in the United States of America. (SB)

9 8 7 6 5 4 3 2 1 SPIN 11301783

Birkhäuser is a part of *Springer Science+Business Media*

www.birkhauser.com

To the memory of my friend
Hans-Joachim Nastold (1929–2004)
and of our teacher
Friedrich Karl Schmidt (1901–1977)

Preface

This book is a slightly extended elaboration of a course on commutative ring theory and plane algebraic curves that I gave several times at the University of Regensburg to students with a basic knowledge of algebra. I thank Richard Belshoff for translating the German lecture notes into English and for preparing the numerous figures of the present text.

As in my book *Introduction to Commutative Algebra and Algebraic Geometry*, this book follows the philosophy that the best way to introduce commutative algebra is to simultaneously present applications in algebraic geometry. This occurs here on a substantially more elementary level than in my earlier book, for we never leave plane geometry, except in occasional notes without proof, as for instance that the abstract Riemann surface of a plane curve is "actually" a smooth curve in a higher-dimensional space. In contrast to other presentations of curve theory, here the algebraic viewpoint stays strongly in the foreground. This is completely different from, for instance, the book of Brieskorn–Knörrer [BK], where the geometric–topological–analytic aspects are particularly stressed, and where there is more emphasis on the history of the subject. Since these things are explained there in great detail, and with many beautiful pictures, I felt relieved of the obligation to go into the topological and analytical connections. In the lectures I recommended to the students that they read the appropriate sections of Brieskorn–Knörrer [BK]. The book by G. Fischer [F] can also serve this purpose.

We will study algebraic curves over an algebraically closed field K. It is not at all clear a priori, but rather to be regarded as a miracle, that there is a close correspondence between the details of the theory of curves over $\mathbb{C}$ and that of curves over an arbitrary algebraically closed field. The parallel between curves over fields of prime characteristic and over fields of characteristic 0 ends somewhat earlier. In the last few decades algebraic curves of prime characteristic made an entrance into coding theory and cryptography, and thus into applied mathematics.

The following are a few ways in which this course differs from other introductions to the theory of plane algebraic curves known to me: Filtered algebras, the associated graded rings, and Rees rings will be used to a great extent, in order to deduce basic facts about intersection theory of plane curves. There will be modern proofs for many classical theorems on this subject. The techniques which we apply are nowadays also standard tools of computer algebra.

Also, a presentation of algebraic residue theory in the affine plane will be given, and its applications to intersection theory will be considered. Many of the theorems proved here about the intersection of two plane curves carry over with relatively minor changes to the case of the intersection of n hypersurfaces in n-dimensional space, or equivalently, to the solution sets of n algebraic equations in n unknowns.

The treatment of the Riemann–Roch theorem and its applications is based on ideas of proofs given by F.K. Schmidt in 1936. His methods of proof are an especially good fit with the presentation given here, which is formulated in the language of filtrations and associated graded rings.

The book contains an introduction to the algebraic classification of plane curve singularities, a subject on which many publications have appeared in recent years and to which references are given. The lectures had to end at some point, and so resolution of singularities was not treated. For this subject I refer to Brieskorn–Knörrer or Fulton [Fu]. Nevertheless I hope that the reader will also get an idea of the problems and some of the methods of higher-dimensional algebraic geometry.

The present work is organized so that the algebraic facts that are used and that go beyond a standard course in algebra are collected together in Appendices A–L, which account for about one-third of the text and are referred to as needed. A list of keywords in the section "Algebraic Foundations" should make clear what parts of algebra are deemed to be well-known to the reader. We always strive to give complete and detailed proofs based on these foundations

My former students Markus Nübler, Lutz Pinkofsky, Ulrich Probst, Wolfgang Rauscher and Alfons Schamberger have written diploma theses in which they have generalized parts of the book. They have contributed to greater clarity and better readability of the text. To them, and to those who have attended my lectures, I owe thanks for their critical comments. My colleague Rolf Waldi who has used the German lecture notes in his seminars deserves thanks for suggesting several improvements.

Regensburg
December 2004

Ernst Kunz

Conventions and Notation

(a) By a *ring* we shall always mean an associative, commutative ring with identity.
(b) For a ring R, let $\operatorname{Spec} R$ be the set of all prime ideals $\mathfrak{p} \neq R$ of R (the *Spectrum of* R). The set of all maximal (minimal) prime ideals will be denoted by $\operatorname{Max} R$ (respectively $\operatorname{Min} R$).
(c) A ring homomorphism $\rho : R \to S$ shall always map the identity of R to the identity of S. We also say that S/R is an *algebra* over R given by ρ. Every ring is a $\mathbb{Z}$-algebra.
(d) For an algebra S over a field K we denote by $\dim_K S$ the dimension of S as a K-vector space.
(e) For a polynomial f in a polynomial algebra $R[X_1, \ldots, X_n]$, we let $\deg f$ stand for the *total degree* of f and $\deg_{X_i} f$ the *degree in* X_i.
(f) If K is a field, $K(X_1, \ldots, X_n)$ denotes the *field of rational functions* in the variables $X_1, \ldots, X_n$ over K (the quotient field of $K[X_1, \ldots, X_n]$).
(g) The minimal elements in the set of all prime ideals containing an ideal I are called the minimal prime divisors of I.

Contents

Introduction to
Plane Algebraic Curves

Part I

Plane Algebraic Curves

1

Affine Algebraic Curves

This section uses only a few concepts and facts from algebra. It assumes a certain familiarity with polynomial rings $K[X_1, \ldots, X_n]$ over a field, in particular that $K[X]$ is a principal ideal domain, and that $K[X_1, \ldots, X_n]$ is a unique factorization domain in general. Also, ideals and quotient rings will be used. Finally, one must know that an algebraically closed field has infinitely many elements.

We will study algebraic curves over an arbitrary algebraically closed field K. Even if one is only interested in curves over $\mathbb{C}$, the investigation of the $\mathbb{Z}$-rational points of curves by "reduction mod p" leads into the theory of curves over fields with prime characteristic p. Such curves also appear in algebraic coding theory (Pretzel [P], Stichtenoth [St]) and cryptography (Koblitz [K], Washington [W]).

$\mathbb{A}^2(K) := K^2$ denotes the affine plane over K, and $K[X, Y]$ the polynomial algebra in the variables X and Y over K. For $f \in K[X, Y]$, we call

$$\mathcal{V}(f) := \{(x, y) \in \mathbb{A}^2(K) \mid f(x, y) = 0\}$$

the *zero set of* f. We set $D(f) := \mathbb{A}^2(K) \setminus \mathcal{V}(f)$ for the set of points where f does not vanish.

Definition 1.1. A subset $\Gamma \subset \mathbb{A}^2(K)$ is called a (plane) *affine algebraic curve* (for short: curve) if there exists a nonconstant polynomial $f \in K[X, Y]$ such that $\Gamma = \mathcal{V}(f)$. We write $\Gamma : f = 0$ for this curve and call $f = 0$ an *equation for* Γ.

If $K_0 \subset K$ is a subring and $\Gamma = \mathcal{V}(f)$ for a nonconstant polynomial $f \in K_0[X, Y]$, we say that Γ *is defined over* K_0 and call $\Gamma_0 := \Gamma \cap K_0^2$ the set of K_0-*rational points of* Γ.

Examples 1.2.

(a) The zero sets of linear polynomials $aX + bY + c = 0$ with $(a, b) \neq (0, 0)$ are called *lines*. If $K_0 \subset K$ is a subfield and $a, b, c \in K_0$, then the line $g : aX + bY + c = 0$ certainly possesses K_0-rational points. Through two different points of $\mathbb{A}^2(K_0)$ there is exactly one line (defined over K_0).

(b) If $\Gamma_1, \ldots, \Gamma_h$ are algebraic curves with equations $f_i = 0$ $(i = 1, \ldots, h)$, then $\Gamma := \cup_{i=1}^h \Gamma_i$ is also an algebraic curve. It is given by the equation $\prod_{i=1}^h f_i = 0$. In particular, the union of finitely many lines is an algebraic curve (see Figure 1.1).

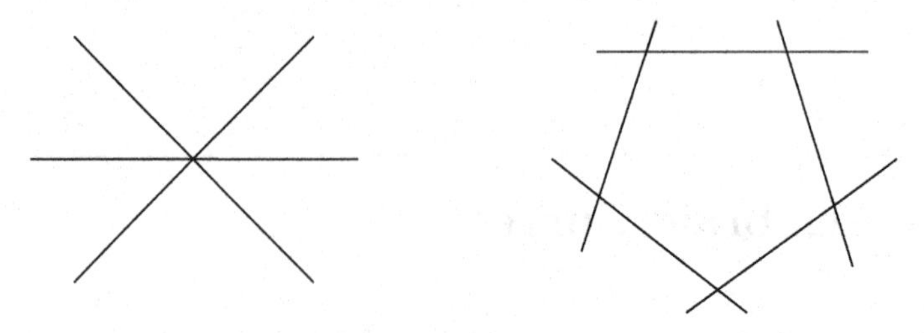

Fig. 1.1. The union of finitely many lines is an algebraic curve.

(c) Let $\Gamma = \mathcal{V}(f)$ with a nonconstant $f \in K[Y]$ (so f does not depend on X). The decomposition of f into linear factors

$$f = c \cdot \prod_{i=1}^{d}(Y - a_i) \qquad (c \in K^* := K \setminus \{0\},\ a_1, \dots, a_d \in K)$$

shows that Γ is the union of lines $g_i : Y - a_i = 0$ parallel to the X-axis.

(d) The zero sets of quadric polynomials

$$f = aX^2 + bXY + cY^2 + dX + eY + g \qquad (a, b, \dots g \in K;\ (a, b, c) \neq (0, 0, 0))$$

are called *quadrics.* In case $K = \mathbb{C}$, $K_0 = \mathbb{R}$ we get the *conic sections,* whose $\mathbb{R}$-rational points are shown in Figures 1.2 through 1.5. Defined

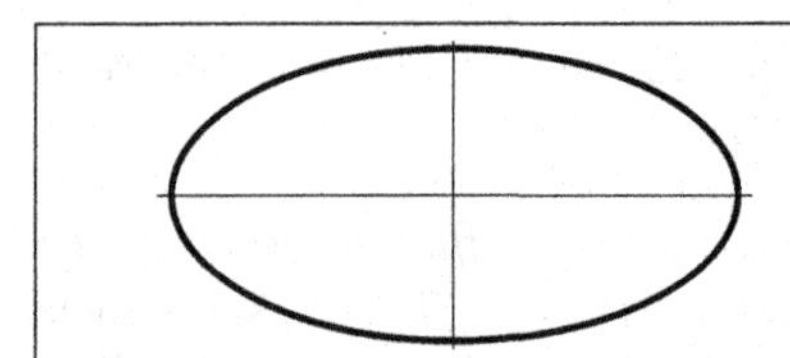

Fig. 1.2. Ellipse: $\frac{X^2}{a^2} + \frac{Y^2}{b^2} = 1$, $(a, b \in \mathbb{R}_+)$

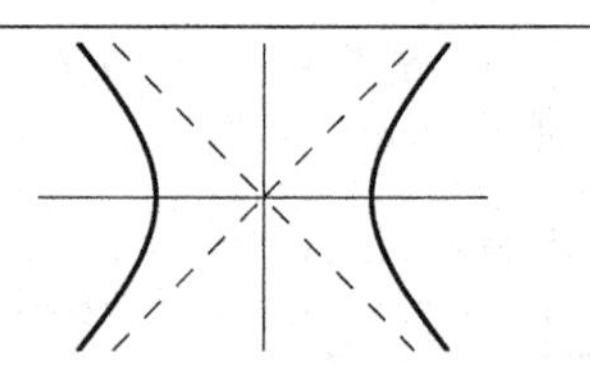

Fig. 1.3. Hyperbola: $\frac{X^2}{a^2} - \frac{Y^2}{b^2} = 1$, $(a, b \in \mathbb{R}_+)$

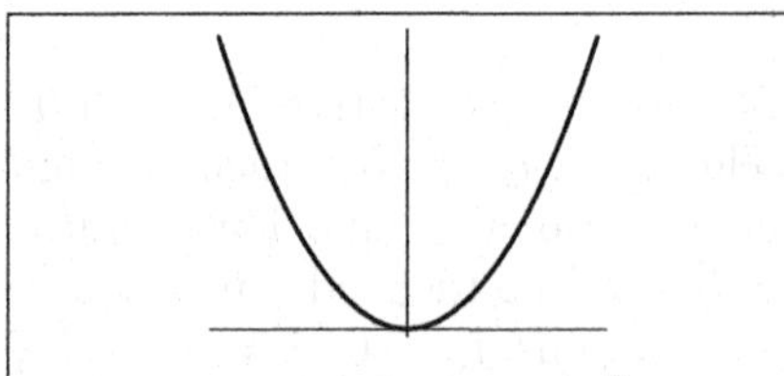

Fig. 1.4. Parabola: $Y = aX^2$, $(a \in \mathbb{R}_+)$

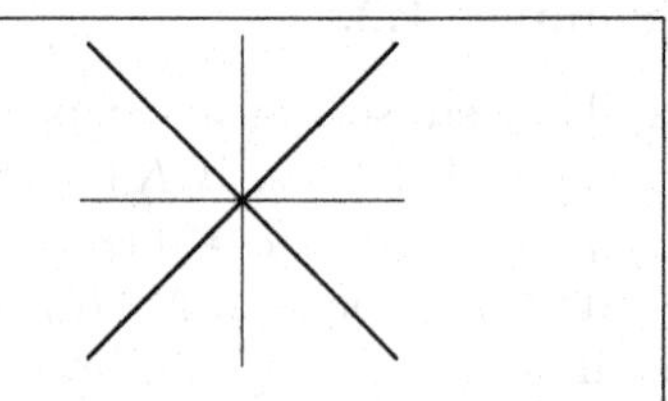

Fig. 1.5. Line pair: $X^2 - Y^2 = 0$

as sections of a cone with a plane, they were thoroughly studied in ancient Greek mathematics. Many centuries later, they became important in Kepler's laws of planetary motion and in Newton's mechanics. Unlike the $\mathbb{R}$-rational points, questions about the $\mathbb{Q}$-rational points of quadrics have, in general, nontrivial answers (cf. Exercises 2–4).

(e) The zero sets of polynomials of degree 3 are called *cubics*. The $\mathbb{R}$-rational points of some prominent cubics are sketched in Figures 1.6 through 1.9. Cubic curves will be discussed in 7.17 and in Chapter 10.

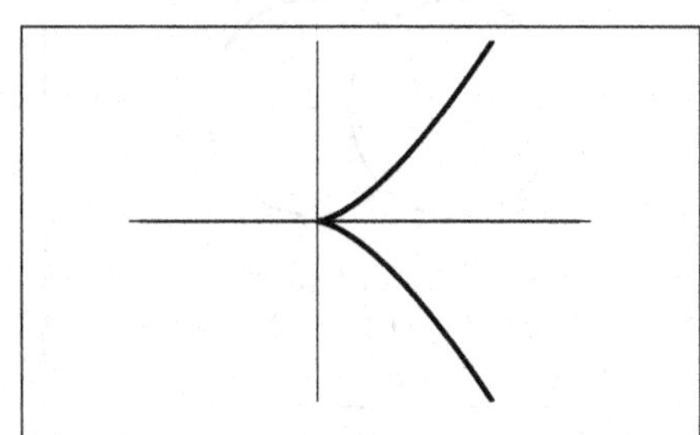

Fig. 1.6. Neil's semicubical parabola: $X^3 - Y^2 = 0$

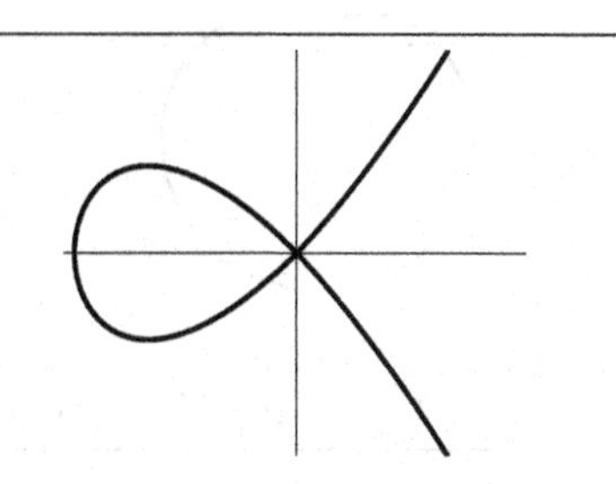

Fig. 1.7. Folium of Descartes: $X^3 + X^2 - Y^2 = 0$

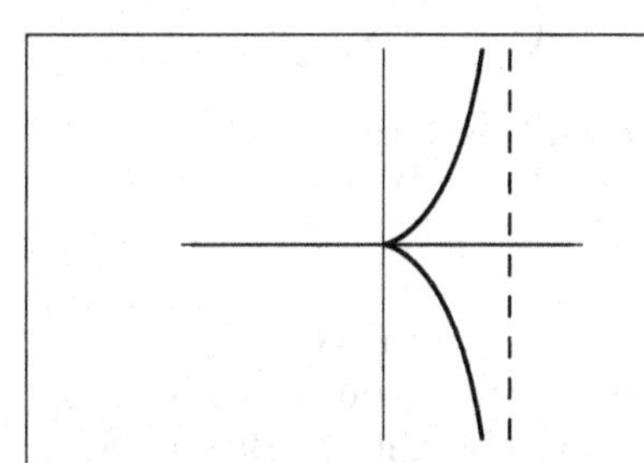

Fig. 1.8. Cissoid of Diocles: $Y^2(1 - X) - X^3 = 0$

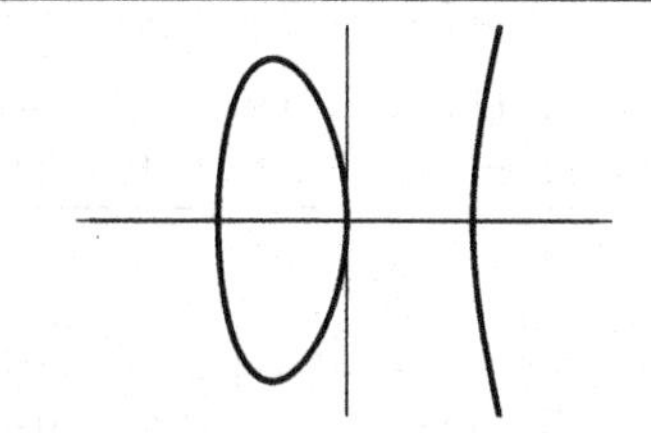

Fig. 1.9. Elliptic curve in Weierstraß normal form ($e_1 < e_2 < e_3$ real): $Y^2 = 4(X - e_1)(X - e_2)(X - e_3)$

(f) Some curves with equations of higher degrees are sketched in Figures 1.10 through 1.15. For the origin of these curves and the others indicated above, one can consult the book by Brieskorn–Knörrer [BK]. See also Xah Lee's "Visual Dictionary of Special Plane Curves" `http://xahlee.org`, and the "Famous Curves Index" at the MacTutor History of Mathematics archive `http://www-history.mcs.st-and.ac.uk/history`.

(g) The *Fermat curve* F_n ($n \geq 3$) is given by the equation $X^n + Y^n = 1$. It is connected with some of the most spectacular successes of curve theory in recent years. *Fermat's last theorem* (1621) asserted that the only $\mathbb{Q}$-rational points on this curve are the obvious ones: $(1, 0)$ and $(0, 1)$ in

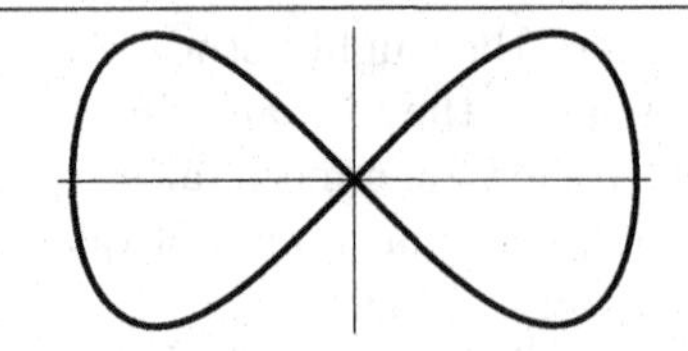

Fig. 1.10. Lemniscate:
$X^2(1-X^2)-Y^2=0$

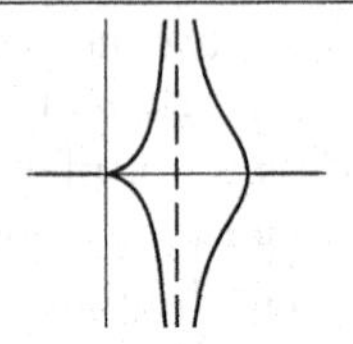

Fig. 1.11. Conchoid of Nichomedes:
$(X^2+Y^2)(X-1)^2-X^2=0$

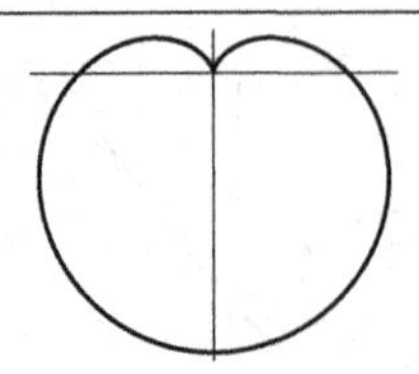

Fig. 1.12. Cardioid:
$(X^2+Y^2+4Y)^2-16(X^2+Y^2)=0$

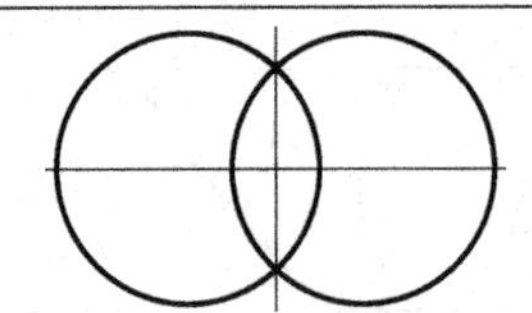

Fig. 1.13. Union of two circles:
$(X^2-4)^2+(Y^2-9)^2+2(X^2+4)(Y^2-9)=0$

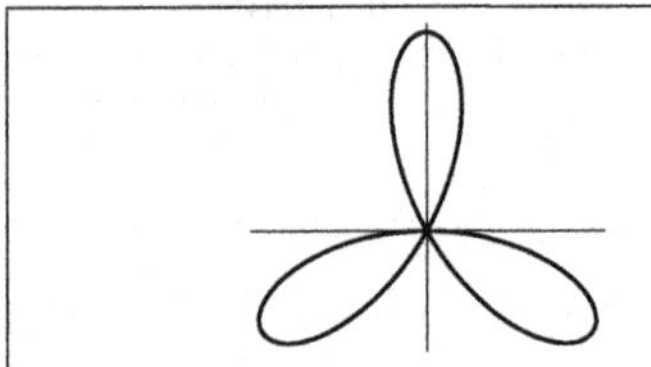

Fig. 1.14. Three-leaf rose:
$(X^2+Y^2)^2+3X^2Y-Y^3=0$

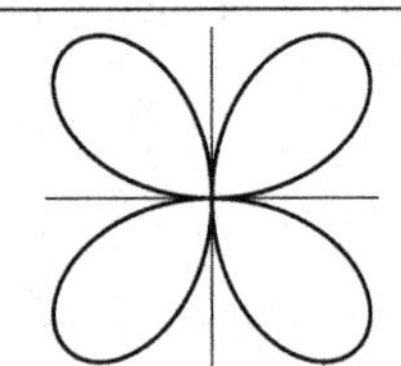

Fig. 1.15. Four-leaf rose:
$(X^2+Y^2)^3-4X^2Y^2=0$

case n is odd; and $(\pm 1, 0)$, $(0, \pm 1)$ in case n is even. G. Faltings [Fa] in 1983 showed that there are only finitely many $\mathbb{Q}$-rational points on F_n, a special case of *Mordell's conjecture* proved by him. In 1986 G. Frey observed that Fermat's last theorem should follow from a conjecture about elliptic curves (the *Shimura–Taniyama theorem*), for which Andrew Wiles (see [Wi], [TW]) gave a proof in 1995, hence also proving Fermat's last theorem. These works are far beyond the scope of the present text. The reader interested in the history of the problem and its solution may enjoy Simon Singh's bestselling book *Fermat's last theorem* [Si].

Having seen some of the multifaceted aspects of algebraic curves, we turn now to the general theory of these curves. The examples $X^2+Y^2=0$ and $X^2+Y^2+1=0$ show that the set of $\mathbb{R}$-rational points of a curve can be finite, or even empty. For points with coordinates in an algebraically closed field, however, this cannot happen.

Theorem 1.3. *Every algebraic curve $\Gamma \subset \mathbb{A}^2(K)$ consists of infinitely many points, and also $\mathbb{A}^2(K) \setminus \Gamma$ is infinite.*

Proof. Let $\Gamma = \mathcal{V}(f)$ with $f = a_0 + a_1X + \cdots + a_pX^p$, where $a_i \in K[Y]$ $(i = 0, \ldots, p)$ and $a_p \neq 0$. If $p = 0$, we are in the situation of Example 1.2 (c) above, and since an algebraically closed field has infinitely many elements, there is nothing more to be shown. Therefore, let $p > 0$. Since a_p has only finitely many zeros in K, there are infinitely many $y \in K$ with $a_p(y) \neq 0$. Then

$$f(X, y) = a_0(y) + a_1(y)X + \cdots + a_p(y)X^p$$

is a nonconstant polynomial in $K[X]$. If $x \in K$ is a zero of this polynomial, then $(x, y) \in \Gamma$; therefore, Γ contains infinitely many points. If $x \in K$ is not a zero, then $(x, y) \in D(f)$, and therefore there are also infinitely many points in $\mathbb{A}^2(K) \setminus \Gamma$.

An important theme in curve theory is the investigation of the intersection of two algebraic curves. Our first instance of this is furnished by the following theorem. It assumes a familiarity with unique factorization domains.

Theorem 1.4. *Let f and g be nonconstant relatively prime polynomials in $K[X, Y]$. Then*

(a) *$\mathcal{V}(f) \cap \mathcal{V}(g)$ is finite. In other words, the system of equations*

$$f(X, Y) = 0, \quad g(X, Y) = 0$$

has only finitely many solutions in $\mathbb{A}^2(K)$.

(b) *The K-algebra $K[X, Y]/(f, g)$ is finite-dimensional.*

For the proof we will use

Lemma 1.5. *Let R be a UFD with quotient field K. If $f, g \in R[X]$ are relatively prime, then they are also relatively prime in $K[X]$, and there exists an element $d \in R \setminus \{0\}$ such that*

$$d = af + bg$$

for some polynomials $a, b \in R[X]$.

Proof. Suppose that $f = \alpha h$, $g = \beta h$ for polynomials $\alpha, \beta, h \in K[X]$, where h is not a constant polynomial. Since any denominators that appear in h may be brought over to α and β, we may assume that $h \in R[X]$. We then write

$$\alpha = \textstyle\sum \alpha_i X^i, \quad \beta = \sum \beta_j X^j \qquad (\alpha_i, \beta_j \in K).$$

Let $\delta \in R \setminus \{0\}$ be the least common denominator for the α_i and β_j. Then we have

$$\delta f = \phi h, \qquad \delta g = \psi h$$

with $\phi := \delta\alpha \in R[X]$, $\psi := \delta\beta \in R[X]$. A prime element of R that divides δ cannot simultaneously divide ϕ and ψ, since δ was chosen to be the least common denominator. It follows that every prime factor of δ must divide h. Consequently, δ is a divisor of h, and there are equations $f = \phi h_1$ and $g = \psi h_1$ for some nonconstant polynomial $h_1 \in R[X]$. This is a contradiction, and therefore f and g are also relatively prime in $K[X]$.

In $K[X]$ we then have an equation

$$1 = Af + Bg \qquad (A, B \in K[X]).$$

Multiplying through by a common denominator for all the coefficients of A and B, we get an equation $d = af + bg$ with $a, b \in R[X]$, and $d \neq 0$.

PROOF OF 1.4:
(a) By Lemma 1.5 we have equations

$$(1) \qquad d_1 = a_1 f + b_1 g, \qquad d_2 = a_2 f + b_2 g,$$

with $d_1 \in K[X] \setminus \{0\}$, $d_2 \in K[Y] \setminus \{0\}$, and $a_i, b_i \in K[X, Y]$ $(i = 1, 2)$. If $(x, y) \in \mathcal{V}(f) \cap \mathcal{V}(g)$, then x is a zero of d_1 and y is a zero of d_2. Therefore, there can be only finitely many $(x, y) \in \mathcal{V}(f) \cap V(g)$.

(b) Suppose the polynomial d_k in (1) has degree m_k $(k = 1, 2)$. Dividing a polynomial $F \in K[X, Y]$ by d_1 using the division algorithm gives us an equation $F = Gd_1 + R_1$, where $G, R_1 \in K[X, Y]$ and $\deg_X R_1 < m_1$. Similarly, we have $R_1 = Hd_2 + R_2$, where $H, R_2 \in K[X, Y]$, $\deg_X R_2 < m_1$, and $\deg_Y R_2 < m_2$. It follows that $F \equiv R_2 \operatorname{mod}(f, g)$. Let ξ, η be the residue classes of X, Y in $A := K[X, Y]/(f, g)$. Then $\{\xi^i \eta^j \mid 0 \leq i < m_1, 0 \leq j < m_2\}$ is a set of generators for A as a K-vector space.

Using Theorem 1.4 one sees, for example, that a line g intersects an algebraic curve Γ in finitely many points or else is completely contained in Γ; for if $\Gamma = \mathcal{V}(f)$, then the linear polynomial g is either a factor of f, or f and g are relatively prime. (In this simple case there is, of course, a direct proof that does not use Theorem 1.4.) The sine curve cannot be the real part of an algebraic curve in $\mathbb{A}^2(\mathbb{C})$ because there are infinitely many points of intersection with the X-axis.

Next we will investigate the question of which polynomials can define a given algebraic curve Γ. Let $f = 0$ be an equation for Γ. We decompose f into a product of powers of irreducible polynomials:

$$f = cf_1^{\alpha_1} \cdots f_h^{\alpha_h} \qquad (c \in K^*, \quad f_i \in K[X, Y] \text{ irreducible}, \quad \alpha_i \in \mathbb{N}_+).$$

Here f_i and f_j are not associates if $i \neq j$.

Definition 1.6. $\mathcal{J}(\Gamma) := \{g \in K[X, Y] \mid g(x, y) = 0 \text{ for all } (x, y) \in \Gamma\}$ is called the *vanishing ideal of Γ*.

Theorem 1.7. *$\mathcal{J}(\Gamma)$ is the principal ideal generated by $f_1 \cdots f_h$.*

Proof. It is clear that $\Gamma = \mathcal{V}(f_1 \cdots f_h) = \mathcal{V}(f_1) \cup \cdots \cup \mathcal{V}(f_h)$ and therefore $f_1 \cdots f_h \in \mathcal{J}(\Gamma)$. If $g \in \mathcal{J}(\Gamma)$, it follows that $\Gamma \subset \mathcal{V}(g)$. Suppose f_j, for some $j \in \{1, \dots, h\}$, were not a divisor of g. Then the set $\mathcal{V}(f_j) = \mathcal{V}(f_j) \cap \mathcal{V}(g)$ would be finite by 1.4. But this cannot happen by 1.3. Therefore $f_1 \cdots f_h$ is a divisor of g and $\mathcal{J}(\Gamma) = (f_1 \cdots f_h)$.

Definition 1.8. Given $\mathcal{J}(\Gamma) = (f)$ with $f \in K[X,Y]$, we call f a *minimal polynomial for Γ*. Its degree is called the *degree of Γ*, and $K[\Gamma] := K[X,Y]/(f)$ is called the (affine) *coordinate ring of Γ*.

The minimal polynomial is uniquely determined by Γ up to a constant factor from K^*, so the degree of Γ is well-defined. Theorem 1.7 shows us how to get a minimal polynomial for Γ given any equation $f = 0$ for Γ. Conversely, it is also clear which polynomials define Γ.

We call a polynomial in $K[X,Y]$ *reduced* if it does not contain the square of an irreducible polynomial as a factor.

From 1.7 we infer the following.

Corollary 1.9. *The algebraic curves $\Gamma \subset \mathbb{A}^2(K)$ are in one-to-one correspondence with the principal ideals of $K[X,Y]$ generated by nonconstant reduced polynomials.*

In the following let $\Gamma \subset \mathbb{A}^2(K)$ be a fixed algebraic curve.

Definition 1.10. Γ is called *irreducible* if whenever $\Gamma = \Gamma_1 \cup \Gamma_2$ for algebraic curves Γ_i $(i = 1,2)$, then $\Gamma = \Gamma_1$ or $\Gamma = \Gamma_2$.

Theorem 1.11. *Let f be a minimal polynomial for Γ. Then Γ is irreducible if and only if f is an irreducible polynomial.*

Proof. Let Γ be irreducible and suppose $f = f_1 f_2$ for some polynomials $f_i \in K[X,Y]$ $(i = 1,2)$. Then $\Gamma = \mathcal{V}(f_1) \cup \mathcal{V}(f_2)$. If f_1 and f_2 were not constant, then we would have $\mathcal{V}(f_1) = \Gamma$ or $\mathcal{V}(f_2) = \Gamma$. But then it would follow that $f_1 \in (f)$ or $f_2 \in (f)$, and this cannot happen, since the f_i are proper factors of f. Therefore, f is an irreducible polynomial.

Conversely, suppose f is irreducible and let $\Gamma = \Gamma_1 \cup \Gamma_2$ be a decomposition of Γ into curves Γ_i $(i = 1,2)$. If f_i is a minimal polynomial for Γ_i, then $f \in \mathcal{J}(\Gamma_i) = (f_i)$, i.e., f is divisible by f_1 (and by f_2). Since f is irreducible, we must have that f is an associate of f_i for some i, and therefore $\Gamma = \Gamma_i$. Hence Γ is irreducible.

Among the examples above one finds many irreducible algebraic curves. One can check, using appropriate irreducibility tests, that their defining polynomials are irreducible.

Corollary 1.12. *The following statements are equivalent:*

(a) Γ *is an irreducible curve.*
(b) $\mathcal{J}(\Gamma)$ *is a prime ideal in* $K[X,Y]$.
(c) $K[\Gamma]$ *is an integral domain.*

The irreducible curves $\Gamma \subset \mathbb{A}^2(K)$ *are in one-to-one correspondence with the principal ideals* $\neq (0),(1)$ *in* $K[X,Y]$ *that are simultaneously prime ideals.*

Theorem 1.13. *Every algebraic curve* Γ *has a unique (up to order) representation*

$$\Gamma = \Gamma_1 \cup \cdots \cup \Gamma_h,$$

where the Γ_i *are irreducible curves* $(i = 1,\ldots,h)$ *corresponding to the decomposition of a minimal polynomial of* Γ *into irreducible factors.*

The proof of the uniqueness starts with an arbitrary representation $\Gamma = \Gamma_1 \cup \cdots \cup \Gamma_h$. If f, respectively f_i, is a minimal polynomial of Γ, respectively Γ_i $(i = 1,\ldots,h)$, then $(f) = (f_1 \cdots f_h)$, because f and $f_1 \cdots f_h$ are reduced polynomials with the same zero set. The f_i are therefore precisely the irreducible factors of f, and as a result, the Γ_i are uniquely determined by Γ.

We call the Γ_i the *irreducible components* of Γ. Theorem 1.4 (a) can now be reformulated to say: Two algebraic curves that have no irreducible components in common intersect in finitely many points.

The previous observations allow us to make the following statements about the prime ideals of $K[X,Y]$.

Theorem 1.14.

(a) *The maximal ideals of* $K[X,Y]$ *are in one-to-one correspondence with the points of* $\mathbb{A}^2(K)$*: Given a point* $P = (a,b) \in \mathbb{A}^2(K)$*, then* $\mathfrak{M}_P := (X-a, Y-b) \in \operatorname{Max} K[X,Y]$*, and every maximal ideal is of this form for a uniquely determined point* $P \in \mathbb{A}^2(K)$.
(b) *The nonmaximal prime ideals* $(\neq (0),(1))$ *of* $K[X,Y]$ *are in one-to-one correspondence with the irreducible curves of* $\mathbb{A}^2(K)$*: These are exactly the principal ideals* (f) *generated by irreducible polynomials.*

Proof. The K-homomorphism $K[X,Y] \to K$, where $X \mapsto a$ and $Y \mapsto b$ is onto and has kernel $\mathfrak{M}_P$. Since $K[X,Y]/\mathfrak{M}_P \cong K$ is a field, $\mathfrak{M}_P$ is a maximal ideal.

Now let $\mathfrak{p} \in \operatorname{Spec} K[X,Y]$, $\mathfrak{p} \neq (0)$. Then $\mathfrak{p}$ contains a nonconstant polynomial and therefore also contains an irreducible polynomial f. If $\mathfrak{p} = (f)$, then $\mathfrak{p}$ is not maximal, for $\mathfrak{p} \subset \mathfrak{M}_P$ for all $P \in \mathcal{V}(f)$ and $\mathcal{V}(f)$ contains infinitely many points P by 1.3.

On the other hand, if $\mathfrak{p}$ is not generated by f, then $\mathfrak{p}$ contains a polynomial g that is not divisible by f. As in the proof of 1.4 we have two equations of the form (1). Since $d_1 \in K[X]$ decomposes into linear factors, $\mathfrak{p}$ contains a polynomial $X - a$ for some $a \in K$. Similarly, $\mathfrak{p}$ contains a polynomial $Y - b$ $(b \in K)$, and it follows that $\mathfrak{p} = (X-a, Y-b)$.

If Γ is an algebraic curve, then the maximal ideals of $K[X,Y]$ that contain $\mathcal{J}(\Gamma)$ are precisely the $\mathfrak{M}_P$ for which $P \in \Gamma$. The other elements of $\operatorname{Spec} K[X,Y]$ that contain an arbitrary Γ are the $\mathcal{J}(\Gamma_i)$, where the Γ_i are the irreducible components of Γ. The coordinate ring $K[\Gamma]$ of Γ "knows" the points of Γ and the irreducible components of Γ:

Corollary 1.15.

(a) $\operatorname{Max} K[\Gamma] = \{\mathfrak{M}_P/\mathcal{J}(\Gamma) \mid P \in \Gamma\}$.
(b) $\operatorname{Spec} K[\Gamma] \setminus \operatorname{Max} K[\Gamma] = \{\mathcal{J}(\Gamma_i)/\mathcal{J}(\Gamma)\}_{i=1,\dots,h}$.

We will see even closer relationships between algebraic curves in $\mathbb{A}^2(K)$ and ideals in $K[X,Y]$ as we learn more about algebraic curves.

Definition 1.16. The *divisor group* $\mathcal{D}$ of $\mathbb{A}^2(K)$ is the free abelian group on the set of all irreducible curves in $\mathbb{A}^2(K)$. Its elements are called *divisors* on $\mathbb{A}^2(K)$.

A divisor D is therefore a (formal) linear combination

$$D = \sum_{\Gamma \text{ irred.}} n_\Gamma \Gamma \qquad (n_\Gamma \in \mathbb{Z}, \quad n_\Gamma \neq 0 \text{ for only finitely many } \Gamma),$$

$\deg D := \sum n_\Gamma \deg \Gamma$ is called the *degree* of the divisor, and D is called *effective* if $n_\Gamma \geq 0$ for all Γ. For such a D we call

$$\operatorname{Supp}(D) := \bigcup_{n_\Gamma > 0} \Gamma$$

the *support of* D. This is an algebraic curve, except when $D = 0$ is the *zero divisor*, i.e., $n_\Gamma = 0$ for all Γ.

One can think of a divisor as an algebraic curve whose irreducible components have certain positive or negative multiplicities (weights) attached. For example, it is sometimes appropriate to say that the equation $X^2 = 0$ represents the Y-axis "counted twice."

If $D = \sum_{i=1}^h n_i \Gamma_i$ is effective, and f_i is a minimal polynomial for Γ_i, then we call $f_1^{n_1} \cdots f_h^{n_h}$ a polynomial for D,

$$\mathcal{J}(D) := (f_1^{n_1} \cdots f_h^{n_h})$$

the *ideal* (*vanishing ideal*) of D, and

$$K[D] := K[X,Y]/\mathcal{J}(D)$$

the *coordinate ring* of D. These concepts generalize the earlier ones introduced for curves.

It is clear that the effective divisors of $\mathbb{A}^2(K)$ are in one-to-one correspondence with the principal ideals $\neq (0)$ in $K[X,Y]$, and the ideal (1) corresponds to the zero divisor. The maximal ideals of $K[D]$ are in one-to-one correspondence with the points of $\operatorname{Supp}(D)$, and the nonmaximal prime ideals $\neq (0), (1)$ are in one-to-one correspondence with the components Γ of D with $n_\Gamma > 0$.

Exercises

1. Let K be an algebraically closed field and $K_0 \subset K$ a subfield. Let $\Gamma \subset \mathbb{A}^2(K)$ be an algebraic curve of degree d and let L be a line that intersects Γ in exactly d points. Assume that Γ and L have minimal polynomials in $K_0[X, Y]$. Show that if $d-1$ of the intersection points are K_0-rational, then all of the intersection points are K_0-rational.
2. Let K be an algebraically closed field of characteristic $\neq 2$ and let $K_0 \subset K$ be a subfield. Show that the K_0-rational points of the curve $\Gamma : X^2+Y^2 = 1$ are $(0, 1)$ and
$$\left(\frac{2t}{t^2+1}, \frac{t^2-1}{t^2+1}\right) \qquad \text{with } t \in K_0, \quad t^2+1 \neq 0.$$
(Consider all lines through $(0, -1)$ that are defined over K_0 and their points of intersection with Γ.)
3. (Diophantus of Alexandria $\sim$ 250 AD.) A triple $(a, b, c) \in \mathbb{Z}^3$ is called "Pythagorean" if $a^2 + b^2 = c^2$. Show, using Exercise 2, that for $\lambda, u, v \in \mathbb{Z}$, the triple $\lambda(2uv, u^2 - v^2, u^2 + v^2)$ is Pythagorean, and for every Pythagorean triple (a, b, c), either (a, b, c) or (b, a, c) can be represented in this way.
4. The curve in $\mathbb{A}^2(K)$ with equation $X^2+Y^2 = 3$ has no $\mathbb{Q}$-rational points.
5. Convince yourself that the curves in 1.2(e) and 1.2(f) really do appear as indicated in the sketches. Also check which of those curves are irreducible.
6. Sketch the following curves.
   (a) $4[X^2 + (Y+1)^2 - 1]^2 + (Y^2 - X^2)(Y+1) = 0$
   (b) $(X^2+Y^2)^5 - 16X^2Y^2(X^2-Y^2)^2 = 0$

2

Projective Algebraic Curves

Besides facts from linear algebra we will use the concept of a homogeneous polynomial; see the beginning of Appendix A. Specifically, Lemma A.3 and Theorem A.4 will play a role.

In studying algebraic curves one has to distinguish between local and global properties. Beautiful global theorems can be obtained by completing affine curves to projective curves by adding "points at infinity." Here we will discuss these "compactifications." A certain familiarity with the geometry of the projective plane will be useful. The historical development of projective geometry is sketched out in Brieskorn–Knörrer [BK]. The modern access to projective geometry comes at the end of a long historical process.

The *projective plane* $\mathbb{P}^2(K)$ over a field K is the set of all lines in K^3 through the origin. The points $P \in \mathbb{P}^2(K)$ will therefore be given by triples $\langle x_0, x_1, x_2 \rangle$, with $(x_0, x_1, x_2) \in K^3$, $(x_0, x_1, x_2) \neq (0,0,0)$, where $\langle x_0, x_1, x_2 \rangle = \langle y_0, y_1, y_2 \rangle$ if and only if $(y_0, y_1, y_2) = \lambda(x_0, x_1, x_2)$ for some $\lambda \in K^*$. The triple (x_0, x_1, x_2) is called a *system of homogeneous coordinates* for $P = \langle x_0, x_1, x_2 \rangle$. Observe that there is no point $\langle 0,0,0 \rangle$ in $\mathbb{P}^2(K)$. Two points $P = \langle x_0, x_1, x_2 \rangle$ and $Q = \langle y_0, y_1, y_2 \rangle$ are distinct if and only if (x_0, x_1, x_2) and (y_0, y_1, y_2) are linearly independent over K.

Generalizing $\mathbb{P}^2(K)$, one can define *n-dimensional projective space* $\mathbb{P}^n(K)$ as the set of all lines in K^{n+1} through the origin. The points of $\mathbb{P}^n(K)$ are the "homogeneous $(n+1)$-tuples" $\langle x_0, \dots, x_n \rangle$ with $(x_0, \dots, x_n) \neq (0, \dots, 0)$. As a special case we have the *projective line* $\mathbb{P}^1(K)$ given by

$$\mathbb{P}^1(K) = \{\langle x_0, x_1 \rangle \mid (x_0, x_1) \in K^2 \setminus (0,0)\}.$$

Still more generally, given any K-vector space V, there is an associated projective space $\mathbb{P}(V)$ defined as the set of all 1-dimensional subspaces of V.

In the following let K again be an algebraically closed field, and let $K[X_0, X_1, X_2]$ be the polynomial algebra over K in the variables X_0, X_1, X_2. If $F \in K[X_0, X_1, X_2]$ is a *homogeneous* polynomial and $P = \langle x_0, x_1, x_2 \rangle$ is a point of $\mathbb{P}^2(K)$, we will call P a *zero* of F if $F(x_0, x_1, x_2) = 0$. If $\deg F = d$, then $F(\lambda X_0, \lambda X_1, \lambda X_2) = \lambda^d F(X_0, X_1, X_2)$ for any $\lambda \in K$, and therefore the condition $F(x_0, x_1, x_2) = 0$ does not depend on the particular choice of homogeneous coordinates for P. So we can then write $F(P) = 0$. The set

$$\mathcal{V}_+(F) := \{P \in \mathbb{P}^2(K) \mid F(P) = 0\}$$

is called the *zero set of* F in $\mathbb{P}^2(K)$.

Definition 2.1. A subset $\Gamma \subset \mathbb{P}^2(K)$ is called a *projective algebraic curve* if there exists a homogeneous polynomial $F \in K[X_0, X_1, X_2]$ with $\deg F > 0$ such that $\Gamma = \mathcal{V}_+(F)$. A polynomial of least degree of this kind is called a *minimal polynomial* for Γ, and its degree is called the *degree of* Γ ($\deg \Gamma$).

We shall see in 2.10 that the minimal polynomial is unique up to multiplication by a constant $\lambda \in K^*$.

If $K_0 \subset K$ is a subring and Γ has a minimal polynomial F with $F \in K_0[X_0, Y_0, Z_0]$, then we say that Γ *is defined over* K_0. The points $P \in \Gamma$ that can be written as $P = \langle x_0, x_1, x_2 \rangle$ with $x_i \in K_0$ are called the K_0*-rational points of* Γ.

Example 2.2. Curves of degree 1 in $\mathbb{P}^2(K)$ are called *projective lines.* These are the solution sets of homogeneous linear equations

$$a_0X_0 + a_1X_1 + a_2X_2 = 0 \qquad (a_0, a_1, a_2) \neq (0, 0, 0).$$

A line uniquely determines its equation up to a constant factor $\lambda \in K^*$. Furthermore, through any two points $P = \langle x_0, x_1, x_2 \rangle$ and $Q = \langle y_0, y_1, y_2 \rangle$ with $P \neq Q$ there is exactly one line g through P and Q, for the system of equations

$$\begin{aligned} a_0x_0 + a_1x_1 + a_2x_2 &= 0, \\ a_0y_0 + a_1y_1 + a_2y_2 &= 0, \end{aligned}$$

has a unique solution $(a_0, a_1, a_2) \neq (0, 0, 0)$ up to a constant factor. The line is then

$$g = \{\langle \lambda(x_0, x_1, x_2) + \mu(y_0, y_1, y_2) \rangle \mid \lambda, \mu \in K \text{ not both } = 0\},$$

which we abbreviate as $g = \lambda P + \mu Q$. Also note that three points $P_i = \langle x_{0i}, x_{1i}, x_{2i} \rangle$ $(i = 1, 2, 3)$ lie on a line whenever (x_{0i}, x_{1i}, x_{2i}) are linearly dependent over K.

Two projective lines always intersect, and the point of intersection is unique if the lines are different. This is clear, because a system of equations

$$\begin{aligned} a_0X_0 + a_1X_1 + a_2X_2 &= 0, \\ b_0X_0 + b_1X_1 + b_2X_2 &= 0, \end{aligned}$$

always has a nontrivial solution (x_0, x_1, x_2) that is unique up to a constant factor if the coefficient matrix has rank 2.

A mapping $c : \mathbb{P}^2(K) \to \mathbb{P}^2(K)$ is called a (projective) *coordinate transformation* if there is a matrix $A \in GL(3, K)$ such that for each point $\langle x_0, x_1, x_2\rangle \in \mathbb{P}^2(K)$,

$$c(\langle x_0, x_1, x_2\rangle) = \langle (x_0, x_1, x_2)A\rangle.$$

The matrix A is uniquely determined by c up to a factor $\lambda \in K^*$: First of all, it is clear that λA defines the same coordinate transformation as A. If $B \in GL(3, K)$ is another matrix that defines c, then BA^{-1} is the matrix of a linear transformation that is an automorphism of K^3 that maps all lines through the origin to themselves; it follows that $B = \lambda A$ for some $\lambda \in K^*$.

One applies coordinate transformations to bring a configuration of points and curves into a clearer position. Let $\Gamma = \mathcal{V}_+(F)$ be a curve, where F is a homogeneous polynomial, and let c be a coordinate transformation with matrix A. Then

$$c(\Gamma) = \mathcal{V}_+(F^A),$$

where (in the above notation)

$$F^A(X_0, X_1, X_2) = F((X_0, X_1, X_2)A^{-1}).$$

Thus F^A is homogeneous with $\deg F^A = \deg F$. A coordinate transformation maps a projective curve to a projective curve of the same degree, and we tend to identify two curves that differ only by a coordinate transformation.

After this summary of facts, which we assume to be known, we come to the "passage from affine to projective."

We have an injection given by

$$i : \mathbb{A}^2(K) \to \mathbb{P}^2(K), \qquad i(x, y) = \langle 1, x, y\rangle,$$

from the affine to the projective plane. We identify $\mathbb{A}^2(K)$ with its image under i. Then $\mathbb{A}^2(K)$ is the complement of the line $X_0 = 0$ in $\mathbb{P}^2(K)$. This line is called the *line at infinity* of $\mathbb{P}^2(K)$; the points of this line are called *points at infinity*, and the points of $\mathbb{A}^2(K)$ are called *points at finite distance*. For $P = \langle 1, x, y\rangle \in \mathbb{P}^2(K)$ we call (x, y) the *affine coordinates* of P.

Given a polynomial $f \in K[X, Y]$ with $\deg f = d$, we can define by

$$\hat{f}(X_0, X_1, X_2) := X_0^d f\left(\frac{X_1}{X_0}, \frac{X_2}{X_0}\right) \tag{1}$$

a homogeneous polynomial $\hat{f} \in K[X_0, X_1, X_2]$ with $\deg \hat{f} = d$. It is called the *homogenization* of f.

Definition 2.3. Let $\Gamma \subset \mathbb{A}^2(K)$ be an algebraic curve with minimal polynomial $f \in K[X, Y]$ and let $\hat{f}$ be the homogenization of f. Then the projective algebraic curve $\hat{\Gamma} = \mathcal{V}_+(F)$ is called the *projective closure* of f.

The curve $\hat{\Gamma}$ depends only on the curve Γ and not on the choice of a minimal polynomial for Γ. By 1.8 this polynomial is uniquely determined by Γ up to a factor $\lambda \in K^*$, and it is obvious that $\widehat{\lambda f} = \lambda \hat{f}$.

We can give the following description for $\hat{f}$: If f is of degree d and

$$f = f_0 + f_1 + \cdots + f_d, \tag{2}$$

where the f_i are homogeneous polynomials of degree i (so in particular $f_d \neq 0$), then

(3)

$$\hat{f} = X_0^d f_0(X_1, X_2) + X_0^{d-1} f_1(X_1, X_2) + \cdots + X_0 f_{d-1}(X_1, X_2) + f_d(X_1, X_2).$$

It follows that

Lemma 2.4. $\Gamma = \hat{\Gamma} \cap \mathbb{A}^2(K)$.

The points of $\hat{\Gamma} \setminus \Gamma$ are called the *points at infinity of* Γ. The next lemma shows how to calculate them.

Lemma 2.5. *Every affine curve Γ of degree d has at least one and at most d points at infinity. These are the points $\langle 0, a, b \rangle$, where (a, b) runs over all the zeros of f_d, where $\Gamma = \mathcal{V}(f)$ and f is written as in (2).*

Proof. $\hat{\Gamma} \setminus \Gamma$ consists of the solutions $\langle x_0, x_1, x_2 \rangle$ to the equation $\hat{f} = 0$ with $x_0 = 0$. By (3) the second assertion of the lemma is satisfied. A homogeneous polynomial f_d of degree d decomposes into d homogeneous linear factors by A.4. The first assertion of the lemma follows from this.

Examples 2.6. (a) For an affine line

$$g : aX + bY + c = 0, \qquad (a, b) \neq (0, 0),$$

the projective closure is given by

$$\hat{g} : cX_0 + aX_1 + bX_2 = 0.$$

The point at infinity on g is $\langle 0, b, -a \rangle$. Two affine lines are then parallel if and only if they meet at infinity, i.e., their points at infinity coincide.

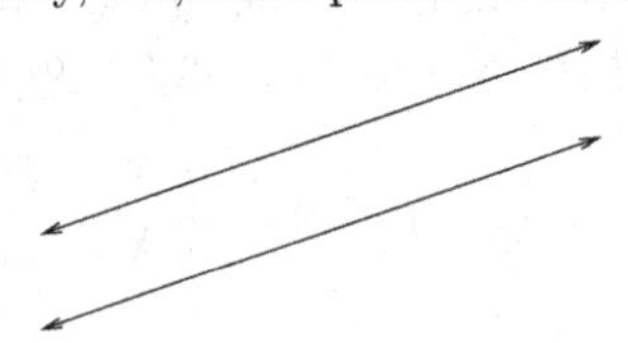

(b) The ellipse $\frac{X^2}{a^2} + \frac{Y^2}{b^2} = 1$ $(a, b \in \mathbb{R}_+)$ has two points at infinity, $\langle 0, a, \pm ib \rangle$, which, however, are not $\mathbb{R}$-rational. The hyperbola $\frac{X^2}{a^2} - \frac{Y^2}{b^2} = 1$ $(a, b \in \mathbb{R}_+)$ has two points at infinity, $\langle 0, a, \pm b \rangle$, which are both $\mathbb{R}$-rational. The parabola $Y = aX^2$ $(a \in \mathbb{R}_+)$ has exactly one point at infinity, namely $\langle 0, 0, 1 \rangle$. All circles $(X - a)^2 + (Y - b)^2 = r^2$ $(a, b, r \in \mathbb{R})$ have the same points $\langle 0, 1, \pm i \rangle$ at infinity.

If $h : a_0X_0 + a_1X_1 + a_2X_2 = 0$ is a projective line different from the line at infinity $h_\infty : X_0 = 0$, then $(a_1, a_2) \neq (0, 0)$, and $h = \hat{g}$, where g is given by the equation $a_1X + a_2Y + a_0 = 0$. Consequently, there is a bijection given by $g \mapsto \hat{g}$ from the set of affine lines to the set of projective lines $\neq h_\infty$. There is a similar result for arbitrary algebraic curves, as we will now see.

For a homogeneous polynomial $F \in K[X_0, X_1, X_2]$ we call the polynomial f in $K[X, Y]$ given by $f(X, Y) = F(1, X, Y)$ the *dehomogenization* of F (with respect to X_0). If X_0 is not a factor of F, then

$$\deg f = \deg F$$

and

$$F = \hat{f},$$

as one sees immediately from equation (3).

Theorem 2.7. *Let Δ be a projective algebraic curve with minimal polynomial F and let $\Gamma := \Delta \cap \mathbb{A}^2(K)$. Then*

(a) *If Δ is not the line at infinity, then Γ is an affine algebraic curve.*
(b) *If Δ does not contain the line at infinity, then the dehomogenization f of F is a minimal polynomial of Γ and*

$$\Delta = \hat{\Gamma}, \text{ its projective closure.}$$

Proof. (a) By the hypotheses on Δ, f is not constant and $\Gamma = \mathcal{V}(f)$ is an affine curve.

(b) We notice first of all that for polynomials $f_1, f_2 \in K[X, Y]$, the formula

$$\widehat{f_1 f_2} = \hat{f}_1 \hat{f}_2 \tag{4}$$

holds, as one easily sees by the definition (1) of homogenization.

Let $f = cf_1^{\alpha_1} \cdots f_h^{\alpha_h}$ be a decomposition of f into irreducible factors ($c \in K^*, \alpha_i \in \mathbb{N}_+, f_i$ irreducible, $f_i \not\sim f_j$ for $i \neq j$). Because X_0 is not a factor of F,

$$F = c\hat{f}_1^{\alpha_1} \cdots \hat{f}_h^{\alpha_h}$$

by the above formula (4), and because F is a minimal polynomial of Δ, we must have $\alpha_1 = \cdots = \alpha_h = 1$. Also, $\Delta = \hat{\Gamma}$ by definition of $\hat{\Gamma}$.

Corollary 2.8. *The affine algebraic curves are in one-to-one correspondence with the projective algebraic curves that do not contain the line h_∞ at infinity. Other than h_∞, the only other projective curves are of the form $\hat{\Gamma} \cup h_\infty$, where Γ is an affine curve.*

Corollary 2.9. *Every projective curve Δ consists of infinitely many points, and also $\mathbb{P}^2(K) \setminus \Delta$ is infinite.*

This follows immediately from the corresponding result for affine curves (1.3).

Under a projective coordinate transformation, the line at infinity $X_0 = 0$ is in general sent to some other line, and conversely, the line at infinity in the new coordinate system came from some other line. By an appropriate choice of a coordinate system, any line can be the line at infinity. More specifically: given four points $P_i \in \mathbb{P}^2(K)$ $(i = 0, 1, 2, 3)$, no three of which lie on a line, there is a unique coordinate transformation c of $\mathbb{P}^2(K)$ such that $c(P_0) = \langle 1, 0, 0\rangle$, $c(P_1) = \langle 0, 1, 0\rangle$, $c(P_2) = \langle 0, 0, 1\rangle$, and $c(P_3) = \langle 1, 1, 1\rangle$. If g is a projective line and $P \neq Q$ are points not on g, then one can always find a projective coordinate transformation that maps g to the line at infinity, maps P to the origin $(0, 0)$ in the affine plane, and maps Q to an arbitrary given point $(a, b) \neq (0, 0)$. We will make frequent use of these facts.

Corollary 2.10. *The minimal polynomial of a projective curve Δ is uniquely determined by Δ up to a constant factor $\neq 0$.*

Proof. If F is a minimal polynomial of Δ and c is a coordinate transformation with matrix A, then F^A is a minimal polynomial of $c(\Gamma)$. We can assume that $c(\Gamma)$ does not contain the line at infinity (in the new coordinate system). Then, by 2.7, F^A is the homogenization of the minimal polynomial of the affine part of $c(\Gamma)$. This is unique up to a constant factor $\neq 0$, and therefore so are F^A and F.

Definition 2.11. The *vanishing ideal* of a projective curve Δ is the ideal $\mathcal{J}_+(\Delta) \subset K[X_0, X_1, X_2]$ generated by all homogeneous polynomials that vanish at all points of Δ.

$\mathcal{J}_+(\Delta)$ is therefore a homogeneous ideal (A.7).

Theorem 2.12. *$\mathcal{J}_+(\Delta)$ is the principal ideal generated by any minimal polynomial of Δ.*

Proof. We can assume that Δ does not contain the line at infinity. Let F be a minimal polynomial for Δ and let $G \in \mathcal{J}_+(\Delta)$ be homogeneous. Write $G = X_0^\alpha H$, where H is not divisible by X_0. Then $h := H(1, X, Y)$ is contained in the vanishing ideal of $\Gamma := \Delta \cap \mathbb{A}^2(K)$. This is generated by $f := F(1, X, Y)$; hence $h = fg$ for some $g \in K[X, Y]$. By (4) we have $H = \hat{h} = \hat{f}\hat{g} = F\hat{g}$, and so $G \in (F)$.

As in the affine case we call a curve $\Delta \subset \mathbb{P}^2(K)$ *irreducible* provided that whenever $\Delta = \Delta_1 \cup \Delta_2$ for projective curves Δ_i $(i = 1, 2)$, then $\Delta = \Delta_1$ or $\Delta = \Delta_2$.

Corollary 2.13. *The following are equivalent:*

(a) *Δ is irreducible.*
(b) *The minimal polynomial of Δ is an irreducible polynomial.*
(c) *$\mathcal{J}_+(\Delta)$ is a (homogeneous) prime ideal.*

Proof. (a) $\Rightarrow$ (b). Let F be a minimal polynomial of Δ and suppose that $F = F_1F_2$ for some $F_i \in K[X_0, X_1, X_2]$ $(i = 1, 2)$. These polynomials are homogeneous by A.3, and we have $\Delta = \mathcal{V}_+(F_1) \cup \mathcal{V}_+(F_2)$. Hence without loss of generality $\Delta = \mathcal{V}_+(F_1)$. Then F_1 is divisible by F, and it follows that F_2 must be constant. Therefore, F is irreducible.

(b) $\Rightarrow$ (c) is clear.

(c) $\Rightarrow$ (a). Let $\Delta = \Delta_1 \cup \Delta_2$, where the curves Δ_i have minimal polynomials F_i $(i = 1, 2)$. For $F := F_1F_2$ we have $F \in \mathcal{J}_+(\Delta)$. Therefore $F_1 \in \mathcal{J}_+(\Delta)$ or $F_2 \in \mathcal{J}_+(\Delta)$, because $\mathcal{J}_+(\Delta)$ is a prime ideal. Then, however, $\Delta \subset \mathcal{V}_+(F) = \Delta_i$ for an $i \in \{1, 2\}$, and therefore $\Delta = \Delta_i$.

Corollary 2.14. (a) (*Decomposition into irreducible components*) *Every projective algebraic curve has a unique (up to order) representation*

$$\Delta = \Delta_1 \cup \cdots \cup \Delta_h,$$

where the Δ_i $(i = 1, \ldots, h)$ are irreducible curves. These are in one-to-one correspondence with the irreducible factors of a minimal polynomial for Δ.

(b) *If Γ is an affine curve with irreducible decomposition*

$$\Gamma = \Gamma_1 \cup \cdots \cup \Gamma_h,$$

then

$$\hat{\Gamma} = \hat{\Gamma}_1 \cup \cdots \cup \hat{\Gamma}_h,$$

and the $\hat{\Gamma}_i$ are the irreducible components of $\hat{\Gamma}$.

This follows by 1.13 and by the previous observation about the affine and projective minimal polynomials.

In 1.14 the prime ideals of $K[X, Y]$ were described. We now do the same for the *homogeneous* prime ideals in $K[X_0, X_1, X_2]$.

Theorem 2.15. *$K[X_0, X_1, X_2]$ has the following homogeneous prime ideals and no others:*

(a) *The zero ideal.*
(b) *The principal ideals (F) generated by irreducible homogeneous polynomials $F \neq 0$. These are in one-to-one correspondence with the irreducible curves in $\mathbb{P}^2(K)$.*
(c) *The ideals $\mathfrak{p}_P := (aX_1 - bX_0, aX_2 - cX_0, bX_2 - cX_1)$, where $P = \langle a, b, c \rangle \in \mathbb{P}^2(K)$. These prime ideals are in one-to-one correspondence with the points of $\mathbb{P}^2(K)$.*
(d) *The homogeneous maximal ideal (X_0, X_1, X_2).*

Proof. Let $\mathfrak{p}$ be a homogeneous prime ideal in $K[X_0, X_1, X_2]$. If $\mathfrak{p} \neq (0)$, then $\mathfrak{p}$ contains an irreducible homogeneous polynomial $F \neq 0$. If $\mathfrak{p} \neq (F)$, then there is a homogeneous polynomial G in $\mathfrak{p}$ that is not divisible by F. According to 1.5 the ideal (F, G) contains, and therefore $\mathfrak{p}$ contains, a homogeneous polynomial $d_1 \neq 0$ in which X_2 does not appear, and a homogeneous polynomial $d_2 \neq 0$ in which X_1 does not appear. Now d_1 and d_2 decompose into homogeneous linear factors by A.4, and so $\mathfrak{p}$ must contain linear polynomials of the form $aX_1 - bX_0 \neq 0$, $a'X_2 - b'X_0 \neq 0$. One sees easily that an ideal of the form $\mathfrak{p}_P$ $(P \in \mathbb{P}^2(K))$ is then also contained in $\mathfrak{p}$. Such a prime ideal is already generated by two homogeneous linear polynomials, and it follows that $K[X_0, X_1, X_2]/\mathfrak{p}_P \cong K[T]$, a polynomial ring in one variable. The only homogeneous prime ideals in this ring are (0) and (T). The preimages of these prime ideals in $K[X_0, X_1, X_2]$ are $\mathfrak{p}_P$ and (X_0, X_1, X_2). Hence we must have $\mathfrak{p} = \mathfrak{p}_P$ or $\mathfrak{p} = (X_0, X_1, X_2)$.

As in the affine case, the *divisor group* of $\mathbb{P}^2(K)$ is defined as the free abelian group on the set of irreducible projective curves Δ. For a divisor $D = \sum n_\Delta \Delta$, the *degree* of D is defined by

$$\deg D := \sum n_\Delta \deg \Delta.$$

D is called *effective* if $n_\Delta \geq 0$ for all Δ. For an effective divisor D we call the Δ with $n_\Delta > 0$ the *components of* D, and

$$\operatorname{Supp}(D) := \bigcup_{n_\Delta > 0} \Delta$$

is called the *support of* D. If a divisor is given in the form $D = \sum_{i=1}^{h} n_i \Delta_i$ and if F_i is a minimal polynomial of Δ_i $(i = 1, \ldots, h)$, then we assign to $\mathcal{D}$ the polynomial $F := \prod_{i=1}^{h} F_i^{n_i}$. Conversely, since the factors of a homogeneous polynomial are themselves homogeneous (A.3), every homogeneous polynomial $F \neq 0$ determines, by decomposition into irreducible factors, a unique effective divisor, which for simplicity we also denote by F. We can choose to think of F as either a homogeneous polynomial $\neq 0$ in $K[X_0, X_1, X_2]$, or as the associated divisor in $\mathbb{P}^2(K)$. From now on, effective divisors in $\mathbb{P}^2(K)$ will be called "curves in $\mathbb{P}^2(K)$." These are the curves we considered earlier, whose irreducible components where furnished with "weights" from $\mathbb{N}$. If Δ is a curve with decomposition into irreducible components $\Delta = \Delta_1 \cup \cdots \cup \Delta_h$, it will be identified with the effective divisor $\Delta_1 + \cdots + \Delta_h$. Divisors of this kind will be called "reduced curves" in the future. These correspond to the reduced homogeneous polynomials $\neq 0$ in $K[X_0, x_1, x_2]$. The curve Δ is *defined* over a subfield $K_0 \subset K$ if an appropriate associated polynomial can be chosen in $K_0[X_0, X_1, X_2]$.

If the line at infinity is not a component of the curve $\Delta = \sum_{i=1}^{h} n_i \Delta_i$ (and this can always be arranged by choosing an appropriate coordinate system), then we call $\Gamma := \sum_{i=1}^{h} n_i \Gamma_i$ with $\Gamma_i := \Delta_i \cap \mathbb{A}^2(K)$ the *affine curve belonging to* Γ. It corresponds to the dehomogenization of the polynomial associated with Δ.

Exercises

1. Determine the points at infinity for the curves in 1.2(e) and 1.2(f).
2. A *projective quadric* is a curve Q (i.e., an effective divisor) in $\mathbb{P}^2(K)$ of degree 2. Show that if char $K \neq 2$, then in an appropriate coordinate system, Q has one of the following equations:
   (a) $X_0^2 + X_1^2 + X_2^2 = 0$ (nonsingular quadric)
   (b) $X_0^2 + X_1^2 = 0$ (pair of lines)
   (c) $X_0^2 = 0$ (a double line)
3. Show that if char $K \neq 2$ and $i := \sqrt{-1}$, then

$$\alpha : \mathbb{P}^1(K) \to \mathbb{P}^2(K), \qquad \alpha(\langle u, v\rangle) = \langle 2uv, u^2 - v^2, i(u^2 + v^2)\rangle,$$

   gives a bijection of the projective line $\mathbb{P}^1(K)$ with the nonsingular quadric (cf. Exercises 2 and 3 in Chapter 1).
4. Let $c : \mathbb{P}^2(K) \to \mathbb{P}^2(K)$ be the coordinate transformation that sends the points

$$\langle 1,0,0\rangle, \ \langle 0,1,0\rangle, \ \langle -1,0,1\rangle, \ \langle 0,1,1\rangle$$

   to the points

$$\langle 1,0,0\rangle, \ \langle 0,1,0\rangle, \ \langle 0,0,1\rangle, \ \langle 1,1,1\rangle.$$

   (a) Determine the equation of the curve

$$X_0^2X_2 - X_0X_1^2 + X_0X_2^2 - 2X_0X_1X_2 - X_1^2X_2 - 2X_1X_2^2 = 0$$

   in the new coordinate system.
   (b) Is the curve irreducible?

3

The Coordinate Ring of an Algebraic Curve and the Intersections of Two Curves

From now on, we assume that the reader is familiar with the material in Appendices A and B. Above all, we will use the methods contained in Appendix B repeatedly. We will also apply the elementary Lemmas D.5 and I.4.

Let F be an algebraic curve in $\mathbb{P}^2(K)$, i.e., F is an effective divisor according to our convention in Chapter 2. At the same time, F denotes a homogeneous polynomial in $K[X_0, X_1, X_2]$ that defines the curve. Instead of $\text{Supp}(F)$ we also write $\mathcal{V}_+(F)$:

$$\text{Supp}(F) = \mathcal{V}_+(F) = \{P \in \mathbb{P}^2(K) \mid F(P) = 0\}.$$

Definition 3.1. The residue class ring $K[F] := K[X_0, X_1, X_2]/(F)$ is called the *projective coordinate ring* of F.

Since F is homogeneous, $K[F]$ is a graded K-algebra (A.7). A coordinate transformation on $\mathbb{P}^2(K)$ with a matrix A defines a K-automorphism of $K[X_0, X_1, X_2]$ given by

$$(X_0, X_1, X_2) \mapsto (X_0, X_1, X_2) \cdot A^{-1}.$$

Under this automorphism, F becomes F^A, which corresponds to a curve in the new coordinate system, and it induces a K-isomorphism

$$K[X_0, X_1, X_2]/(F) \cong K[X_0, X_1, X_2]/(F^A).$$

The coordinate ring is therefore independent—up to K-isomorphism—of the choice of coordinates.

We can, and will, choose the coordinates such that X_0 does not divide F. The curve f in $\mathbb{A}^2(K) = \mathbb{P}^2(K) \setminus \mathcal{V}_+(X_0)$ corresponding to F is then the dehomogenization f of F, i.e., f is the polynomial in $K[X, Y]$ with

$$f(X, Y) = F(1, X, Y).$$

We have already introduced the affine coordinate ring $K[f] = K[X, Y]/(f)$ in Chapter 1. A point $P = (a, b) \in \mathbb{A}^2(K)$ belongs to $\text{Supp}(f) = \mathcal{V}(f)$ if and only if $f \in \mathfrak{M}_P = (X - a, Y - b)$. In this case we denote by $\mathfrak{m}_P := \mathfrak{M}_P/(f)$

the image of $\mathfrak{M}_P$ in $K[f]$. Then $\mathfrak{m}_P \in \operatorname{Max} K[f]$, and every maximal ideal of $K[f]$ is of this form for a uniquely determined $P \in \operatorname{Supp}(f)$.

If g is another curve and ψ is the residue class of the polynomial g in $K[f]$, then there is a one-to-one correspondence between the maximal ideals $\mathfrak{m}$ of $K[f]$ with $g \in \mathfrak{m}$ and the points of $\mathcal{V}(f) \cap \mathcal{V}(g)$. The question, "How many points of intersection do the curves have?" (or equivalently, "How many solutions does the system of equations $f = 0$, $g = 0$ have?") can be reformulated as, "How many maximal ideals are there in $K[f]$ that contain ψ?"

We now endow $K[X,Y]$ with its degree filtration $\mathcal{G}$, and $K[f]$ with the corresponding residue class filtration $\mathcal{F}$ induced by $\mathcal{G}$ (see Appendix B). According to B.4(a), $K[X_0, X_1, X_2]$ can be interpreted as the Rees algebra $\mathcal{R}_\mathcal{G} K[X,Y]$, and the homogenization of a polynomial from $K[X,Y]$ in the sense of Appendix B is the usual one. Since F is the homogenization f^* of f, it follows from B.8 and B.12 that:

Remark 3.2. $K[F] \cong \mathcal{R}_\mathcal{F} K[f]$.

Proof.

$$\begin{aligned} K[F] &= K[X_0, X_1, X_2]/(F) = \mathcal{R}_\mathcal{G} K[X,Y]/(f^*) \\ &\cong \mathcal{R}_\mathcal{F}(K[X,Y]/(f)) = \mathcal{R}_\mathcal{F} K[f]. \end{aligned}$$

One can deduce the following immediately from B.5.

Remark 3.3. The image x_0 of X_0 in $K[F]$ is not a zerodivisor, $K[x_0]$ is a polynomial algebra, and there are K-isomorphisms

$$\begin{aligned} K[f] &\cong K[F]/(x_0 - 1), \\ \operatorname{gr}_\mathcal{F} K[f] &\cong K[F]/(x_0). \end{aligned}$$

If f_p is the homogeneous component of largest degree of f (therefore the leading form $L_\mathcal{G} f$ using the degree filtration), then B.8 and B.12 also yield

Remark 3.4. $\operatorname{gr}_\mathcal{F} K[f] \cong K[X,Y]/(f_p)$.

Since f_p describes the points at infinity of f, $\operatorname{gr}_\mathcal{F} K[f]$ has something to do with the points at infinity. By A.12(a) we know the Hilbert function of $\operatorname{gr}_\mathcal{F} K[f]$. We have:

Remark 3.5.

$$\dim_K \mathcal{F}_k/\mathcal{F}_{k-1} = \begin{cases} k+1 & \text{for } 0 \le k < p, \\ p & \text{for } p \le k. \end{cases}$$

Now let two curves F, G in $\mathbb{P}^2(K)$ be given, and assume that they have no common components.

Definition 3.6. $K[F \cap G] := K[X_0, X_1, X_2]/(F, G)$ is called the *projective coordinate ring* of the intersection of F and G.

If Ψ denotes the image of G in $K[F]$, and Φ the image of F in $K[G]$, then by the Noether isomorphism theorem,

$$K[F \cap G] \cong K[F]/(\Psi) \cong K[G]/(\Phi),$$

where Φ is not a zerodivisor on $K[G]$ and Ψ is not a zerodivisor on $K[F]$.

Now let f and g be two affine curves with $\deg f =: p$, $\deg g =: q$, and assume that they have no common components.

Definition 3.7. $K[f \cap g] := K[X,Y]/(f,g)$ is called the *affine coordinate ring* of the intersection of f and g.

We have $K[f \cap g] \cong K[f]/(\psi) \cong K[g]/(\phi)$, where ψ denotes the image of g in $K[f]$ and ϕ the image of f in $K[g]$. Since f and g are relatively prime, ϕ and ψ are not zerodivisors in their respective rings.

From what was said above, the points of $\mathcal{V}(f) \cap \mathcal{V}(g)$ are in one-to-one correspondence with the maximal ideals of $K[f \cap g]$. Also, $K[f \cap g]$ is a finite-dimensional K-algebra by 1.4. How big is its dimension? If f and g do not have any points at infinity in common, the observations about Rees algebras and associated graded algebras in Appendix B give us the answer immediately, which we will now show.

We will denote by $\mathcal{F}$ the residue class filtration induced on $K[f \cap g]$ by the degree filtration $\mathcal{G}$ of $K[X,Y]$. Let $f_p = L_{\mathcal{G}} f$ and $g_q = L_{\mathcal{G}} g$ be the homogeneous components of largest degree of f and g.

By 2.5, to say that f and g have no points at infinity in common is equivalent to saying that f_p and g_p are relatively prime. In concrete cases one can decide using the Euclidean algorithm whether this condition is satisfied, without being forced to calculate the points at infinity explicitly (Exercise 4). That f_p and g_q are relatively prime is equivalent to the statement that f_p is not a zerodivisor modulo g_q, and g_q is not a zerodivisor modulo f_p.

Let F and G be the projective curves associated with f and g, that is, the homogenizations of f and g in $K[X_0, X_1, X_2]$. We have $\deg F = p$ and $\deg G = q$. Again, B.5, B.8, and B.12 are applicable. We get the following.

Theorem 3.8. *Suppose f and g have no points at infinity in common. Then*

(a) $K[F \cap G] \cong \mathcal{R}_{\mathcal{F}} K[f \cap g]$.
(b) *The image x_0 of X_0 in $K[F \cap G]$ is not a zerodivisor, $K[x_0]$ is a polynomial algebra, and we have*

$$K[f \cap g] \cong K[F \cap G]/(x_0 - 1),$$
$$\operatorname{gr}_{\mathcal{F}} K[f \cap g] \cong K[F \cap G]/(x_0) \cong K[X,Y]/(f_p, g_q).$$

By A.12(b), $\dim_K K[F \cap G]/(f_p, g_q) = p \cdot q$. Therefore, B.6 gives the answer to the above question:

Theorem 3.9. *If f and g have no points at infinity in common, then*

(a) $\dim_K K[f \cap g] = p \cdot q$.
(b) $K[F \cap G]$ *is a free* $K[x_0]$*-module of rank* $p \cdot q$.

Now let F and G be two arbitrary curves in $\mathbb{P}^2(K)$ with $\deg F = p > 0$, $\deg G = q > 0$, and with no common components (as polynomials they are relatively prime). Then $\mathcal{V}_+(F) \cap \mathcal{V}_+(G)$ consists of only finitely many affine points (by 1.4), and since one of the curves does not contain the line at infinity, $\mathcal{V}_+(F) \cap \mathcal{V}_+(G)$ also contains only finitely many points at infinity. We can choose the coordinate system so that all the points in $\mathcal{V}_+(F) \cap \mathcal{V}_+(G)$ are points at finite distance. Then $\mathcal{V}_+(F) \cap \mathcal{V}_+(G) = \mathcal{V}(f) \cap \mathcal{V}(g)$, where f and g are the affine curves associated with F and G, and we are in the situation of Theorems 3.8 and 3.9. We then get the following.

Corollary 3.10. *Assume that F and G have no common components. Then the intersection $\mathcal{V}_+(F) \cap \mathcal{V}_+(G)$ contains at least one and at most $p \cdot q$ points.*

Proof. Since $K[f \cap g]$ is not the zero-algebra, it has at least one maximal ideal. This corresponds to a point in $\mathcal{V}(f) \cap \mathcal{V}(g)$. From the elementary lemma D.5 and 3.9(a) it follows that $K[f \cap g]$ has at most $p \cdot q$ maximal ideals. Hence $\mathcal{V}(f) \cap \mathcal{V}(g)$ contains at most $p \cdot q$ points.

The statement of the corollary is a weak form of Bézout's theorem, which we will discuss in detail later (See 5.7). In the projective plane not only do two lines always intersect, but also any two curves of positive degree always intersect. The corollary could also be formulated in this way: A system of equations

$$F(X_0, X_1, X_2) = 0, \qquad G(X_0, X_1, X_2) = 0,$$

with relatively prime homogeneous polynomials F and G of degrees p and q has at least one and at most $p \cdot q$ solutions in $\mathbb{P}^2(K)$.

For the remainder of this section we assume, as in 3.8 and 3.9, that f and g have no points at infinity in common. We strive now to find a few more precise statements about the structure of the projective coordinate ring $K[F \cap G]$ and the affine coordinate ring $K[f \cap g]$. These follow from theorems in Appendix B, and they also allow later geometric applications.

Let $B := \operatorname{gr}_{\mathcal{F}} K[f \cap g]$. By A.12(b), we know the Hilbert function of B exactly. If $B_k := \operatorname{gr}^k_{\mathcal{F}} K[f \cap g]$ is the homogeneous component of degree k and if, without loss of generality, $p \leq q$, then by A.12(b) and 3.8(b),

$$\chi_B(k) = \dim_K B_k = \begin{cases} k+1, & \text{if } 0 \leq k < p, \\ p, & \text{if } p \leq k < q, \\ p+q-k-1, & \text{if } q \leq k < p+q-1, \\ 0, & \text{if } p+q-1 \leq k. \end{cases}$$

An application of B.6 then gives

Theorem 3.11. (a) *As a* $K[x_0]$*-module,* $K[F \cap G]$ *has a basis* $\{s_1, \dots, s_{p \cdot q}\}$ *of homogeneous elements* s_i $(i = 1, \dots, p \cdot q)$*, where*

$$0 = \deg s_1 \le \deg s_2 \le \cdots \le \deg s_{p \cdot q} = p + q - 2$$

and

$$\deg s_i + \deg s_{p \cdot q - i} = p + q - 2$$

for $i = 1, \dots, p \cdot q$*. For each* $k \in \{0, \dots, p+q-2\}$ *there are exactly* $\chi_B(k)$ *basis elements in* $\{s_1, \dots, s_{p \cdot q}\}$ *of degree* k*.*

(b) $K[f \cap g]$ *has a* K*-basis* $\{\bar{s}_1, \dots, \bar{s}_{p \cdot q}\}$*, where*

$$0 = \operatorname{ord}_{\mathcal{F}} \bar{s}_1 \le \operatorname{ord}_{\mathcal{F}} \bar{s}_2 \le \cdots \le \operatorname{ord}_{\mathcal{F}} \bar{s}_{p \cdot q} = p + q - 2,$$

$$\operatorname{ord}_{\mathcal{F}} \bar{s}_i + \operatorname{ord}_{\mathcal{F}} \bar{s}_{p \cdot q - i} = p + q - 2 \quad (i = 1, \dots, p \cdot q),$$

and where for each $k \in \{1, \dots, p + q - 2\}$*, exactly* $\chi_B(k)$ *basis elements have order* k*. Furthermore,*

$$\mathcal{F}_k = \mathcal{F}_{p+q-2} \quad \textit{for each } k \ge p + q - 2.$$

The degree $(p + q - 2)$ component of B is especially interesting. It is 1-dimensional, and we will find a basis for it. To do that, we will write

$$f_p = c_{11} X + c_{12} Y,$$
$$g_q = c_{21} X + c_{22} Y,$$

with homogeneous $c_{ij} \in K[X, Y]$. Then $\det(c_{ij})$ is homogeneous of degree $p + q - 2$ and the image Δ of this determinant in B is in any case contained in B_{p+q-2}. One can check easily (I.4) that Δ depends only on f_p and g_q, and not on any particular choice of the c_{ij}. We can view Δ as a generalization of the Jacobian determinant of f_p and g_q with respect to X, Y.

Examples 3.12.

(a) Decompose f_p and g_q into homogeneous linear factors (A.4):

$$f_p = \prod_{i=1}^{p} (a_i X + b_i Y) = \frac{f_p}{a_i X + b_i Y} \cdot (a_i X + b_i Y),$$

$$g_q = \prod_{j=1}^{q} (c_j X + d_j Y) = \frac{g_q}{c_j X + d_j Y} \cdot (c_j X + d_j Y).$$

Then Δ is the image of

$$(a_i d_j - b_i c_j) \cdot \frac{f_p \cdot g_q}{(a_i X + b_i Y)(c_j X + d_j Y)}$$

in B, and this does not depend on i and j.

(b) By the Euler relation (Appendix A, formula (A2)), we have

$$p \cdot f_p = \frac{\partial f_p}{\partial X} \cdot X + \frac{\partial f_p}{\partial Y} \cdot Y,$$
$$q \cdot g_q = \frac{\partial g_q}{\partial X} \cdot X + \frac{\partial g_q}{\partial Y} \cdot Y.$$

Denote the image of the Jacobian determinant $\frac{\partial(f_p, g_q)}{\partial(X,Y)}$ in B by j, so that

$$j = p \cdot q \cdot \Delta.$$

If the characteristic of K does not divide $p \cdot q$, then

$$\Delta = \frac{1}{p \cdot q} \cdot j.$$

Now let $B_+ := \oplus_{k>0} B_k$ be the homogeneous maximal ideal of B.

Definition 3.13. $\mathfrak{S}(B) := \{z \in B \mid B_+ \cdot z = 0\}$ is called the *socle* of B.

One can check that $\mathfrak{S}(B)$ is a homogeneous ideal of B, and $B_{p+q-2} \subset \mathfrak{S}(B)$, because $B_+ \cdot B_{p+q-2} = 0$, since $B_k = 0$ for $k > p + q - 2$.

Theorem 3.14. *We have* $\mathfrak{S}(B) = B_{p+q-2} = K \cdot \Delta$. *In particular,* $\mathfrak{S}(B)$ *is a* 1*-dimensional* K*-vector space and* $\Delta \neq 0$. *If the characteristic of* K *does not divide* $p \cdot q$*, then the Jacobian determinant* j *is nonzero and* $\mathfrak{S}(B) = K \cdot j$.

Proof. It remains to show that every homogeneous element $\eta \in \mathfrak{S}(B)$ is contained in (Δ). This follows from I.5, but we give here a simpler, more direct proof. Let $H \in K[X,Y]$ be a homogeneous preimage for η. As in 3.12(a), write $f_p = \Phi \cdot L_1$, $g_q = \Psi \cdot L_2$, with homogeneous linear factors $L_1 = aX + bY$, $L_2 = cX + dY$, and homogeneous polynomials Φ, $\Psi \in K[X,Y]$. Then

$$\Delta = (ad - bc) \cdot \phi \cdot \psi,$$

where ϕ and ψ are the images of Φ and Ψ in B. Since f_p and g_q are relatively prime, $ad - bc \neq 0$.

By the hypothesis on η, we have $(X,Y) \cdot H \subset (f_p, g_q)$, and therefore we have equations

$$L_1 H = R_1 \Phi L_1 + R_2 \Psi L_2,$$
$$L_2 H = S_1 \Phi L_1 + S_2 \Psi L_2,$$

with homogeneous R_i, $S_i \in K[X,Y]$ $(i = 1, 2)$. Since L_1 is not a divisor of Ψ, and L_2 is not a divisor of Φ, we get

$$H = R_1 \Phi + R_2^* \Psi L_2 \qquad (R_2^* := L_1^{-1} R_2),$$
$$H = S_1^* \Phi L_1 + S_2 \Psi \qquad (S_1^* := L_2^{-1} S_1),$$

and therefore

$$\Phi \cdot (R_1 - S_1^* L_1) = \Psi \cdot (S_2 - R_2^* L_2).$$

Since Φ and Ψ are relatively prime, it follows that

$$R_1 - S_1^* L_1 = T \cdot \Psi$$

for some homogeneous $T \in K[X, Y]$, and

$$H = T \cdot \Phi \cdot \Psi + S_1^* \cdot f_p + R_2^* \cdot g_q.$$

Hence $\eta \in (\phi \cdot \psi) = (\Delta)$.

Corollary 3.15. *Let $d \in K[f \cap g]$ be an element with $L_{\mathcal{F}} d = \Delta$. Then*

$$K[f \cap g] = K \cdot d \oplus \mathcal{F}_{p+q-3}.$$

If Char K *does not divide $p \cdot q$, this statement is true if one replaces d by the image J of the Jacobian determinant $\frac{\partial(f_p, g_q)}{\partial(X,Y)}$ in $K[f \cap g]$.*

The formula for $K[f \cap g]$ follows from 3.14 and B.6. For the last statement, observe that $L_{\mathcal{F}} J$ is the image of $\frac{\partial(f_p, g_q)}{\partial(X,Y)}$ in B.

Exercises

1. Let K be a field. Give a basis for the K-algebra

$$K[X, Y]/(f, g),$$

   where $f = X^4 - Y^4 + X$, $g = X^2Y^3 - X + 1$.
2. Determine the solutions in $\mathbb{P}^2(\mathbb{C})$ for the following systems of equations, and illustrate the situation with a sketch.
   (a)

$$X_1^2 + X_2^2 - X_0^2 = 0,$$
$$(X_1^2 + X_2^2)^3 - \lambda X_0^2 X_1^2 X_2^2 = 0 \qquad (\lambda \in \mathbb{C}).$$

   (b)

$$X_1 X_2 (X_1^2 - X_2^2) = 0,$$
$$(X_1 + 2X_2)((X_1 + X_2)^2 - X_0^2) = 0.$$

3. Let F and G be reduced curves in $\mathbb{P}^2(K)$, all of whose components are lines. For $P \in \mathbb{P}^2(K)$, let $m_P(F) = m$ if exactly m components of F contain the point P. Give a rule for the number of intersection points of F and G involving $m_P(F)$ and $m_P(G)$.
4. How many points at infinity do the curves f and g have in common, where

$$f = X^5 - X^3Y^2 + X^2Y^3 - 2XY^4 - 2Y^5 + XY$$
$$g = X^4 + X^3Y - X^2Y^2 - 2XY^3 - 2Y^4 + Y^2?$$

4

Rational Functions on Algebraic Curves

Besides the coordinate ring, the ring of rational functions on an algebraic curve is another invariant that can be used to study and to classify algebraic curves. This section uses Appendix C on rings of quotients and Appendix D on the Chinese remainder theorem.

The *field of rational functions* on $\mathbb{P}^2(K)$ is the set of all quotients

$$\frac{\phi}{\psi} \in K(X_0, X_1, X_2),$$

where ϕ, $\psi \in K[X_0, X_1, X_2]$ are relatively prime, homogeneous of the same degree, and $\psi \neq 0$. It is clear that these fractions form a subfield of $K(X_0, X_1, X_2)$. For a point $P = \langle x_0, x_1, x_2 \rangle \in \mathbb{P}^2(K)$ with $\psi(P) \neq 0$,

$$\frac{\phi(x_0, x_1, x_2)}{\psi(x_0, x_1, x_2)}$$

is independent of the choice of homogeneous coordinates for P. Thus $\frac{\phi}{\psi}$ gives in fact a function

$$r : \mathbb{P}^2(K) \setminus \mathcal{V}_+(\psi) \to K \qquad \left(P \mapsto \frac{\phi(P)}{\psi(P)}\right)$$

that vanishes on $\mathcal{V}_+(\phi)$. We call r a rational function on the projective plane, whose domain of definition, $\mathrm{Def}(r)$, is $\mathbb{P}^2(K) \setminus \mathcal{V}_+(\psi)$. We call ψ the *pole divisor* and ϕ the *zero divisor* of r. The difference $\phi - \psi$ in the divisor group $\mathcal{D}$ of $\mathbb{P}^2(K)$ is called the *principal divisor* belonging to r. The principal divisors of rational functions on $\mathbb{P}^2(K)$ form a subgroup $\mathcal{H}$ of $\mathcal{D}$. The residue class group $\mathrm{Cl}(\mathbb{P}^2) = \mathcal{D}/\mathcal{H}$ is called the *divisor class group* of $\mathbb{P}^2(K)$. Using the degree of a divisor, it is easy to show that

$$\mathrm{Cl}(\mathbb{P}^2) \cong (\mathbb{Z}, +).$$

We shall write $\mathcal{R}(\mathbb{P}^2)$ for the field of rational functions on $\mathbb{P}^2(K)$. The field K is embedded into $\mathcal{R}(\mathbb{P}^2)$ as the field of constant functions. Recall that for $\frac{\phi}{\psi} \in \mathcal{R}(\mathbb{P}^2)$ we assume ϕ and ψ to be relatively prime.

For $P \in \mathbb{P}^2(K)$, we denote by $\mathcal{O}_P \subset \mathcal{R}(\mathbb{P}^2)$ the subring of all rational functions whose domain of definition contains P. That is,

$$\mathcal{O}_P = \left\{ \frac{\phi}{\psi} \in \mathcal{R}(\mathbb{P}^2) \mid \psi(P) \neq 0 \right\}.$$

We call $\mathcal{O}_P$ the *local ring* of P on $\mathbb{P}^2(K)$. In the language of appendix C, the ring $\mathcal{O}_P$ is the homogeneous localization of $K[X_0, X_1, X_2]$ at

$$\mathfrak{p}_P := (\{\psi \in K[X_0, X_1, X_2] \mid \psi \text{ is homogeneous and } \psi(P) = 0\}),$$

the prime ideal corresponding to P:

$$\mathcal{O}_P = K[X_0, X_1, X_2]_{(\mathfrak{p}_P)}. \tag{1}$$

The maximal ideal of $\mathcal{O}_P$ is

$$\mathfrak{m}_P = \left\{ \frac{\phi}{\psi} \in \mathcal{O}_P \mid \phi(P) = 0 \right\}.$$

Let a coordinate transformation $c : \mathbb{P}^2(K) \to \mathbb{P}^2(K)$ be given by a matrix A. We will use the notation introduced in Chapter 2. From c we get a K-automorphism

$$\gamma : \mathcal{R}(\mathbb{P}^2) \to \mathcal{R}(\mathbb{P}^2) \qquad \left(\frac{\phi}{\psi} \mapsto \frac{\phi^A}{\psi^A} \right),$$

which one can describe as follows: Every rational function r on $\mathbb{P}^2(K)$ will be mapped to $r \circ c^{-1}$ by γ. In particular, γ induces, for each $P \in \mathbb{P}^2(K)$, a K-isomorphism

$$\gamma_P : \mathcal{O}_P \to \mathcal{O}_{c(P)}.$$

We come now to the "affine description" of the field $\mathcal{R}(\mathbb{P}^2)$ and the ring $\mathcal{O}_P$.

Lemma 4.1. *Dehomogenization gives a K-isomorphism*

$$\rho : \mathcal{R}(\mathbb{P}^2) \xrightarrow{\cong} K(X, Y) \qquad \left(\frac{\phi}{\psi} \mapsto \frac{\phi(1, X, Y)}{\psi(1, X, Y)} \right).$$

If $P = (a, b)$ is a point at finite distance and $\mathfrak{M}_P = (X - a, Y - b)$ its maximal ideal in $K[X, Y]$, then ρ induces a K-isomorphism

$$\mathcal{O}_P \xrightarrow{\cong} K[X, Y]_{\mathfrak{M}_P}$$

onto the localization of $K[X, Y]$ with respect to $\mathfrak{M}_P$.

Proof. The mapping ρ is well-defined and is a K-homomorphism by the rules for calculating with fractions. Since ρ is obviously injective, it needs only to be shown that ρ is surjective. For $f, g \in K[X, Y]$, $g \neq 0$, let ϕ and ψ be the homogenizations of f and g in $K[X_0, X_1, X_2]$. If $\deg \phi \leq \deg \psi$, then

$$\rho\left(\frac{X_0^{\deg \psi - \deg \phi} \cdot \phi}{\psi} \right) = \frac{f}{g},$$

and otherwise,

$$\rho\left(\frac{\phi}{X_0^{\deg\phi-\deg\psi}\cdot\psi}\right)=\frac{f}{g}.$$

Since $\frac{\phi}{\psi}\in\mathcal{O}_P$, we have $\psi(1,a,b)\neq 0$, hence $\psi(1,X,Y)\notin\mathfrak{M}_P$. Therefore it is clear that $\mathcal{O}_P$ is mapped onto $K[X,Y]_{\mathfrak{M}_P}$ by ρ.

The elements of $K(X,Y)$ can, in an obvious way, be thought of as functions on $\mathbb{A}^2(K)$, and ρ assigns to each rational function on $\mathbb{P}^2(K)$ its restriction to $\mathbb{A}^2(K)$. We call

$$\mathcal{R}(\mathbb{A}^2)=K(X,Y)$$

the *field of rational functions* on $\mathbb{A}^2(K)$. For $P\in\mathbb{A}^2(K)$,

$$\mathcal{O}'_P:=K[X,Y]_{\mathfrak{M}_P}$$

is the subring of all functions from $\mathcal{R}(\mathbb{A}^2)$ that are defined at P.

From a different point of view one can interpret a rational function $r=\frac{\phi}{\psi}\in\mathcal{R}(\mathbb{P}^2)$ as a function from $\mathbb{P}^2(K)\setminus(\mathcal{V}_+(\phi)\cap\mathcal{V}_+(\psi))$ to $\mathbb{P}^1(K)$ in which each point $P\in\mathbb{P}^2(K)\setminus(\mathcal{V}_+(\phi)\cap\mathcal{V}_+(\psi))$ is assigned to the point $\langle\psi(P),\phi(P)\rangle\in\mathbb{P}^1(K)$. We denote this mapping also by r. Since ϕ and ψ are relatively prime by hypothesis, $\mathcal{V}_+(\phi)\cap\mathcal{V}_+(\psi)$ is finite by 3.10, and therefore r is not defined only on a finite set, the *set of indeterminate points* of r. Furthermore, r is either constant or surjective, for if $\deg\phi=\deg\psi>0$, and $\langle a,b\rangle\in\mathbb{P}^1(K)$ with $a\neq 0$ and $b\neq 0$ is given, then the equation $a\phi-b\psi=0$ has a solution $P\notin\mathcal{V}_+(\phi)\cap\mathcal{V}_+(\psi)$ by 2.9 and 3.10, and it follows that $\langle a,b\rangle=\langle\psi(P),\phi(P)\rangle$. It is also clear that $\langle 1,0\rangle$ and $\langle 0,1\rangle$ belong to the image of r.

Let F be a curve in $\mathbb{P}^2(K)$ of positive degree. Then

$$\mathcal{O}_F:=\left\{\frac{\phi}{\psi}\in\mathcal{R}(\mathbb{P}^2)\mid F\text{ and }\psi\text{ are relatively prime}\right\}$$

is a subring of $\mathcal{R}(\mathbb{P}^2)$, and

$$I_F:=\left\{\frac{\phi}{\psi}\in\mathcal{O}_F\mid\phi\in(F)\right\}$$

is an ideal of $\mathcal{O}_F$. The ring $\mathcal{O}_F$ consists of precisely the rational functions that are defined on $\mathcal{V}_+(F)$ up to a finite set of exceptions, and I_F consists of the functions that vanish on $\mathcal{V}_+(F)$.

Definition 4.2. The residue class ring $\mathcal{R}(F):=\mathcal{O}_F/I_F$ is called the *ring of rational functions on* F.

Each residue class $\frac{\phi}{\psi}+I_F$ defines a function $\mathcal{V}_+(F)\setminus\mathcal{V}_+(\psi)\to K$ by restricting $\frac{\phi}{\psi}$ to $\mathcal{V}_+(F)$. Different representatives of the residue class agree with the function on the intersection of the domains of definition, and therefore each residue class defines a function on the union of the domains of definition, the "rational function associated with the residue class." Different residue classes can yield the same function. However, this does not occur for reduced curves, as the following lemma shows.

We call a subset F^* of $\mathcal{V}_+(F)$ *dense* in $\mathcal{V}_+(F)$ if F^* contains infinitely many points from each irreducible component of F. The domain of definition of a function in $\mathcal{R}(F)$ has finite complement in $\mathcal{V}_+(F)$, so in particular it is dense in $\mathcal{V}_+(F)$.

Lemma 4.3. *Let F be reduced. If r, $\bar{r} \in \mathcal{R}(\mathbb{P}^2)$ agree on a dense subset of $\mathcal{V}_+(F)$, then*

$$r|_{\mathcal{V}_+(F)\cap \mathrm{Def}(r)\cap \mathrm{Def}(\bar{r})} = \bar{r}|_{\mathcal{V}_+(F)\cap \mathrm{Def}(r)\cap \mathrm{Def}(\bar{r})}.$$

Proof. Let $r = \frac{\phi}{\psi}$, $\tilde{r} = \frac{\tilde{\phi}}{\tilde{\psi}}$. By hypothesis, $\phi\tilde{\psi} - \tilde{\phi}\psi$ vanishes on a dense subset of $\mathcal{V}_+(F)$. By 3.10, every irreducible factor of F must then be a divisor of $\phi\tilde{\psi} - \tilde{\phi}\psi$. Because F is reduced, F itself must divide $\phi\tilde{\psi} - \tilde{\phi}\psi$, and therefore there exists a homogeneous $A \in K[X_0, X_1, X_2]$ such that

$$\frac{\phi}{\psi} - \frac{\tilde{\phi}}{\tilde{\psi}} = \frac{AF}{\psi\tilde{\psi}} \in I_F.$$

Using this lemma, one can identify, for a reduced curve, an element of $\mathcal{R}(F)$ with the function defined by it. Similarly, one can consider such a function as a mapping to $\mathbb{P}^1(K)$, which is defined at all but finitely many points of $\mathcal{V}_+(F)$. If a function $r \in \mathcal{R}(F)$ is represented by $\frac{\phi}{\psi}$ and $\phi(P) \neq 0$, $\psi(P) = 0$, then $r(P)$ is the point at infinity $\langle 0, 1\rangle$ of $\mathbb{P}^1(K)$. The "poles" of r will be mapped to the point at infinity of the projective line, and only these poles will be mapped to the point at infinity. On the other hand, if P is an indeterminate point of $\frac{\phi}{\psi}$, then r does not assign any function value in $\mathbb{P}^1(K)$, but changing to another representative of the rational function may yield a function value.

The following theorem gives an affine description of the ring of rational functions of a curve F.

Theorem 4.4. *Suppose X_0 is not a component of F, and f is the affine curve associated with F; suppose $K[f]$ is its coordinate ring, and $Q(K[f])$ is the full ring of quotients of $K[f]$. Then there is a K-isomorphism*

$$\mathcal{R}(F) \cong Q(K[f]).$$

Proof. Dehomogenization $\frac{\phi}{\psi} \mapsto \frac{\phi(1,X,Y)}{\psi(1,X,Y)}$ defines a K-isomorphism from $\mathcal{O}_F$ to the ring of all quotients $\frac{\phi}{\psi} \in K(X,Y)$ such that ϕ, $\psi \in K[X,Y]$ and $\gcd(f,\phi) = 1$. This is the localization of $K[X,Y]$ at the set N of all elements ψ of nonzerodivisors mod (f). The above isomorphism maps the ideal I_F to the principal ideal (f), and hence there is an induced K-isomorphism

$$\mathcal{R}(F) = \mathcal{O}_F/I_F \xrightarrow{\cong} K[X,Y]_N/(f).$$

On account of the permutability of localization and quotient rings (C.8), there is also a K-isomorphism

$$K[X,Y]_N/(f) \xrightarrow{\cong} (K[X,Y]/(f))_{\bar{N}} = K[f]_{\bar{N}},$$

where $\bar{N}$ is the set of all nonzerodivisors of $K[f]$. This proves the theorem.

Again one can assign to an element of $Q(K[f])$ a function on $\mathcal{V}(f)$, namely the restriction of the corresponding function from $\mathcal{V}_+(F)$ to $\mathcal{V}(f)$. We therefore call $\mathcal{R}(f) := Q(K[f])$ the ring of *rational functions of* f. If f is reduced, one can even identify the elements of $\mathcal{R}(f)$ with their associated functions.

Since $K[f]$ is a subring of $\mathcal{R}(f)$, the elements of the affine coordinate ring $K[f]$ are in particular assigned to functions on $\mathcal{V}(f)$ that are defined on all of $\mathcal{V}(f)$. As an example, the residue classes x and y of X and Y in $K[f]$ are the "coordinate functions" that assign to each point $P \in \mathcal{V}(f)$ its X-coordinate, respectively its Y-coordinate. In contrast, the coordinate ring $K[F]$ of a projective curve F is not a ring of functions on $\mathcal{V}_+(F)$.

Let $f = c \cdot f_1^{\alpha_1} \cdot \dots \cdot f_h^{\alpha_h}$ ($c \in K^*$, $\alpha_i \in \mathbb{N}_+$) be the decomposition of f into irreducible factors f_i, and let $\bar{f}_i$ be the residue class of f_i in $K[f]$ ($i = 1, \dots, h$). Then the principal ideals $(\bar{f}_i)$ are all the minimal prime ideals of $K[f]$ (1.15) and we have $\bar{N} = K[f] \setminus \bigcup_{i=1}^h (\bar{f}_i)$. Thus $K[f]_{\bar{N}}$ is by C.9 a ring with only finitely many prime ideals $\mathfrak{p}_i = (\bar{f}_i)K[f]_{\bar{N}}$, and by the Chinese remainder theorem (D.3) it follows that $Q(K[f]) = K[f]_{\bar{N}}$ is the direct product of its localizations at these prime ideals. We have $(K[f]_{\bar{N}})_{\mathfrak{p}_i} \cong K[f]_{(\bar{f}_i)}$ by C.8. But then, however,

$$\begin{aligned} K[f]_{(\bar{f}_i)} &\cong K[X,Y]_{(f_i)}/(f)K[X,Y]_{(f_i)} \\ &= K[X,Y]_{(f_i)}/(f_i^{\alpha_i})K[X,Y]_{(f_i)} \cong Q(K[f_i^{\alpha_i}]), \end{aligned}$$

and hence there is a K-isomorphism

$$Q(K[f]) \cong Q(K[f_1^{\alpha_1}]) \times \dots \times Q(K[f_h^{\alpha_h}]).$$

Therefore, in the projective case we have the following.

Theorem 4.5. *If $F = cF_1^{\alpha_1} \cdots F_h^{\alpha_h}$ is the decomposition of F into irreducible factors, then there is a K-isomorphism*

$$\mathcal{R}(F) \cong \mathcal{R}(F_1^{\alpha_1}) \times \dots \times \mathcal{R}(F_h^{\alpha_h}).$$

If F is reduced, that is, if $\alpha_1 = \dots = \alpha_h = 1$, it is not hard to see that the above isomorphism assigns to each $r \in \mathcal{R}(F)$ the system $(r|_{F_1}, \dots, r|_{F_h})$ of restrictions to the irreducible components of F.

Corollary 4.6. *The curve F is irreducible if and only if $\mathcal{R}(F)$ is a field. This field is then K-isomorphic to $Q(K[f])$. If x and y denote the residue classes of X and Y in $K[f]$, then $\mathcal{R}(F) \cong K(x,y)$. In this case x (without loss of generality) is transcendental over K, and $\mathcal{R}(F)$ is a separable algebraic extension of $K(x)$.*

Proof. We need to establish only the last statement. We cannot have both x and y algebraic over K, for otherwise, f must divide the minimal polynomials for x and y over K. These are polynomials in which only X, respectively only Y, appears. Then f must be a constant, a contradiction.

Both partial derivatives $\frac{\partial f}{\partial X}$ and $\frac{\partial f}{\partial Y}$ cannot vanish, for otherwise, f would be a polynomial in X^p and Y^p, where $p := \mathrm{Char} K > 0$. Since K is algebraically closed, f would be a pth power, hence certainly not irreducible. So we can, without loss of generality, let $\frac{\partial f}{\partial Y} \neq 0$. Then x is transcendental over K, and y is separable algebraic over $K(x)$.

Let L be an extension field of K. If there is an element $x \in L$ that is transcendental over K, and L is finite algebraic over $K(x)$, we call L/K an *algebraic function field of one variable.* By the corollary, the function fields $\mathcal{R}(F)$ of irreducible curves are such fields. A theorem from field theory says that every algebraic function field of one variable over an algebraically closed field K is of the form $L = K(x,y)$, where x is transcendental over K and y is algebraic over K (a generalization of the theorem of the primitive element).

Theorem 4.7. *Every algebraic function field L/K is K-isomorphic to the field $\mathcal{R}(F)$ of rational functions of a suitably chosen irreducible algebraic curve F in $\mathbb{P}^2(K)$.*

Proof. Write $L = K(x,y)$ as indicated and consider the minimal polynomial $\phi \in K(x)[Y]$ of y over $K(x)$. Multiplying ϕ by a common denominator for the coefficients from $K(x)$, we get an irreducible polynomial $f \in K[X,Y]$, and we have $L \cong Q(K[X,Y]/(f))$. The homogenization of f gives the desired curve F.

In the situation of the theorem, one calls F a *plane projective model* for the function field L/K. One studies these field extensions as they arise as the function fields of projective algebraic curves. Conversely, these function fields lead to a classification of irreducible curves.

Definition 4.8.

(a) Two curves F and G in $\mathbb{P}^2(K)$ are called *birationally equivalent* if there is a K-isomorphism $\mathcal{R}(F) \cong \mathcal{R}(G)$.
(b) An irreducible curve F is called *rational* if there is a K-isomorphism $\mathcal{R}(F) \cong K(T)$ with the quotient field $K(T)$ of the polynomial ring $K[T]$.

These concepts extend analogously to affine curves. Of course, curves that arise through coordinate transformations are birationally equivalent, but birational equivalence is a weaker condition than projective equivalence. Later, we will give a more geometrical interpretation of birational equivalence.

One tries to classify curves up to birational equivalence, which for irreducible curves is equivalent with the classification of algebraic function fields of one variable up to isomorphism. Since projective lines are certainly rational, a curve is rational if and only if it is birationally equivalent to a line. We will later show that this is the case if and only if the curve has a "rational parametrization." (See 8.5.)

Example 4.9. *Irreducible quadrics are rational* (Char $K \neq 2$). We can assume after a coordinate transformation that we are dealing with the quadric $Q = X_1^2 + X_2^2 - X_0^2$. This is given affinely as $q := X^2 + Y^2 - 1$. In Chapter 1, Exercise 2, it was shown that the points of q are $(1, 0)$ and those given by

$$\left(\frac{2t}{t^2+1}, \frac{t^2-1}{t^2+1}\right) \qquad (t \in K, \quad t^2 + 1 \neq 0).$$

It follows immediately by substitution that q is contained in the kernel of the K-homomorphism

$$\alpha : K[X, Y] \to K(T) \qquad \left(\alpha(X) = \frac{2T}{T^2+1}, \quad \alpha(Y) = \frac{T^2-1}{T^2+1}\right),$$

and because q is irreducible, we must then have $\ker \alpha = (q)$. We then get an injective K-homomorphism $K[q] = K[X, Y]/(q) \hookrightarrow K(T)$, and therefore also a K-homomorphism $\mathcal{R}(Q) = Q(K[q]) \hookrightarrow K(T)$. Since

$$\frac{\alpha(Y)+1}{\alpha(X)} = \left(\frac{T^2-1}{T^2+1} + 1\right) \frac{T^2+1}{2T} = T,$$

the homomorphism is also surjective and is therefore an isomorphism.

Exercises

1. Suppose an irreducible affine algebraic curve is given by an equation $f_{n+1} + f_n = 0$, where $f_i \in K[X, Y]$ is homogeneous of degree i ($i = n, n+1$; $n \in \mathbb{N}$). Show that the curve is rational. (This shows the rationality of some of the curves in Figures 1.6–1.15).
2. Let f be the lemniscate with equation

$$(X^2 + Y^2)^2 = \alpha(X^2 - Y^2) \qquad (\alpha \in K^*)$$

   and let $x, y \in K[f]$ be the associated coordinate functions. Prove the rationality of f by showing that $\mathcal{R}(f) = K(t)$ with $t := \frac{x^2+y^2}{x-y}$.

5

Intersection Multiplicity and Intersection Cycle of Two Curves

Two projective curves of degrees p and q that have no common components intersect in at least one and at most pq points (3.10).

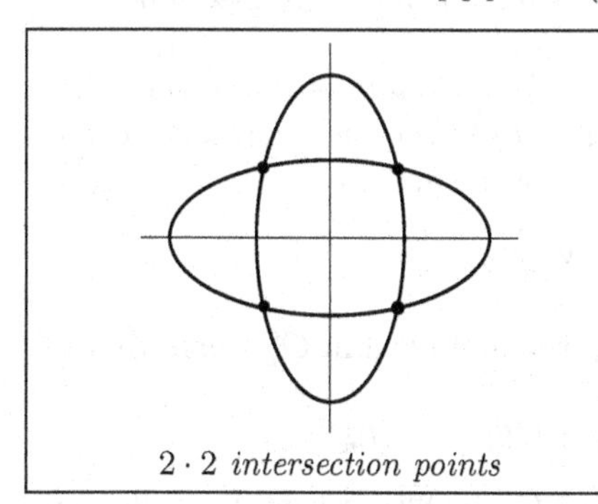

$2 \cdot 2$ *intersection points*

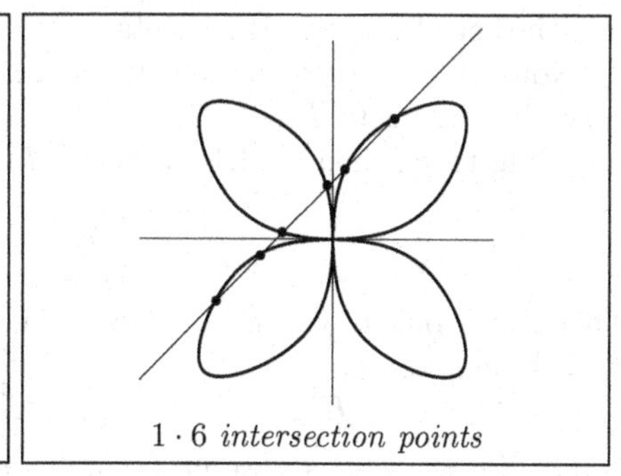

$1 \cdot 6$ *intersection points*

When the number of intersection points is less than that maximum number pq, it is because certain intersection points "coincide," or that the curves intersect in some points "to a higher order":

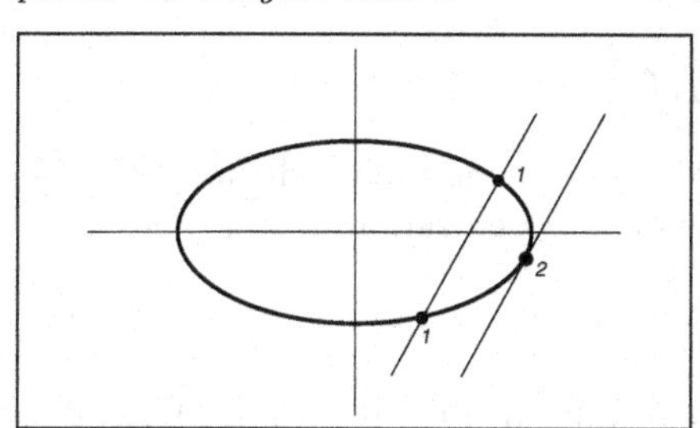

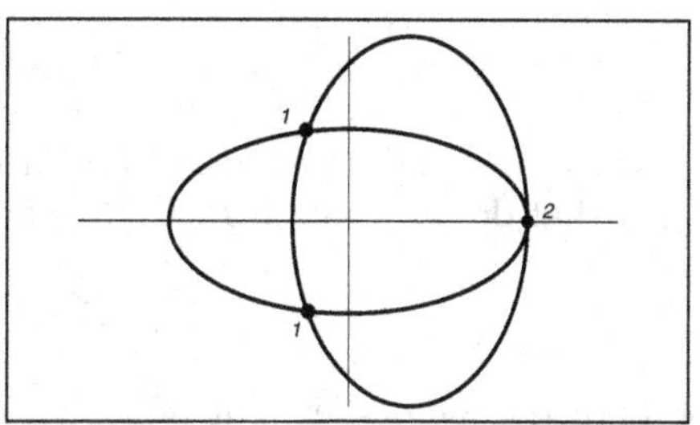

We will see in this chapter that one can assign a "multiplicity" to the intersection points of two curves in such a way that when one counts the points "with multiplicity," the number of intersection points is exactly pq. But to do that requires some preparations.

Let $\mathcal{O}_P$ be the local ring of a point P in $\mathbb{P}^2(K)$ and let $\mathfrak{m}_P$ be its maximal ideal. For a curve F in $\mathbb{P}^2(K)$ we call

$$I(F)_P := \left\{ \frac{\phi}{\psi} \in \mathcal{O}_P \mid \phi \in (F) \right\}$$

the *ideal of F in $\mathcal{O}_P$*. For curves $F_1, \ldots, F_m$ in $\mathbb{P}^2(K)$ we set

$$\mathcal{O}_{F_1\cap\cdots\cap F_m,P} := \mathcal{O}_P/(I(F_1)_P + \cdots + I(F_m)_P).$$

If $P \notin \mathcal{V}_+(F_1)\cap\cdots\cap\mathcal{V}_+(F_m)$, then this is the zero ring, because at least one of the ideals $I(F_j)_P$ equals $\mathcal{O}_P$. Conversely, if $P \in \mathcal{V}_+(F_1)\cap\cdots\cap\mathcal{V}_+(F_m)$, then $\mathcal{O}_{F_1\cap\cdots\cap F_m,P} \neq 0$. The set of points of $\mathcal{V}_+(F_1)\cap\cdots\cap\mathcal{V}_+(F_m)$ together with the local rings $\mathcal{O}_{F_1\cap\cdots\cap F_m,P}$ is called the *intersection scheme* $F_1\cap\cdots\cap F_m$ of the curves $F_1,\ldots,F_m$, and $\mathcal{O}_{F_1\cap\cdots\cap F_m,P}$ is called the *local ring of the point* P *on the intersection scheme*. In particular, this also defines the local ring $\mathcal{O}_{F,P}$ of a point P on a curve F. For $P \in \mathcal{V}_+(F_1)\cap\cdots\cap\mathcal{V}_+(F_m)$ we think of the local ring as being "attached" at the point P. The intersection scheme contains much more information about the behavior of the intersections than just the set of intersection points.

Now let $P = (a,b)$ be a point at finite distance, and let f_j be the affine curve belonging to F_j. If $\mathfrak{M}_P = (X-a, Y-b)$ is the maximal ideal corresponding to P, then by 4.1 there is a K-isomorphism

$$\mathcal{O}_P \xrightarrow{\sim} \mathcal{O}'_P := K[X,Y]_{\mathfrak{M}_P}.$$

This isomorphism will map $I(F_j)_P$ to the principal ideal in $\mathcal{O}'_P$ generated by f_j. Setting

$$K[f_1\cap\cdots\cap f_m] := K[X,Y]/(f_1,\ldots,f_m)$$

and denoting the image of $\mathfrak{M}_P$ in $K[f_1\cap\cdots\cap f_m]$ by $\overline{\mathfrak{M}}_P$, we get the following due to the commutativity of rings of quotients and residue class rings (C.8).

Theorem 5.1. *There is a* K*-isomorphism*

$$\mathcal{O}_{F_1\cap\cdots\cap F_m,P} \cong K[f_1\cap\cdots\cap f_m]_{\overline{\mathfrak{M}}_P}.$$

Let F_j^* be the curve obtained from F_j by omitting the irreducible components that do not contain P. If f_j^* is the dehomogenization of F_j^*, then

$$f_j\,\mathcal{O}'_P = f_j^*\,\mathcal{O}'_P,$$

because the factors left out of f_j become units in $\mathcal{O}'_P$. By 5.1 we have the following.

Corollary 5.2.

$$\mathcal{O}_{F_1\cap\cdots\cap F_m,P} \cong \mathcal{O}_{F_1^*\cap\cdots\cap F_m^*,P}.$$

Hence we get the following finiteness theorem.

Theorem 5.3. *Let* $m \geq 2$ *and suppose two of the curves from* $\{F_1,\ldots,F_m\}$ *do not have irreducible components in common containing* P*. Then*

$$\dim_K \mathcal{O}_{F_1\cap\cdots\cap F_m,P} < \infty.$$

Proof. We can assume that F_1 and F_2 do not have irreducible components in common containing P. Since $\mathcal{O}_{F_1\cap\cdots\cap F_m,P}$ is a homomorphic image of $\mathcal{O}_{F_1\cap F_2,P}$, it is enough to show that $\dim_K \mathcal{O}_{F_1\cap F_2,P} < \infty$. By 5.2 we can assume that F_1 and F_2 have no components in common at all. Then, however, $K[f_1 \cap f_2]$ is a finite-dimensional K-algebra by 1.4(b). By the Chinese remainder theorem D.4, we have that $K[f_1\cap f_2]_{\overline{\mathfrak{M}}_P}$ is a direct factor of $K[f_1 \cap f_2]$. Therefore, we also have that $\mathcal{O}_{F_1\cap F_2,P} \cong K[f_1 \cap f_2]_{\overline{\mathfrak{M}}_P}$ is finite-dimensional over K.

The remark below follows immediately from the definition of $\mathcal{O}_{F_1\cap\cdots\cap F_m,P}$.

Remark 5.4. (a) $\dim_K \mathcal{O}_{F_1\cap\cdots\cap F_m,P} = 0$ if and only if

$$P \notin \mathcal{V}_+(F_1) \cap \cdots \cap \mathcal{V}_+(F_m).$$

(b) $\dim_K \mathcal{O}_{F_1\cap\cdots\cap F_m,P} = 1$ if and only if

$$\mathfrak{m}_P = I(F_1)_P + \cdots + I(F_m)_P.$$

Definition 5.5. For two curves F_1, F_2 in $\mathbb{P}^2(K)$ we call

$$\mu_P(F_1, F_2) := \dim_K \mathcal{O}_{F_1\cap F_2,P}$$

the *intersection multiplicity* of F_1 and F_2 at the point P. If $\mu_P(F_1, F_2) = \mu$, we say that P is a *μ-fold point* of $F_1 \cap F_2$. The intersection multiplicity $\mu_P(f_1, f_2)$ of two affine curves f_1, f_2 is defined analogously.

This definition is somewhat abstract, but we will see by and by that this concept does possess the geometric properties that we want. A major advantage of the definition lies in the fact that it is very easy to see the independence of the choice of coordinates. A coordinate transformation $c : \mathbb{P}^2(K) \to \mathbb{P}^2(K)$ given by a matrix A induces a K-isomorphism

$$\mathcal{O}_P \xrightarrow{\sim} \mathcal{O}_{c(P)} \qquad \left(\frac{\phi}{\psi} \mapsto \frac{\phi^A}{\psi^A}\right)$$

with the property that for each curve F the ideal $I(F)_P$ is mapped to $I(c(F))_{c(P)}$. Therefore, $\mathcal{O}_{F_1\cap\cdots\cap F_m,P}$ and $\mathcal{O}_{c(F_1)\cap\cdots\cap c(F_m),c(P)}$ are K-isomorphic, and so in particular, they have the same dimension.

By 5.4(a) we have $\mu_P(F_1, F_2) = 0$ precisely when $P \notin \mathcal{V}_+(F_1) \cap \mathcal{V}_+(F_2)$. Furthermore, $\mu_P(F_1, F_2) = 1$ precisely when

$$\mathfrak{m}_P = I(F_1)_P + I(F_2)_P.$$

We will see later (see 7.6 and 7.7) that this condition is satisfied if and only if F_1 and F_2 intersect "transversally" at P:

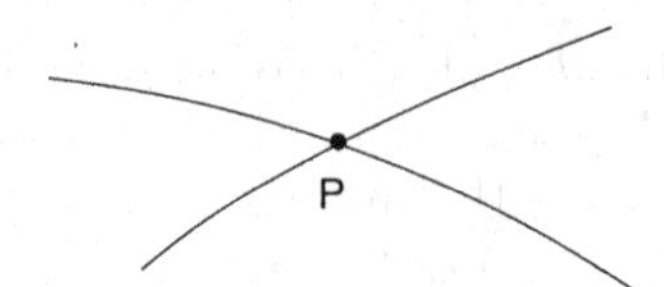

By 5.3, we have $\mu_P(F_1, F_2) < \infty$ if F_1 and F_2 do not have a common irreducible component containing P. On the other hand, if they do possess such a common component F, then $I(F_1)_P + I(F_2)_P \subset I(F)_P$, and $\mathcal{O}_{F_1 \cap F_2, P}$ has $\mathcal{O}_{F,P}$ as a homomorphic image. This is an integral domain with quotient field $\mathcal{R}(F)$, and so it cannot be finite-dimensional as a K-algebra. Therefore, $\mu_P(F_1, F_2) = \infty$.

To study global questions about the intersections of two projective curves it is convenient to introduce the following concept.

Definition 5.6. A *cycle* Z in $\mathbb{P}^2(K)$ is an element of the free abelian group on the set of all points of $\mathbb{P}^2(K)$:

$$Z = \sum_{P \in \mathbb{P}^2(K)} m_P \cdot P \qquad (m_P \in \mathbb{Z},\ m_P \neq 0 \text{ for only finitely many } P).$$

We define $\deg Z := \sum m_P$, the *degree of* Z. For curves F_1, F_2 in $\mathbb{P}^2(K)$ that have no common components, we call

$$F_1 \star F_2 = \sum_{P \in \mathbb{P}^2(K)} \mu_P(F_1, F_2) \cdot P$$

the *intersection cycle* of F_1 and F_2.

Here the notation $F_1 \star F_2$ has nothing to do with the product of the two homogeneous polynomials. The intersection cycle describes the intersection of F_1 and F_2 by indicating the points of $\mathcal{V}_+(F_1) \cap \mathcal{V}_+(F_2)$ and their associated intersection multiplicities. It contains less information than the intersection scheme $F_1 \cap F_2$, but more than $\mathcal{V}_+(F_1) \cap \mathcal{V}_+(F_2)$. We have now arrived at the main theorem of this chapter, whose proof is quite easy by the above.

Bézout's Theorem 5.7. *For two curves F_1 and F_2 in $\mathbb{P}^2(K)$ with no common component we have*

$$\deg(F_1 \star F_2) = \deg F_1 \cdot \deg F_2.$$

Two curves F_1, F_2 in $\mathbb{P}^2(K)$ always intersect in $\deg F_1 \cdot \deg F_2$ points if F_1 and F_2 have no common component and their intersection points are counted with the appropriate intersection multiplicities.

Proof. We can assume that

$$\mathcal{V}_+(F_1) \cap \mathcal{V}_+(F_2) = \{P_1, \ldots, P_r\}$$

has only points at finite distance. If f_1 and f_2 are the corresponding affine curves, then by 3.9(a) we have $\dim_K K[f_1 \cap f_2] = p \cdot q$. Moreover, $\operatorname{Max} K[f_1 \cap f_2] = \{\overline{\mathfrak{M}}_{P_1}, \dots, \overline{\mathfrak{M}}_{P_r}\}$. Therefore, by the Chinese remainder theorem,

$$K[f_1 \cap f_2] = K[f_1 \cap f_2]_{\overline{\mathfrak{M}}_{P_1}} \times \cdots \times K[f_1 \cap f_2]_{\overline{\mathfrak{M}}_{P_r}}, \tag{1}$$

and hence

$$\deg F_1 \cdot \deg F_2 = p \cdot q = \sum_{i=1}^{r} \mu_{P_i}(F_1, F_2) = \deg(F_1 \star F_2).$$

Next we show the *additivity of intersection multiplicities and intersection cycles.*

Theorem 5.8. *Let F, G, and H be curves in $\mathbb{P}^2(K)$. Denote by $F + G$ the sum of the divisors F and G, i.e., the curve corresponding to $F \cdot G$. If $F + G$ and H have no common component containing P, then*

$$\mu_P(F + G, H) = \mu_P(F, H) + \mu_P(G, H).$$

Proof. By 5.2 we can assume that $F + G$ and H have no components in common at all. Let $P = (a, b)$ be an affine point and let f, g, h be the affine curves corresponding to F, G, H. Then $f \cdot g$ is a polynomial in $K[X, Y]$ corresponding to the divisor $F + G$. We denote the residue class of f, respectively g, in $K[h] = K[X, Y]/(h)$ by ϕ and ψ. Since $f \cdot g$ and h are relatively prime, ϕ and ψ are not zero divisors in $K[h]$ and also $\phi \cdot \psi$ is not a zero divisor. We will denote by $\mathfrak{m}_P$ the maximal ideal in $K[h]$ corresponding to P and set $R := K[h]_{\mathfrak{m}_P}$. Because of the commutativity of the formation of rings of quotients and residue classes (C.8), there are K-isomorphisms

$$\mathcal{O}_P / I(F + G)_P + I(H)_P \cong R/(\phi \cdot \psi),$$

$$\mathcal{O}_P / I(F)_P + I(H)_P \cong R/(\phi), \qquad \mathcal{O}_P / I(G)_P + I(H)_P \cong R/(\psi).$$

The statement of the theorem then follows from the definition of intersection multiplicity and by the following lemma.

Lemma 5.9. *Let R be a K-algebra and let ϕ, $\psi \in R$. If ϕ is a nonzerodivisor on R and if $\dim_K R/(\phi \cdot \psi) < \infty$, then*

$$\dim_K R/(\phi \cdot \psi) = \dim_K R/(\phi) + \dim_K R/(\psi).$$

Proof. We have $(\phi \cdot \psi) \subset (\phi) \subset R$ and hence

$$\dim_K R/(\phi \cdot \psi) = \dim_K R/(\phi) + \dim_K (\phi)/(\phi \cdot \psi).$$

Since ϕ is a nonzerodivisor on R, multiplication by ϕ yields a K-isomorphism $R \xrightarrow{\sim} (\phi)$, under which (ψ) is mapped to $(\phi \cdot \psi)$. Therefore,

$$\dim_K (\phi)/(\phi \cdot \psi) = \dim_K R/(\psi).$$

Corollary 5.10. *Suppose $F+G$ and H have no common components. Then*

$$(F+G)\star H = F\star H + G\star H.$$

Theorem 5.8 implies in particular that $\mu_P(F,H) > 1$ if a multiple component containing P appears in F (or in G).

Corollary 5.11. *Under the assumptions of Bézout's theorem, suppose that $\mathcal{V}_+(F_1)\cap\mathcal{V}_+(F_2)$ is a set of points at finite distance, and let f_1, f_2 be the affine curves corresponding to F_1, F_2. Then the following are equivalent:*

(a) *$\mathcal{V}_+(F_1)\cap\mathcal{V}_+(F_2)$ consists of $\rho := \deg F_1\cdot\deg F_2$ distinct points.*
(b) *F_1 and F_2 are reduced curves and $K[f_1\cap f_2]$ is a direct product of ρ copies of the field K:*

$$K[f_1\cap f_2] = \prod_{i=1}^{\rho} K.$$

Proof. By Bézout's theorem, we have (a) precisely when $\mu_P(F_1,F_2)=1$ for all $P\in\mathcal{V}_+(F_1)\cap\mathcal{V}_+(F_2)$, i.e., when $\mathcal{O}_{F_1\cap F_2,P}\cong K$. By the above remark, F_1 and F_2 are reduced. Since $\mathcal{V}_+(F_1)\cap\mathcal{V}_+(F_2)=\mathcal{V}_+(f_1)\cap\mathcal{V}_+(f_2)$, it follows from (1) that condition (a) holds if and only if $K[f_1\cap f_2]=\prod_{i=1}^{\rho}K$.

It will later turn out that the conclusions of the corollary occur exactly when F_1 and F_2 are reduced curves that "intersect transversally everywhere," as was suggested by the figures at the beginning of this chapter.

Many theorems about algebraic curves deal with the *interpolation problem*: An algebraic curve of a certain degree is to pass through some given points in the plane, where the behavior of the curve at the points (maybe its direction) is prescribed. We consider here the case in which the given points are just the intersection points of two curves F_1 and F_2 without common components.

Definition 5.12. Let G be another curve. We say that $F_1\cap F_2$ is a *subscheme* of G if

$$\dim_K \mathcal{O}_{F_1\cap F_2\cap G,P} = \mu_P(F_1,F_2) \tag{2}$$

for all points $P\in\mathcal{V}_+(F_1)\cap\mathcal{V}_+(F_2)$.

Condition (2) is equivalent to the condition that for all $P\in\mathcal{V}_+(F_1)\cap\mathcal{V}_+(F_2)$,

$$I(G)_P\subset I(F_1)_P+I(F_2)_P,\ \text{and thus}\ \mathcal{O}_{F_1\cap F_2,P}=\mathcal{O}_{F_1\cap F_2\cap G,P}. \tag{3}$$

If $\mu_P(F_1,F_2)=1$, this is the same as saying that $P\in V_+(G)$, and therefore that G "goes through" P.

Lemma 5.13. *Suppose $\mathcal{V}_+(F_1)\cap\mathcal{V}_+(F_2)$ consists of points at finite distance, and f_1, f_2, and g are the polynomials in $K[X,Y]$ corresponding to F_1, F_2, and G. Then $F_1\cap F_2$ is a subscheme of G if and only if $g\in(f_1,f_2)$.*

Proof. Let ψ be the image of g in $K[f_1 \cap f_2]$. Condition (3) is then equivalent to saying that the image of ψ in $K[f_1 \cap f_2]_{\overline{\mathfrak{m}}_P}$ vanishes, where $\overline{\mathfrak{m}}_P$ is the maximal ideal in $K[f_1 \cap f_2]$ corresponding to P. By the Chinese remainder theorem (1) it follows that (3) holds for all $P \in \mathcal{V}_+(F_1) \cap \mathcal{V}_+(F_2)$ if and only if $\psi = 0$, i.e., $g \in (f_1, f_2)$.

Fundamental Theorem of Max Noether 5.14. *The following are equivalent:*

(a) *$F_1 \cap F_2$ is a subscheme of G.*
(b) *$G \in (F_1, F_2)$.*
(c) *There are homogeneous polynomials $A, B \in K[X_0, X_1, X_2]$ with $\deg A = \deg G - \deg F_1$ and $\deg B = \deg G - \deg F_2$ such that*

$$G = A \cdot F_1 + B \cdot F_2.$$

Proof. We can assume that we are in the situation of the lemma. If $G \in (F_1, F_2)$, then by dehomogenization, $g \in (f_1, f_2)$. Now suppose conversely that this condition is satisfied. Write $g = af_1 + bf_2$ with $a, b \in K[X, Y]$. Then this yields by homogenization an equation of the form

$$X_0^\nu G = A \cdot F_1 + B \cdot F_2,$$

where $\nu \in \mathbb{N}$ and $A, B \in K[X_0, X_1, X_2]$ are homogeneous polynomials. By 3.8(b) X_0 is not a zerodivisor modulo (F_1, F_2). Therefore, it follows that $G \in (F_1, F_2)$. Hence by Lemma 5.13, statements (a) and (c) of the theorem are equivalent. Now (b) $\Rightarrow$ (c) follows, because the polynomials F_1, F_2, and G are homogeneous, and (c) $\Rightarrow$ (b) is trivial.

Corollary 5.15. *Suppose G and F_1 have no common components. If $F_1 \cap F_2$ is a subscheme of G, then there is a curve H satisfying $\deg H = \deg G - \deg F_2$ and such that*

$$G \star F_1 = H \star F_1 + F_2 \star F_1.$$

Proof. Choose an equation as in 5.14(c) and set $H := B$. We may assume that we are in the situation of Lemma 5.13 and denote by a and h the dehomogenizations of A and H. Then

$$\begin{aligned}\mu_P(G, F_1) &= \dim_K \mathcal{O}'_P/(g, f_1) = \dim_K \mathcal{O}'_P/(af_1 + hf_2, f_1)\\ &= \dim_K \mathcal{O}'_P/(hf_2, f_1) = \mu_P(H + F_2, F_1) = \mu_P(H, F_1) + \mu_P(F_2, F_1)\end{aligned}$$

for all $P \in \mathbb{P}^2(K)$, and the statement follows.

Examples 5.16.

(a) In the situation of 5.14, suppose $\mathcal{V}_+(F_1) \cap \mathcal{V}_+(F_2)$ consists of $\deg F_1 \cdot \deg F_2$ different points. The curve G passes through all of these points if and only if $G \in (F_1, F_2)$.

(b) Suppose $F_1 \cap F_2$ is a subscheme of a curve G and $Z := \mathcal{V}_+(F_1) \cap \mathcal{V}_+(G)$ consists of $\deg F_1 \cdot \deg G$ different points. Then there is a curve H with $\deg H = \deg G - \deg F_2$ that passes through the points of Z not in $\mathcal{V}_+(F_1) \cap \mathcal{V}_+(F_2)$.

(c) Suppose two cubic curves intersect in exactly 9 points, and 6 of these lie on a quadric. Then the remaining 3 intersection points lie on a line.

(d) *Pascal's theorem* ($\sim$ 1639) Suppose the two cubic curves in (c) are each the union of three different lines. Then we have the situation shown in the following figure:

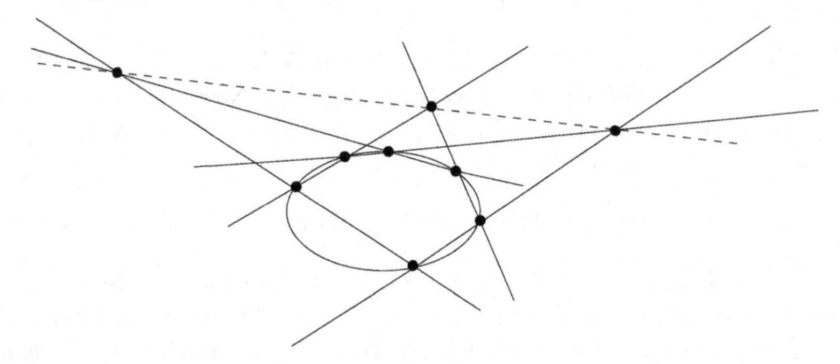

It follows that the three points not lying on the quadric lie on a line.

Usually, Pascal's theorem is formulated by saying that if one chooses 6 points on a quadric so that there is a hexagon inscribed in the quadric, as in the figure, then the opposite sides meet in collinear points. In the special case that the quadric is the union of two lines, Pascal's theorem was known in ancient times (Pappus's theorem). See the following figure.

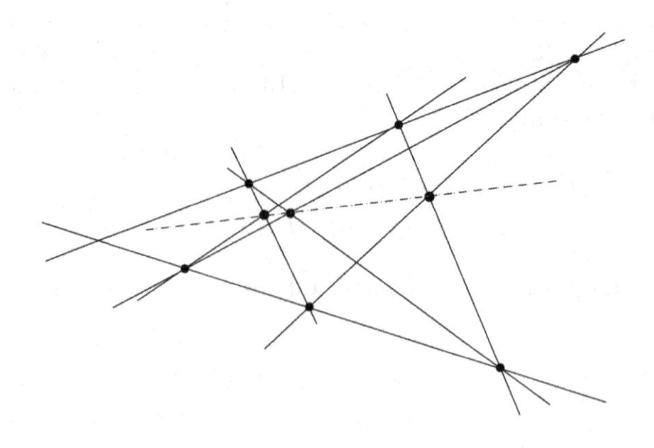

More effort is needed in order to get the following theorem. Our method of proof rests on the results about filtered algebras.

Cayley–Bacharach Theorem 5.17. *Suppose two curves F_1, F_2 in $\mathbb{P}^2(K)$ have no common components. Let $\deg F_1 =: p$, $\deg F_2 =: q$, and let G be another curve with $\deg G =: h < p+q-2$. If*

$$(4) \qquad \sum_{P\in\mathcal{V}_+(F_1)\cap\mathcal{V}_+(F_2)} \dim_K \mathcal{O}_{F_1\cap F_2\cap G,P} \geq (p-1)(q-1)+h+1,$$

then $F_1 \cap F_2$ is a subscheme of G.

Proof. We will use a result from Chapter 3 on the structure of the coordinate ring of the intersections of two curves. As usual, we assume that there are no points at infinity in $\mathcal{V}_+(F_1)\cap\mathcal{V}_+(F_2)$, and that f_1, f_2, and g are the polynomials in $K[X,Y]$ corresponding to F_1, F_2, and G. We set

$$A := K[f_1 \cap f_2] = K[X,Y]/(f_1,f_2)$$

and

$$B := \operatorname{gr}_{\mathcal{F}} A = K[X,Y]/(L_{\mathcal{F}}f_1, L_{\mathcal{F}}f_2),$$

where $\mathcal{F}$ denotes both the degree filtration on $K[X,Y]$ and the induced filtration on the ring $K[f_1\cap f_2]$.

Let γ be the image of g in A. By 5.13, we must show that γ vanishes, under the assumptions of the theorem. Suppose it were the case that $\gamma \neq 0$. Then also $\gamma^0 := L_{\mathcal{F}}\gamma \neq 0$, and this is a homogeneous element of B of degree $< p+q-2$.

By 3.14, B_{p+q-2} is the socle of B. If ξ and η are the images of X and Y in B, then $(\xi,\eta)\cdot\gamma^0 \neq 0$, since $\gamma \notin B_{p+q-2}$. There is thus an element $\alpha_1 \in B_1$ with $\alpha_1\cdot\gamma^0 \neq 0$. By induction we get the following: For $i=0,\dots,p+q-2-h$, there are elements $\alpha_i \in B_i$ ($\alpha_0 = 1$) with $\alpha_i \cdot \gamma^0 \neq 0$. Since these elements have different degrees, they are linearly independent over K. Hence, $\dim_K(\gamma) \geq p+q-1-h$ and $\dim_K A/(\gamma) \leq p\cdot q-(p+q-1-h) = (p-1)(q-1)+h$.

We now apply the Chinese remainder theorem (1), and considering the images of γ in the local rings $K[f_1\cap f_2]_{\overline{\mathfrak{M}}_P}$, we see that the hypothesis (4) of the theorem says that $\dim_K A/(\gamma) \geq (p-1)(q-1)+h+1$. This contradiction shows that we must have $\gamma = 0$.

Examples 5.18. (a) Under the hypotheses of 5.17 suppose $\mathcal{V}_+(F_1)\cap\mathcal{V}_+(F_2)$ consists of $p\cdot q$ different points and $\mathcal{V}_+(G)$ contains $(p-1)(q-1)+h+1$ of these. Then $\mathcal{V}_+(G)$ contains all $p\cdot q$ points of $\mathcal{V}_+(F_1)\cap\mathcal{V}_+(F_2)$.

We recall how we can understand these statements as theorems about systems of algebraic equations: Suppose we are given a system of equations

$$(5) \qquad \begin{aligned} F_1(X_0,X_1,X_2) &= 0,\\ F_2(X_0,X_1,X_2) &= 0,\\ G(X_0,X_1,X_2) &= 0. \end{aligned}$$

Let F_1, F_2, and G be homogeneous of degrees p, q, and $h < p+q-2$. Suppose the system $F_1 = F_2 = 0$ has exactly $p \cdot q$ different solutions P_i in $\mathbb{P}^2(K)$, and $(p-1)(q-1)+h+1$ of these are also solutions of the equation $G = 0$. Then all the P_i are solutions of (5).

(b) Let $p = q = h = n \geq 3$. Then the condition $h < p+q-2$ is satisfied. If $\mathcal{V}_+(F_1) \cap \mathcal{V}_+(F_2)$ consists of n^2 different points and if $(n-1)^2+n+1 = n(n-1)+2$ of these are contained in $\mathcal{V}_+(G)$, then all n^2 points are contained in $\mathcal{V}_+(G)$.

(c) If two cubic curves intersect in 9 different points and another cubic curve contains 8 of these intersection points, then it contains all nine. This is the special case of (b) in which $n = 3$. One can also deduce 5.16(c) and Pascal's theorem from this.

Exercises

1. Let A be an algebra over a field K with the following properties:
   (a) A is a noetherian ring.
   (b) A has exactly one prime ideal $\mathfrak{m}$ (and is therefore local).
   (c) The composite map $K \to A \twoheadrightarrow A/\mathfrak{m}$ is bijective.
   Show that $\dim_K A < \infty$. (Use the fact that $\mathfrak{m}$ is finitely generated and by C.12 consists of nilpotent elements of A.)
2. Let K be an algebraically closed field. A *0-dimensional subscheme* of $\mathbb{A}^2(K)$ is a system $Z = (P_1, \ldots, P_t; A_1, \ldots, A_t)$, where $P_1, \ldots, P_t$ are distinct points in $\mathbb{A}^2(K)$ and $A_1, \ldots, A_t$ are K-algebras with the properties (a)–(c) of Exercise 1, and where the maximal ideal of each A_i is generated by (at most) two elements. We call

$$A := A_1 \times \cdots \times A_t$$

   the *affine algebra* of Z. Also, we call $\sum_{i=1}^t \dim_K A_i \cdot P_i$ the *cycle of* Z and $\dim_K A$ the *degree of* Z.
   (a) Show that there is a K-isomorphism

$$A \cong K[X,Y]/I,$$

   where I is an ideal of the polynomial ring $K[X,Y]$.
   (b) Conversely, assign a 0-dimensional subscheme of $\mathbb{A}^2(K)$ to each finite-dimensional K-algebra of the form $K[X,Y]/I$.
3. Let $F, G \in \mathbb{R}[X_0, X_1, X_2]$ be homogeneous polynomials. For a point $P = \langle x_0, x_1, x_2 \rangle \in \mathbb{P}^2(\mathbb{C})$, denote by $\overline{P} := \langle \overline{x}_0, \overline{x}_1, \overline{x}_2 \rangle$ the complex conjugate point of P ($\overline{x}_i$ is the conjugate of the complex number x_i).
   (a) Show that $\mu_P(F,G) = \mu_{\overline{P}}(F,G)$.
   (b) Conclude that if F and G have odd degree, then the system of equations

$$F(X_0, X_1, X_2) = 0, \qquad G(X_0, X_1, X_2) = 0,$$

   has a solution in $\mathbb{P}^2(\mathbb{R})$.

4. A theorem of Newton (1704): An affine curve f of degree d will intersect a line g in d points $P_1, \ldots, P_d$, where these points can coincide if the intersection multiplicity is > 1. Suppose d does not divide the characteristic of K. The *centroid* of $f \cap g$ is then the point $P^{(g)} := \frac{1}{d}\sum_{i=1}^{d} P_i$, where the sum is constructed using the vector addition of K^2. Show that if a band of parallel lines passes through g, then all the centroids $P^{(g)}$ lie on a line (referred to by Newton as a "diameter" of f). (Hint: One can assume that g is given by $Y = 0$. Write $f = \phi_0 X^d + \phi_1 X^{d-1} + \cdots + \phi_d \quad (\phi_i \in K[Y])$. Consider ϕ_0 and, most importantly, ϕ_1.)
5. A theorem of Maclaurin (1748): Under the hypotheses of Exercise 4, suppose $(0,0) \notin \mathrm{Supp}(f)$. Let g be a line through $(0,0)$ and $P_i = (x_i, y_i)$ $(i = 1, \ldots, d)$. The *harmonic center* of $f \cap g$ is the point

$$H^{(g)} = (x^{(g)}, y^{(g)}) \text{ with } x^{(g)} := d\left(\sum_{i=1}^{d} x_i^{-1}\right)^{-1}, \quad y^{(g)} := d\left(\sum_{i=1}^{d} y_i^{-1}\right)^{-1}$$

if g is not the X-axis or the Y-axis (otherwise, $H^{(g)} = (d(\sum x_i^{-1})^{-1}, 0)$, respectively $H^{(g)} = (0, d(\sum y_i^{-1})^{-1})$). Show that if g runs over all lines through the origin that intersect f in d affine points, then the points $H^{(g)}$ lie on a line. (Hint: For this theorem consider ϕ_{d-1} and ϕ_d.)
6. Determine $\mu_P(f, g)$ if

$$f = (X^2 + Y^2)^3 - 4X^2Y^2,$$
$$g = (X^2 + Y^2)^3 - X^2Y^2,$$

and if $P = (0,0)$.

7. Prove the *converse of Pascal's theorem*: Suppose the intersection points of the 3 pairs of opposite sides of a hexagon lie on a line. Then the vertices of the hexagon lie on a quadric.

6

Regular and Singular Points of Algebraic Curves. Tangents

A point on an algebraic curve is either "simple" or "singular." At a simple point the curve is "smooth." In general, a point on a curve is assigned a "multiplicity" that indicates how many times it has to be counted as a point of the curve. The "tangents" of a curve will also be explained. One can decide whether a point is simple or singular with the help of the local ring at the point. The facts from Appendix E on Noetherian rings and discrete valuation rings will play a role in this chapter. Toward the end, some theorems from Appendix F on integral ring extensions will also be needed.

Definition 6.1. For a curve F in $\mathbb{P}^2(K)$ and a point $P \in \mathbb{P}^2(K)$ we call

$$m_P(F) := \text{Min}\{\mu_P(F,G) \mid G \text{ is a line through } P\}$$

the *multiplicity of* P *on* F (or the multiplicity of F at P).

In the following, F will always be a curve in $\mathbb{P}^2(K)$. By 5.4(a) it is clear that $m_P(F) = 0$ if and only if $P \notin \mathcal{V}_+(F)$. Since there is always a line G through P that is not a component of F, we have $\mu_P(F,G) < \infty$ and hence $m_P(F) < \infty$.

Definition 6.2. Let $P \in \mathcal{V}_+(F)$ and let G be a line through P. If

$$\mu_P(F,G) > m_P(F),$$

we call G *a tangent to* F *at* P.

Observe that the concepts "multiplicity" and "tangent" by their very definition are independent of the choice of coordinates. The following theorem gives a practical way to explicitly determine the multiplicity and the tangents. We suppose that $P = (0,0)$ is the affine origin and f is the affine curve corresponding to F. Denote by Lf the leading form of f for the (X,Y)-adic filtration of $K[X,Y]$, i.e., the homogeneous component of lowest degree with respect to the standard grading of the polynomial ring. In the following let $\deg Lf$ be the degree of Lf with respect to this grading, and not, as in Appendix B, the negative of this degree.

Theorem 6.3. *Under these hypotheses we have*

(a) $m_P(F) = \deg Lf$.
(b) *If* $\deg Lf =: m > 0$ *and* $Lf = \prod_{j=1}^{m}(a_j X - b_j Y)$ *is the decomposition of* Lf *into linear factors, then the lines*

$$t_j : a_j X - b_j Y = 0$$

are all the tangents to F *at* P.

Proof. Statement (a) is trivial for $\deg Lf = 0$. We can therefore assume that $m > 0$ and $P \in \mathcal{V}_+(F)$. We will show that for a line G through P,

$$\mu_P(F,G) \begin{cases} = m, & \text{if } G \notin \{t_1, \dots, t_m\}, \\ > m, & \text{if } G \in \{t_1, \dots, t_m\}, \end{cases} \tag{1}$$

and then the theorem follows.

Let G be given by the equation $aX - bY = 0$, where without loss of generality $b \neq 0$. Then the equation can be put in the form $Y = aX$. With $\mathfrak{M}_P := (X, Y)$, we then have

$$\mu_P(F,G) = \dim_K K[X,Y]_{\mathfrak{M}_P}/(f, Y - aX) = \dim_K K[X]_{(X)}/(f(X, aX))$$

by 5.1, and we have

$$f(X, aX) = X^m f_m(1, a) + X^{m+1} f_{m+1}(1, a) + \cdots + X^d f_d(1, a),$$

where $f = f_m + \cdots + f_d$ $(d \geq m)$ is the decomposition of f into homogeneous components. In particular, $Lf = f_m$. Here $K[X]_{(X)}$ is a discrete valuation ring whose maximal ideal is generated by X. Denote by ν the corresponding discrete valuation, and write

$$f(X, aX) = X^m \cdot [f_m(1, a) + X f_{m+1}(1, a) + \cdots + X^{d-m} f_d(1, a)].$$

Then we see that

$$\nu(F(X, aX)) \begin{cases} = m, & \text{in case } f_m(1, a) \neq 0, \\ > m, & \text{otherwise}, \end{cases} \tag{2}$$

because the expression in the square brackets [ ] is a unit in $K[X]_{(X)}$ if and only if $f_m(1, a) \neq 0$. The second case of (2) occurs precisely when $Y - aX$ is a divisor of Lf, that is, when $G \in \{t_1, \dots, t_m\}$. By E.13, however, we have

$$\dim_K K[X]_{(X)}/(f(X, aX)) = \nu(f(X, aX)).$$

This shows that (1) holds, and the proof is complete.

Corollary 6.4. *At a point of multiplicity* $m > 0$ *on an algebraic curve there is at least one and there are at most* m *tangents.*

Corollary 6.5. *Let $F = \prod_{i=1}^{t} F_i^{n_i}$, where the F_i are irreducible curves and $n_i \in \mathbb{N}_+$ for $i = 1, \ldots, t$. Then for every point $P \in \mathbb{P}^2(K)$,*

$$m_P(F) = \sum_{i=1}^{t} n_i \cdot m_P(F_i).$$

Proof. We can assume that $P = (0,0)$ and that the line at infinity is not a component of F. The affine polynomial f of F then has a factorization $f = \prod_{i=1}^{t} f_i^{n_i}$, where f_i is an irreducible polynomial corresponding to F_i. Because $Lf = \prod_{i=1}^{t} (Lf_i)^{n_i}$, the result follows from 6.3(a).

Under the assumptions of 6.3, one can consider Lf as an affine curve whose irreducible components are lines through P. This is called the *tangent cone* of F at $P = (0,0)$. By translation, the tangent cone is defined at any point of F. If we decompose Lf into a product of powers of linear factors that are pairwise not associates

$$Lf = c \cdot \prod_{i=1}^{\rho} (a_i X - b_i Y)^{\nu_i} \qquad (c \in K^*, \quad (a_i, b_i) \in K^2),$$

then the $a_i X - b_i Y$ define the distinct tangents at P, and ν_i gives the multiplicity with which the tangent should be counted. There are always m tangents at a point of multiplicity m when these tangents are counted with multiplicity.

Examples.

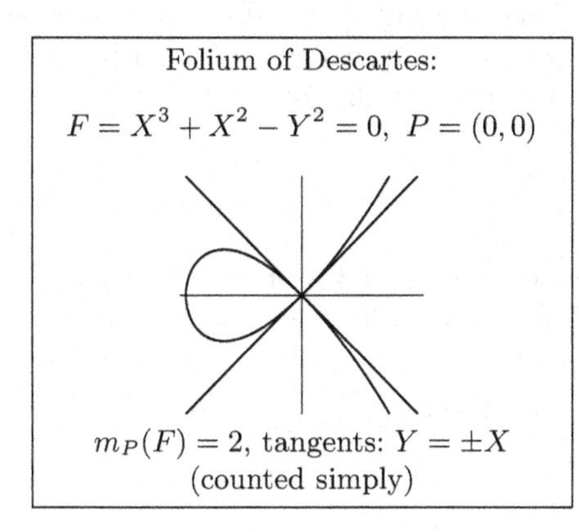

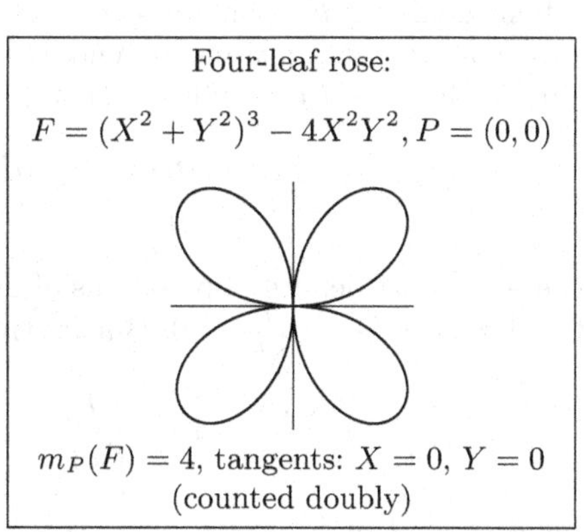

Definition 6.6. A point $P \in \mathcal{V}_+(F)$ is called a *simple* (or *regular*) point of F if $m_P(F) = 1$. In this case one says that F is *smooth* (or regular) at P. If $m_P(F) > 1$, then P is called a *multiple* (*singular*) *point* or a *singularity* of F. A curve that has no singularities is called *smooth* (*nonsingular*). The set of all singular points is denoted by $\operatorname{Sing}(F)$, and $\operatorname{Reg}(F)$ denotes the set of all simple points of F.

By 6.4 a curve has a uniquely determined tangent at a simple point.

Corollary 6.7.

(a) *A simple point of* F *does not lie on two distinct components of* F *and also does not lie on a multiple component of* F.
(b) *Every smooth projective curve is irreducible.*

Proof. (a) follows immediately from the formula in 6.5. A smooth curve can not have a multiple component by (a). If it had two distinct components, these components would have to intersect (Bézout), and each intersection point would be singular. Hence we have (b).

The following theorem will enable us to calculate singularities.

Jacobian Criterion 6.8. *For* $P = \langle x_0, x_1, x_2 \rangle \in \mathcal{V}_+(F)$, *write* $\frac{\partial F}{\partial x_i} := \frac{\partial F}{\partial X_i}(x_0, x_1, x_2)$ *for* $i = 0, 1, 2$. *Then* $P \in \mathrm{Sing}(F)$ *if and only if*

$$\frac{\partial F}{\partial x_0} = \frac{\partial F}{\partial x_1} = \frac{\partial F}{\partial x_2} = 0.$$

Proof. Without loss of generality we can take $x_0 = 1$. Consider the Taylor series of F at $(1, x_1, x_2)$:

$$(3)\quad F = F(1,x_1,x_2) + \frac{\partial F}{\partial x_0}\cdot(X_0 - 1) + \frac{\partial F}{\partial x_1}\cdot(X_1 - x_1) + \frac{\partial F}{\partial x_2}\cdot(X_2 - x_2) + \cdots .$$

Dehomogenizing F with respect to X_0, and setting $X := X_1 - x_1$, $Y := X_2 - x_2$, we get an affine polynomial corresponding to F in a coordinate system with $P = (0,0)$. Since $F(1, x_1, x_2) = 0$, it has the form

$$\frac{\partial F}{\partial x_1}\cdot X + \frac{\partial F}{\partial x_2}\cdot Y + \cdots ,$$

where the dots denote polynomials of degree > 1. By 6.3 we have $m_P(F) > 1$ if and only if $\frac{\partial F}{\partial x_1} = \frac{\partial F}{\partial x_2} = 0$. But by the Euler relation

$$\frac{\partial F}{\partial x_0}\cdot 1 + \frac{\partial F}{\partial x_1}\cdot x_1 + \frac{\partial F}{\partial x_2}\cdot x_2 = (\deg f)\cdot F(1, x_1, x_2) = 0,$$

this is equivalent to $\frac{\partial F}{\partial x_0} = \frac{\partial F}{\partial x_1} = \frac{\partial F}{\partial x_2} = 0$.

Corollary 6.9. *A reduced algebraic curve has only finitely many singularities.*

Proof. Let F be a reduced curve with irreducible components F_i $(i = 1, \dots, t)$. For $P \in \mathbb{P}^2(K)$ we have

$$m_P(F) = \sum_{i=1}^{t} m_P(F_i)$$

by 6.5, and it follows that

$$\text{(4)} \qquad \operatorname{Sing}(F) = \bigcup_{i=1}^{t} \operatorname{Sing}(F_i) \cup \bigcup_{i \neq j} (\mathcal{V}_+(F_i) \cap \mathcal{V}_+(F_j)).$$

Therefore, we must show only the finiteness of $\operatorname{Sing}(F_i)$, since $\mathcal{V}_+(F_i) \cap \mathcal{V}_+(F_j)$ (for $i \neq j$) is finite by Bézout.

So suppose F is irreducible of degree $d > 0$. It cannot be the case that all of the partial derivatives $\frac{\partial F}{\partial X_i}$ vanish ($i = 0, 1, 2$); otherwise, F would be of the form $F = H^p$, for some polynomial H, where $p > 0$ is the characteristic of K; then F would not be irreducible. If some $\frac{\partial F}{\partial X_0} \neq 0$, then $\deg \frac{\partial F}{\partial X_0} = d - 1$. Since F and $\frac{\partial F}{\partial X_0}$ are relatively prime, the system of equations $F = 0$, $\frac{\partial F}{\partial X_0} = 0$ has at most $d(d-1)$ solutions by Bézout. By the Jacobian criterion, $\operatorname{Sing}(F)$ is therefore finite.

Example. The curves $F_n := X_1^n + X_2^n - X_0^n$ ($n \in \mathbb{N}_+$) are smooth if n is not divisible by the characteristic of K. The partial derivatives of F_n then vanish only for $X_0 = X_1 = X_2 = 0$; hence they vanish at no point in the projective plane. In particular, F_n is irreducible. Conversely, if $\operatorname{Char} K$ is a divisor of n, then all points of F_n are singular.

The following theorem shows that the tangents introduced here are for curves defined over $\mathbb{R}$ a generalization of the tangents studied in analysis.

Theorem 6.10 (Tangents at regular points).

(a) *Let $P = (x, y)$ be a regular point of an affine curve f. Then the tangent line to f at P is given by the equation*

$$\frac{\partial f}{\partial x} \cdot (X - x) + \frac{\partial f}{\partial y} \cdot (Y - y) = 0,$$

where $\frac{\partial f}{\partial x} := \frac{\partial f}{\partial X}(x, y)$ and $\frac{\partial f}{\partial y} := \frac{\partial f}{\partial Y}(x, y)$.

(b) *Let $P = \langle x_0, x_1, x_2 \rangle$ be a regular point of a projective curve F. Then the tangent to F at P is given by*

$$\frac{\partial f}{\partial x_0} \cdot X_0 + \frac{\partial f}{\partial x_1} \cdot X_1 + \frac{\partial f}{\partial x_2} \cdot X_2 = 0.$$

Proof. (a) follows from the description of tangents in 6.3(b), since the translation that maps (x, y) to $(0, 0)$ sends $\frac{\partial f}{\partial x} \cdot (X - x) + \frac{\partial f}{\partial y} \cdot (Y - y)$ to the leading form of the polynomial describing the affine curve.

(b) We may suppose without loss of generality that $x_0 \neq 0$ and let $f(X, Y) := F(1, X, Y)$. In affine coordinates the tangent is given by

$$\frac{\partial f}{\partial x} \cdot (X - x) + \frac{\partial f}{\partial y} \cdot (Y - y) = 0 \qquad \text{with } x := \frac{x_1}{x_0}, \quad y := \frac{x_2}{x_0},$$

according to (a), and therefore is given projectively by

$$\frac{\partial f}{\partial x}\cdot X_1+\frac{\partial f}{\partial y}\cdot X_2-\left(\frac{\partial f}{\partial x}x+\frac{\partial f}{\partial y}y\right)\cdot X_0=0. \tag{5}$$

Now $\frac{\partial f}{\partial x}=\frac{\partial F}{\partial X_1}(1,\frac{x_1}{x_0},\frac{x_2}{x_0})$, $\frac{\partial f}{\partial y}=\frac{\partial F}{\partial X_2}(1,\frac{x_1}{x_0},\frac{x_2}{x_0})$, and by the Euler relation we have

$$\begin{aligned}\frac{\partial f}{\partial x}\cdot x+\frac{\partial f}{\partial y}\cdot y&=\frac{\partial F}{\partial X_1}\left(1,\frac{x_1}{x_0},\frac{x_2}{x_0}\right)\cdot\frac{x_1}{x_0}+\frac{\partial F}{\partial X_2}\left(1,\frac{x_1}{x_0},\frac{x_2}{x_0}\right)\cdot\frac{x_2}{x_0}\\&=-\frac{\partial F}{\partial X_0}\left(1,\frac{x_1}{x_0},\frac{x_2}{x_0}\right).\end{aligned}$$

Observing that the partial derivatives are homogeneous of degree $\deg F-1$, we get the equations for the tangents that we want by multiplying through by $x_0^{\deg F-1}$.

Next we want to characterize the regular points P using the local rings $\mathcal{O}_{F,P}$. In the future we will always denote the maximal ideal of $\mathcal{O}_{F,P}$ by $\mathfrak{m}_{F,P}$. If $P=(a,b)$ is an affine point and f the polynomial in $K[X,Y]$ corresponding to F, then

$$\mathcal{O}_{F,P}\cong K[f]_{\overline{\mathfrak{M}}_P},$$

where $K[f]=K[X,Y]/(f)$ and $\overline{\mathfrak{M}}_P$ is the image of the maximal ideal $\mathfrak{M}_P\subset K[X,Y]$. By 1.15 and C.10 we know the prime ideals of $\mathcal{O}_{F,P}$: besides the maximal ideal $\mathfrak{m}_{F,P}$, the ring $\mathcal{O}_{F,P}$ has only finitely many minimal prime ideals, which are in one-to-one correspondence with the components of F that pass through P. In particular, $\mathcal{O}_{F,P}$ is a one-dimensional local ring (E.10(b)). Since $\mathfrak{M}_P=(X-a,Y-b)$, it follows that $\mathfrak{m}_P$ is also generated by two elements, and so we have $\operatorname{edim}\mathcal{O}_{F,P}=2$ or 1.

Regularity Criterion 6.11. *For $P\in\mathcal{V}_+(F)$ the following are equivalent:*

(a) *P is a regular point of F.*
(b) *$\mathcal{O}_{F,P}$ is a discrete valuation ring.*
(c) $\operatorname{edim}\mathcal{O}_{F,P}=1$.

Proof. (a) ⇒ (b). We can assume that $P=(0,0)$, and then we have to show that $K[f]_{\overline{\mathfrak{M}}_P}$ is a discrete valuation ring. Let $f=cf_1^{n_1}\cdots f_h^{n_h}$ be a decomposition of f into irreducible polynomials ($c\in K^*$, $n_i>0$). Since P is a regular point of F, we have $f_i(P)=0$ for exactly one $i\in\{1,\dots,h\}$, and moreover, for this i we have $n_i=1$ (by 6.7(a)). Therefore,

$$(f)\cdot K[X,Y]_{\mathfrak{M}_P}=(f_i)\cdot K[X,Y]_{\mathfrak{M}_P}.$$

Because $m_P(F)=1$, we have $\deg Lf=\deg Lf_i=1$. We can choose the coordinate system so that $Lf_i=Y$; i.e., the tangent to F at P is the X-axis. Then f_i is of the form

$$(6) \quad f_i = \phi_1 \cdot X - \phi_2 \cdot Y, \quad \phi_1,\ \phi_2 \in K[X,Y], \quad \phi_1(0,0) = 0,\ \phi_2(0,0) \neq 0.$$

If ξ, η are the residue classes of X, Y in $K[f]_{\overline{\mathfrak{M}}_P}$, then the maximal ideal $\mathfrak{m} = \mathfrak{m}_{F,P}$ of this ring will be generated by ξ and η, because X and Y generate the ideal $\mathfrak{M}_P$. Since the image of ϕ_2 in $K[f]_{\overline{\mathfrak{M}}_P}$ is a unit, from (6) we get an equation $\eta = r \cdot \xi$ in $K[f]_{\overline{\mathfrak{M}}_P}$ (for some $r \in K[f]_{\overline{\mathfrak{M}}_P}$), and therefore $\mathfrak{m}$ is a principal ideal. Since $\mathcal{O}_{F,P}$ is a 1-dimensional local ring, hence not a field, we have $\xi \neq 0$ and therefore $\operatorname{edim} \mathcal{O}_{F,P} = 1$. We have shown that $\mathcal{O}_{F,P}$ is a discrete valuation ring.

The proof of (b) $\Rightarrow$ (c) is trivial. To prove that (c) implies (a) it is enough to show that $\operatorname{edim} \mathcal{O}_{F,P} = 2$ whenever P is a singularity of F. In this case $\deg Lf \geq 2$, and therefore $f \in (X^2, XY, Y^2) = \mathfrak{M}_P^2$. It follows that

$$\mathfrak{m}/\mathfrak{m}^2 \cong (X,Y)K[X,Y]_{\mathfrak{M}_P}/(X^2, XY, Y^2)K[X,Y]_{\mathfrak{M}_P}.$$

It is easy to see that the ideal $(X,Y)K[X,Y]_{\mathfrak{M}_P}$ is not a principal ideal. By Nakayama's lemma (E.1), $\mathfrak{m}/\mathfrak{m}^2$ is then a K-vector space of dimension 2, and a second application of the lemma shows that $\mathfrak{m}$ is also not a principal ideal.

If F is an irreducible curve, its function field $\mathcal{R}(F)$ contains all of the local rings $\mathcal{O}_{F,P}$, where $P \in \mathcal{V}_+(F)$. For if P is a point at finite distance (an affine point), f the polynomial in $K[X,Y]$ corresponding to F, and $K[f] = K[X,Y]/(f) = K[x,y]$, then $\mathcal{R}(F) = K(x,y)$ and $\mathcal{O}_{F,P}$ is by 5.1 a localization of $K[x,y]$, so it is certainly a subring of $K(x,y)$. It consists just of the rational functions $r \in \mathcal{R}(F)$ for which $P \in \operatorname{Def}(r)$. If P is a smooth point of F, then $\mathcal{O}_{F,P}$ is a discrete valuation ring with field of fractions $\mathcal{R}(F)$ and $K \subset \mathcal{O}_{F,P}$.

In general, we call a discrete valuation ring R with $Q(R) = \mathcal{R}(F)$ and $K \subset R$ a *discrete valuation ring of* $\mathcal{R}(F)/K$. The set $\mathfrak{X}(F)$ of all discrete valuation rings of $\mathcal{R}(F)/K$ is called the *abstract Riemann surface of* $\mathcal{R}(F)/K$. Theorem 6.11 shows that $\mathcal{O}_{F,P}$ belongs to $\mathfrak{X}(F)$ if and only if P is a regular point of F. We now want to investigate more precisely the behavior with respect to F of the abstract Riemann surface of $\mathcal{R}(F)/K$.

Theorem 6.12. *Let F be irreducible. Then*

(a) *To each $R \in \mathfrak{X}(F)$ there is exactly one $P \in \mathcal{V}_+(F)$ such that $\mathcal{O}_{F,P} \subset R$ and $\mathfrak{m}_{F,P} = \mathfrak{m} \cap \mathcal{O}_{F,P}$, where $\mathfrak{m}$ is the maximal ideal of R. There is therefore a natural mapping*

$$\pi : \mathfrak{X}(F) \to \mathcal{V}_+(F) \qquad (R \mapsto P).$$

(b) *The mapping π is surjective. For $P \in \operatorname{Reg}(F)$, the set $\pi^{-1}(P)$ consists of only one "point" R, namely $R = \mathcal{O}_{F,P}$. For $P \in \operatorname{Sing}(F)$, $\pi^{-1}(P)$ is finite.*

Proof. (a) We write $\mathcal{R}(F) = K(x,y)$, where $K[x,y]$ is the affine coordinate ring of F with respect to the line at infinity $X_0 = 0$. If we choose instead

$X_1 = 0$ as the line at infinity, then $K[\frac{1}{x}, \frac{y}{x}]$ is the corresponding coordinate ring, and for $X_2 = 0$ we get the ring $K[\frac{x}{y}, \frac{1}{y}]$.

Now let $v_R : \mathcal{R}(F) \to \mathbb{Z} \cup \infty$ be the valuation belonging to R. If $v_R(x) \geq 0$ and $v_R(y) \geq 0$, then $K[x, y] \subset R$ and $\mathfrak{m} \cap K[x, y]$ is a prime ideal of $K[x, y]$. It cannot be the zero ideal, for in that case we would have $K(x, y) \subset R$. Therefore, $\mathfrak{m} \cap K[x, y] = \mathfrak{m}_P$ is the maximal ideal of some $P \in \mathcal{V}_+(F)$, and we have

$$\mathcal{O}_{F,P} = K[x, y]_{\mathfrak{m}_P} \subset R, \qquad \mathfrak{m} \cap \mathcal{O}_{F,P} = \mathfrak{m}_{F,P}.$$

In case $v_R(x) \geq 0 > v_R(y)$, then $K[\frac{x}{y}, \frac{1}{y}] \subset R$. On the other hand, if $v_R(x) \leq v_R(y) < 0$, then $K[\frac{1}{x}, \frac{y}{x}] \subset R$. In each case, one finds as above a point $P \in \mathcal{V}_+(F)$ with $\mathcal{O}_{F,P} \subset R$ and $\mathfrak{m}_{F,P} = \mathfrak{m} \cap \mathcal{O}_{F,P}$.

Suppose there were another such point $P' \in \mathcal{V}_+(F)$. By a suitable choice of coordinates, both P and P' lie in the affine plane complementary to $X_0 = 0$. Then $\mathfrak{m}_{P'} = K[x, y] \cap \mathfrak{m} = \mathfrak{m}_P$ and hence $P' = P$. This shows the existence of the mapping π.

(b) For $P \in \text{Reg}(F)$ we have $\mathcal{O}_{F,P} \in \mathfrak{X}(F)$. Since a discrete valuation ring is a maximal subring of its field of fractions (E.14), there is only one $R \in \mathfrak{X}(F)$ with $\mathcal{O}_{F,P} \subset R$, namely $R = \mathcal{O}_{F,P}$. Hence $\pi^{-1}(P)$ consists of exactly one point in this case.

We still have to consider the singular points of F. Let $P \in \text{Sing}(F)$. By a suitable choice of coordinate system, we may assume that $P = (0, 0)$, that the polynomial $f \in K[X, Y]$ corresponding to F is monic as a polynomial in Y, and that $\frac{\partial f}{\partial Y} \neq 0$ (4.6). Then $K[f] = K[x, y]$ is integral over $K[x]$ and is a finite $K[x]$-module. If S is the integral closure of $K[x]$ in $\mathcal{R}(F)$, then $K[f] \subset S$. Also, $\mathcal{R}(F)$ is separable algebraic over $K(x)$. By F.7 it follows that S is finitely generated as a $K[x]$-module. Then S is also finitely generated as a $K[f]$-module.

Let $\overline{\mathcal{O}}_{F,P}$ be the integral closure of $\mathcal{O}_{F,P}$ in $\mathcal{R}(F)$. Since $\mathcal{O}_{F,P}$ is a localization of $K[x, y]$, the ring $\overline{\mathcal{O}}_{F,P}$ is the localization of S at the same set of denominators (F.11(a)), and in particular, $\overline{\mathcal{O}}_{F,P}$ is a finitely generated $\mathcal{O}_{F,P}$-module. By F.10(b) respectively F.10(a) it has at least one and at most finitely many maximal ideals $\mathfrak{M}$, and by F.10(b) these all lie over $\mathfrak{m}_{F,P}$, i.e., $\mathfrak{M} \cap \mathcal{O}_{F,P} = \mathfrak{m}_{F,P}$ for all $\mathfrak{M} \in \text{Max}(\overline{\mathcal{O}}_{F,P})$. Furthermore, by F.10(b) respectively F.10(a) and F.8, $(\overline{\mathcal{O}}_{F,P})_{\mathfrak{M}}$ is a discrete valuation ring, and hence an element of $\mathfrak{X}(F)$. This shows that the mapping π is surjective.

Finally, let $R \in \mathfrak{X}(F)$ be an arbitrary element with $\pi(R) = P$, and let $\mathfrak{m}$ be the maximal ideal of R. We will prove that $\overline{\mathcal{O}}_{F,P} \subset R$. For $z \in \overline{\mathcal{O}}_{F,P}$, we must show that $v_R(z) \geq 0$, where v_R denotes the valuation belonging to R. Let

$$z^n + a_1 z^{n-1} + \cdots + a_n = 0$$

be an equation of integral dependence for z over $\mathcal{O}_{F,P}$. From $a_i \in \mathcal{O}_{F,P} \subset R$, it follows that $v_R(a_i) \geq 0$ $(i = 1, \ldots, n)$. If $v_R(z) < 0$, then we would have

$$v_R(z^n) = \text{Min}\{v_R(z^n),\ v_R(a_i z^{n-i}) \mid i = 1, \ldots, n\},$$

and by Rule (c′) in Appendix E, we would have $\infty = v_R(0) = n \cdot v_R(z)$, a contradiction.

Since $\mathcal{O}_{F,P} \subset \overline{\mathcal{O}}_{F,P} \subset R$, we have that $\mathfrak{M} := \mathfrak{m} \cap \overline{\mathcal{O}}_{F,P}$ is one of the maximal ideals of $\overline{\mathcal{O}}_{F,P}$ lying over $\mathfrak{m}_{F,P}$, and $(\overline{\mathcal{O}}_{F,P})_{\mathfrak{M}} = R$, since $(\overline{\mathcal{O}}_{F,P})_{\mathfrak{M}}$ is itself a discrete valuation ring, as we have already shown. Thus, R is the localization of $\overline{\mathcal{O}}_{F,P}$ at one of the finitely many maximal ideals of this ring; i.e., $\pi^{-1}(P)$ is finite.

Corollary 6.13. *If F is smooth, then $\pi : \mathfrak{X} \to \mathcal{V}_+(F)$ is bijective.*

Remark. It can be shown that in general, $\mathfrak{X}(F)$ is the set of all local rings of a smooth curve C in a higher-dimensional projective space, and F appears as a plane curve under a suitable mapping of C to the plane. If, under this mapping, several points of C have the same image, then the outcome of this is a singularity.

The proof of 6.12 has shown that the $R \in \mathfrak{X}(F)$ are exactly the localizations of the rings $\overline{\mathcal{O}}_{F,P}$ $(P \in \mathcal{V}_+(F))$ at their maximal ideals. It is therefore clear that each $R \in \mathfrak{X}(F)$ has K as its residue field up to isomorphism. If v_R is the valuation associated with R, and $r \in \mathcal{R}(F)$ is a rational function, then we call $v_R(r)$ the *order* of r at the zero R. If $v_R(r) > 0$, then we call R a *zero of order* $v_R(r)$ of r; when $v_R(r) < 0$, then we call R a *pole of order* $-v_R(r)$ of r. If $R = \mathcal{O}_{F,P}$ for a regular point P of F, then we also write v_P for v_R and speak of the order of r at the point P.

To each $r \in \mathcal{R}(F)$ one can assign a function on $\mathfrak{X}(F)$ as follows: If $v_R(r) \geq 0$, let $r(R)$ be the image of r under the canonical epimorphism $R \to K$. For $R = \mathcal{O}_{F,P}$ this amounts to the value of the function at the zero P, as we have explained. Looking at the $r \in \mathcal{R}(F)$ as functions on $\mathfrak{X}(F)$ has several advantages: r has an order at each point of $\mathfrak{X}(F)$. Therefore, $R \in \mathfrak{X}(F)$ is a pole of r precisely when R is a zero of r^{-1}. Also, $v_R(r \cdot s) = v_R(r) + v_R(s)$ for all $r, s \in \mathcal{R}(F)$.

Theorem 6.14. *A nonzero rational function $r \in \mathcal{R}(F)$ has only finitely many zeros and poles on $\mathfrak{X}(F)$.*

Proof. Let f be the affine curve belonging to F. We first consider the case $r \in K[f]$. Then r belongs to the local rings $\mathcal{O}_{F,P}$ for every point P of F at finite distance, and r has no poles at the points of $\pi^{-1}(P)$. Let $g \in K[X, Y]$ be a polynomial representing r. Since $r \neq 0$, g is not divisible by f. The affine curves g and f intersect in only finitely many points; i.e., $r \in \mathfrak{m}_{F,P}$ for only finitely many points $P \in \mathcal{V}(f)$; i.e., r has only finitely many zeros at finite distance. Since F has only finitely many points at infinity, each of which has only finitely many π-preimages $R \in \mathfrak{X}(F)$, altogether r has only finitely many zeros and poles.

An arbitrary function in $\mathcal{R}(F)$ is of the form $\frac{r}{s}$, where $r, s \in K[f]$ and $s \neq 0$. It is then clear that any nonzero such function has only finitely many zeros and poles.

As in the case of the projective plane, we define the *divisor group of* $\mathfrak{X}(F)$ (also called the divisor group of $\mathcal{R}(F)/K$) to be the free abelian group on the set of points of $\mathfrak{X}(F)$. The degree of a divisor $D = \sum_{R \in \mathfrak{X}(F)} n_R \cdot R$ is the integer $\sum_{R \in \mathfrak{X}(F)} n_R$. Theorem 6.14 shows that to each function $r \in \mathcal{R}(F) \setminus \{0\}$ there is a *principal divisor*

$$(r) := \sum_{R \in \mathfrak{X}(F)} v_R(r) \cdot R,$$

since we have $v_R(r) \neq 0$ only for finitely many $R \in \mathfrak{X}(F)$. The principal divisors form a subgroup of the divisor group. Its residue class group is called the *divisor class group*, and it is an important invariant of $\mathfrak{X}(F)$ (or of the function field $\mathcal{R}(F)/K$ or of the curve F). This divisor class group is in no way as simple as the divisor class group of the projective plane described at the beginning of Chapter 4. Birationally equivalent curves (see 4.8(a)) in $\mathbb{P}^2(K)$ obviously have isomorphic divisor class groups.

Exercises

1. Determine the real and complex singularities of the projective closures of the curves in 1.2(e) and 1.2(f). Also calculate the multiplicities and the tangents at the singular points. Do this also for the limaçon (snail) of Pascal $(X^2 + Y^2 + 2Y)^2 - (X^2 + Y^2) = 0$.

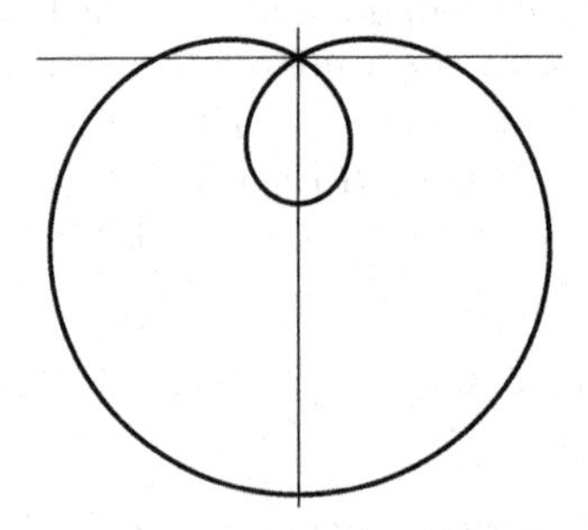

2. Let $\mathcal{O}_{F,P}$ be the local ring of a regular point P of an algebraic curve F. Show that there exists an injective K-algebra homomorphism from $\mathcal{O}_{F,P}$ to the algebra $K[[t]]$ of all formal power series in one variable t over K. (Use the fact that $\mathcal{O}_{F,P}$ is a discrete valuation ring and "expand" the elements of $\mathcal{O}_{F,P}$ using a prime element of the power series ring.)
3. Let $f = f_1 \cdots f_h$ be a reduced affine curve. Show that f has no singularities if and only if the canonical homomorphism

$$K[f] \to K[f_1] \times \cdots \times K[f_h]$$

is bijective and the $K[f_i]$ are integrally closed in their field of fractions.

7

More on Intersection Theory. Applications

In the last section we introduced the multiplicity of a point on an algebraic curve. Using multiplicities we can make more precise statements about the nature of the intersections of two curves than was possible so far. We will also present some further applications of Bézout's theorem.

First we shall give a description of intersection multiplicity by means of valuations. Let F be a curve in $\mathbb{P}^2(K)$ and P a point of F such that the local ring $\mathcal{O}_{F,P}$ is an integral domain. This is the same as saying that P lies on only one irreducible component of F. Let $R_1, \ldots, R_h \in \mathfrak{X}(F)$ be the discrete valuation rings that lie over $\mathcal{O}_{F,P}$ (i.e., for which $\pi(R_i) = P$ using the mapping π of 6.12). Furthermore, let ν_{R_i} be the valuation belonging to R_i $(i = 1, \ldots, h)$. The R_i are precisely the localizations of the integral closures $\overline{\mathcal{O}}_{F,P}$ of $\mathcal{O}_{F,P}$ in $\mathcal{R}(F)$ at their maximal ideals.

Lemma 7.1. *The K-vector space $\overline{\mathcal{O}}_{F,P}/\mathcal{O}_{F,P}$ is finite-dimensional.*

Proof. $\overline{\mathcal{O}}_{F,P}$ is finitely generated as an $\mathcal{O}_{F,P}$-module, and both rings have the same quotient field. Hence there are elements $\omega_i = \frac{a_i}{b}$ $(i = 1, \ldots, n)$ in $\overline{\mathcal{O}}_{F,P}$ with $a_i, b \in \mathcal{O}_{F,P}$, $b \neq 0$, such that

$$\overline{\mathcal{O}}_{F,P} = \sum_{i=1}^{n} \mathcal{O}_{F,P} \cdot \omega_i.$$

Then $b\overline{\mathcal{O}}_{F,P} \subset \mathcal{O}_{F,P} \subset \overline{\mathcal{O}}_{F,P}$.

If b is a unit in $\mathcal{O}_{F,P}$, then $\overline{\mathcal{O}}_{F,P} = \mathcal{O}_{F,P}$, and there is nothing more to show. If b is not a unit of $\mathcal{O}_{F,P}$, then $\overline{\mathcal{O}}_{F,P}/b\overline{\mathcal{O}}_{F,P}$ is a K-algebra with only finitely many prime ideals, namely the images of the maximal ideals of $\overline{\mathcal{O}}_{F,P}$. By D.3,

$$\overline{\mathcal{O}}_{F,P}/b\overline{\mathcal{O}}_{F,P} \cong R_1/(b) \times \cdots \times R_h/(b),$$

and the $R_i/(b)$ are finite-dimensional K-algebras (E.13). Then $\overline{\mathcal{O}}_{F,P}/b\overline{\mathcal{O}}_{F,P}$, and hence also $\overline{\mathcal{O}}_{F,P}/\mathcal{O}_{F,P}$ is a finite-dimensional K-vector space.

Under the hypotheses of 7.1, let G be a curve that has no component that contains the point P in common with F. Let $I(G)_P \subset \mathcal{O}_P$ be the principal ideal belonging to G (Chapter 5).

Theorem 7.2. *Suppose the image of $I(G)_P$ in $\mathcal{O}_{F,P}$ is generated by γ. Then*

$$\mu_P(F,G) = \sum_{i=1}^{h} \nu_{R_i}(\gamma).$$

Proof. We have $\mu_P(F,G) = \dim_K \mathcal{O}_{F,P}/(\gamma)$ by the definition of intersection multiplicity. Consider the diagram

$$\begin{array}{ccc} \gamma\mathcal{O}_{F,P} & \hookrightarrow & \mathcal{O}_{F,P} \\ \downarrow & & \downarrow \\ \gamma\overline{\mathcal{O}}_{F,P} & \hookrightarrow & \overline{\mathcal{O}}_{F,P} \end{array}$$

Using 7.1 we get

$$\dim_K \frac{\overline{\mathcal{O}}_{F,P}}{\mathcal{O}_{F,P}} + \dim_K \frac{\mathcal{O}_{F,P}}{(\gamma)} = \dim_K \frac{\overline{\mathcal{O}}_{F,P}}{(\gamma)} + \dim_K \frac{\gamma\overline{\mathcal{O}}_{F,P}}{\gamma\mathcal{O}_{F,P}}.$$

Since γ is not a zero divisor of $\overline{\mathcal{O}}_{F,P}$, the K-vector spaces $\overline{\mathcal{O}}_{F,P}/\mathcal{O}_{F,P}$ and $\gamma\overline{\mathcal{O}}_{F,P}/\gamma\mathcal{O}_{F,P}$ are isomorphic, and so

$$\mu_P(F,G) = \dim_K \overline{\mathcal{O}}_{F,P}/(\gamma).$$

As in the proof of 7.1 we have

$$\overline{\mathcal{O}}_{F,P}/(\gamma) \cong R_1/(\gamma) \times \cdots \times R_h/(\gamma),$$

and then from E.13,

$$\dim_K \overline{\mathcal{O}}_{F,P}/(\gamma) = \sum_{i=1}^{h} \dim_K(R_i/(\gamma)) = \sum_{i=1}^{h} \nu_{R_i}(\gamma).$$

Theorem 7.3. *Let F be an irreducible curve and $r \in \mathcal{R}(F) \setminus \{0\}$. Then the principal divisor of r on $\mathfrak{X}(F)$ is of degree 0. In other words: The function r has exactly as many zeros as poles when these are counted with their respective orders. If $a \in K$ and $r \neq a$, then r has also as many a-places as poles.*

Proof. Represent r by a rational function $\frac{\Phi}{\Psi}$ on $\mathbb{P}^2(K)$. Suppose $\deg\Phi = \deg\Psi =: q$, $\deg F =: p$, and Φ, Ψ are as always relatively prime. Now, F is not a divisor of Ψ and also not a divisor of Φ, since $r \neq 0$. Therefore, one can choose the line at infinity in such a way that $\Phi(P) \neq 0$ and $\Psi(P) \neq 0$ for every point P of F at infinity. If P is any of these points, then r is a unit in $\mathcal{O}_{F,P}$.

Let $\phi, \psi, f \in K[X,Y]$ be the dehomogenizations of Φ, Ψ, F, and $\overline{\phi}, \overline{\psi}$ the canonical images of ϕ, ψ in $K[f]$. Then $r = \frac{\overline{\phi}}{\overline{\psi}}$, and by 7.2 and Bézout's theorem we have for the degree of the divisor (r),

$$\deg(r) = \deg(\overline{\phi}) - \deg(\overline{\psi}) = \sum_{R \in \mathfrak{X}(F)} \nu_R(\overline{\phi}) - \sum_{R \in \mathfrak{X}(F)} \nu_R(\overline{\psi})$$
$$= \sum_{P \in \mathcal{V}_+(F)} \mu_P(F, \Phi) - \sum_{P \in \mathcal{V}_+(F)} \mu_P(F, \Psi) = p \cdot q - p \cdot q = 0.$$

Since the a-places of r are the zeros of $r - a$ and since $r - a$ has the same poles as r, the statement about a-places has been shown.

In particular, a rational function r that has no pole on $\mathfrak{X}(F)$ must be a constant function.

The next theorem describes the relationship between intersection multiplicity and multiplicity.

Theorem 7.4. *For two curves F, G in $\mathbb{P}^2(K)$ and a point $P \in \mathbb{P}^2(K)$ we always have*

$$\mu_P(F, G) \geq m_P(F) \cdot m_P(G).$$

Equality holds precisely when F and G have no tangent lines in common at the point P.

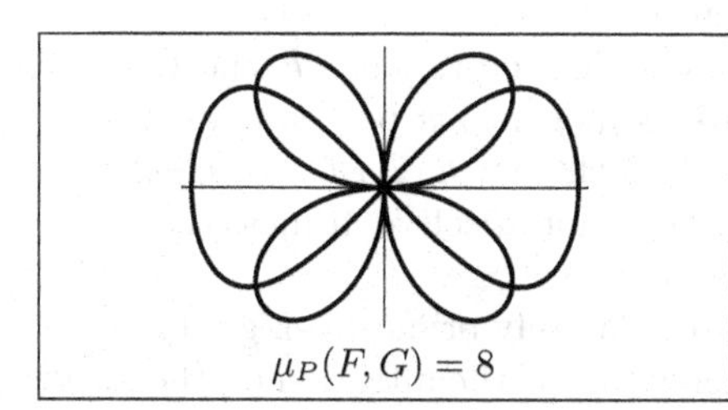

$\mu_P(F, G) = 8$

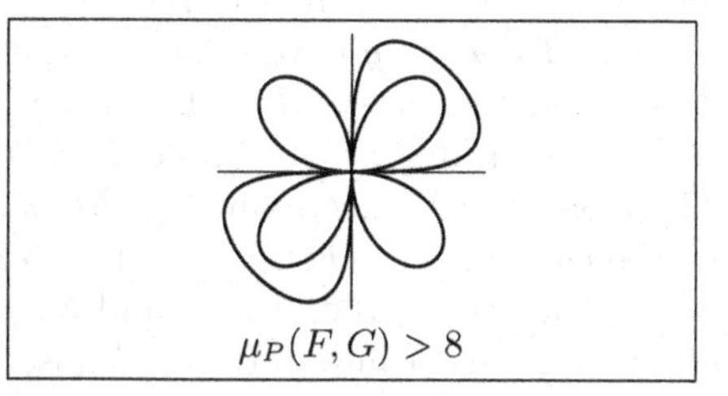

$\mu_P(F, G) > 8$

Proof. We may assume without loss of generality that $P = (0, 0)$ and that F and G have no common component that contains P. Let f and g be the affine curves corresponding to F and G. Denote by $\mathfrak{M} := (X, Y)$ the maximal ideal of $K[X, Y]$ belonging to P and let $\mathcal{F}$ be the $\mathfrak{M}$-adic filtration on $K[X, Y]$. Then by 6.3(b) (after a change of sign),

$$m_P(F) = \deg L_{\mathcal{F}} f, \quad m_P(G) = \deg L_{\mathcal{F}} g.$$

Set $m := m_P(F)$ and $n := m_P(G)$. Using 5.1 and the definition of intersection multiplicity it follows that

$$\begin{aligned} \mu_P(F, G) &= \dim_K K[X, Y]_{\mathfrak{M}}/(f, g) \geq \dim_K K[X, Y]/(f, g, \mathfrak{M}^{m+n}) \\ &= \dim_K K[X, Y]/\mathfrak{M}^{m+n} - \dim_K (f, g, \mathfrak{M}^{m+n})/\mathfrak{M}^{m+n} \\ &= \binom{m+n+1}{2} - \dim_K ((f, g, \mathfrak{M}^{m+n})/\mathfrak{M}^{m+n}). \end{aligned} \tag{1}$$

The K-linear mapping

$$\begin{array}{rcl}\alpha : K[X,Y] \times K[X,Y] & \to & (f,g,\mathfrak{M}^{m+n})/\mathfrak{M}^{m+n}, \\ (a,b) & \mapsto & af + bg + \mathfrak{M}^{m+n},\end{array}$$

is surjective. Also, $\alpha(a,b) = 0$ if $a \in \mathfrak{M}^n$ and $b \in \mathfrak{M}^m$. Hence there is an induced surjection

$$\overline{\alpha} : K[X,Y]/\mathfrak{M}^n \times K[X,Y]/\mathfrak{M}^m \to (f,g,\mathfrak{M}^{m+n})/\mathfrak{M}^{m+n},$$

and we get

$$\begin{aligned}\dim_K((f,g,\mathfrak{M}^{m+n})/\mathfrak{M}^{m+n}) &\leq \dim_K K[X,Y]/\mathfrak{M}^n + \dim_K K[X,Y]/\mathfrak{M}^m \\ &= \binom{n+1}{2} + \binom{m+1}{2}.\end{aligned}$$

By (1) this implies that

$$\mu_P(F,G) \geq \binom{m+n+1}{2} - \binom{n+1}{2} - \binom{m+1}{2} = m \cdot n,$$

and the first part of the theorem has been proved.

If $L_{\mathcal{F}}f$ and $L_{\mathcal{F}}g$ have a nonconstant common factor; hence F and G have a common tangent at P. Then there are homogeneous polynomials $a, b \in K[X,Y]$ with $\deg a = n-1$, $\deg b = m-1$, such that $a \cdot L_{\mathcal{F}}f + b \cdot L_{\mathcal{F}}g = 0$. Therefore, $(a,b) \in \ker \alpha$, and it follows that $\overline{\alpha}$ is not injective. It must therefore be the case that $\mu_P(F,G) > m_P(F) \cdot m_P(G)$.

On the other hand, if $L_{\mathcal{F}}f$ and $L_{\mathcal{F}}g$ are relatively prime, then $\operatorname{gr}_{\mathcal{F}}(f,g) = (L_{\mathcal{F}}f, L_{\mathcal{F}}g)$ according to B.12. We also denote by $\mathcal{F}$ the filtration on the local ring $K[X,Y]_{\mathfrak{M}}$ with respect to its maximal ideal. It is easy to see (C.14) that

$$\operatorname{gr}_{\mathcal{F}} K[X,Y]_{\mathfrak{M}} \cong \operatorname{gr}_{\mathcal{F}} K[X,Y] \cong K[X,Y]$$

in a natural way, and that $L_{\mathcal{F}}f, L_{\mathcal{F}}g$ are also the leading forms of f and g as elements of $K[X,Y]_{\mathfrak{M}}$. By B.6 the vector spaces $K[X,Y]_{\mathfrak{M}}/(f,g)$ and $\operatorname{gr}_{\overline{\mathcal{F}}} K[X,Y]_{\mathfrak{M}}/(f,g))$ have the same K-dimension, where $\overline{\mathcal{F}}$ denotes the induced residue class ring filtration. By B.8,

$$\operatorname{gr}_{\overline{\mathcal{F}}} K[X,Y]_{\mathfrak{M}}/(f,g) = \operatorname{gr}_{\mathcal{F}} K[X,Y]_{\mathfrak{M}}/\operatorname{gr}_{\mathcal{F}}(f,g),$$

and it follows that

$$\mu_P(F,G) = \dim_K(K[X,Y]_{\mathfrak{M}}/(f,g)) = \dim_K(K[X,Y]/(L_{\mathcal{F}}f, L_{\mathcal{F}}g).$$

From equation (4) in Appendix A we see that the last dimension is $m \cdot n$, and therefore $\mu_P(F,G) = m_P(F) \cdot m_P(G)$.

Corollary 7.5. *If F and G have no components in common, then*

$$\deg F \cdot \deg G \geq \sum_{P \in \mathbb{P}^2(K)} m_P(F) \cdot m_P(G).$$

Equality holds if and only if F and G have no tangent lines in common at every point of intersection.

Definition 7.6. We say that F and G *intersect transversally at P* if P is a regular point of F and of G, and if the tangents to F and G at P are different.

Corollary 7.7. *Two curves F and G intersect transversally at P if and only if $\mu_P(F, G) = 1$. If P is a point at finite distance, and f and g are the affine curves corresponding to F and G, then F and G intersect transversally at P if and only if the Jacobian determinant $\frac{\partial(f,g)}{\partial(X,Y)}$ does not vanish at the point P.*

Proof. The first statement follows immediately from Theorem 7.4. The nonvanishing of the Jacobian determinant at the point P is equivalent to saying that the leading forms $L_{\mathcal{F}} f$ and $L_{\mathcal{F}} g$ are of degree 1 and are linearly independent, i.e., that $m_P(F) = m_P(G) = 1$ and that the tangents to F and G at P are different.

Corollary 7.8. *Suppose F and G have no components in common. Then $\mathcal{V}_+(F) \cap \mathcal{V}_+(G)$ consists of $\deg F \cdot \deg G$ distinct points if and only if F and G intersect transversally at all their points of intersection.*

Our goal now is to give a sharpening of Bézout's theorem that will allow us to give a more precise count of the number of singularities of a reduced algebraic curve.

Theorem 7.9. *Let F be an irreducible curve in $\mathbb{P}^2(K)$ of degree d, and let $\operatorname{Sing}(F) = \{P_1, \ldots, P_s\}$. Then*

$$\sum_{i=1}^{s} m_{P_i}(F) \cdot (m_{P_i}(F) - 1) \leq d(d-1).$$

Proof. We may assume that the points P_i are at finite distance. Let $f \in K[X,Y]$ be the dehomogenization of F. Since f is irreducible, both partial derivatives of f cannot vanish, for otherwise, K would be a field of characteristic $p > 0$ and f would be a pth power.

Suppose then $\frac{\partial f}{\partial X} \neq 0$. Then also $\frac{\partial F}{\partial X_1} \neq 0$ and hence $\deg \frac{\partial F}{\partial X_1} = d - 1$. Furthermore, F and $\frac{\partial F}{\partial X_1}$ have no components in common. By Bézout and 7.4 we have

$$\deg F \cdot \deg \frac{\partial F}{\partial X_1} = d(d-1) = \sum_P \mu_P\left(F, \frac{\partial F}{\partial X_1}\right) \geq \sum_P m_P(F) \cdot m_P\left(\frac{\partial F}{\partial X_1}\right).$$

Let $P = (0,0)$. Then $m_P(F) = \deg L_{\mathcal{F}} f$ and $m_P(\frac{\partial F}{\partial X_1}) = \deg L_{\mathcal{F}} \frac{\partial f}{\partial X}$. Because $\deg \frac{\partial F}{\partial X_1} = d - 1$, we have $m_P(\frac{\partial F}{\partial X_1}) \geq m_P(F) - 1$, and it follows that

$$\deg F \cdot \deg \frac{\partial F}{\partial X_1} \geq \sum_P m_P(F) \cdot (m_P(F) - 1).$$

Since only the singularities of F contribute to the sum, the theorem is proved.

Corollary 7.10. *A reduced curve of degree d has at most $\binom{d}{2}$ singularities.*

Proof. For irreducible curves F this follows immediately from 7.9, since $m_P(F) \geq 2$ for each $P \in \operatorname{Sing}(F)$. Suppose F_1 and F_2 are two reduced curves with no common components for which the statement has already been shown, and let $F := F_1 \cdot F_2$.

Denote by s the number of singularities of F, let $d_i := \deg F_i$ $(i = 1, 2)$, and let $d := \deg F = d_1 + d_2$. Using equation (4) of the proof of 6.9 we get

$$s \leq \binom{d_1}{2} + \binom{d_2}{2} + d_1 d_2 = \frac{(d_1 + d_2)^2 - (d_1 + d_2)}{2} = \binom{d}{2}.$$

Now for an arbitrary reduced curve, the statement of the corollary follows by factoring and induction.

Examples 7.11.

(a) A reduced quadric has at most one singularity: In fact, for the line pair, the point of intersection is the singularity.
(b) A reduced cubic can have 3 singularities of multiplicity 2,

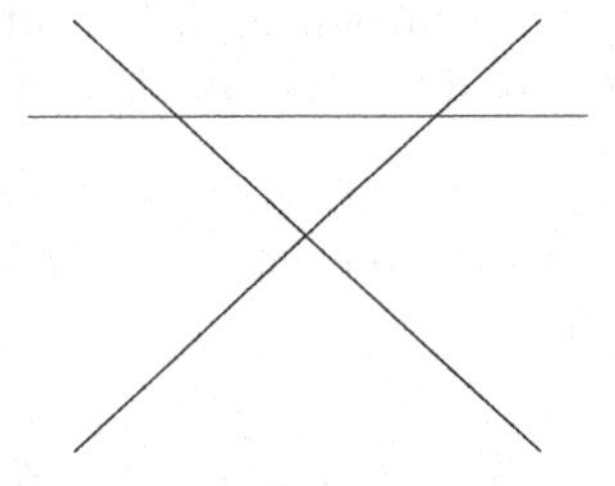

or one singularity of multiplicity 3.

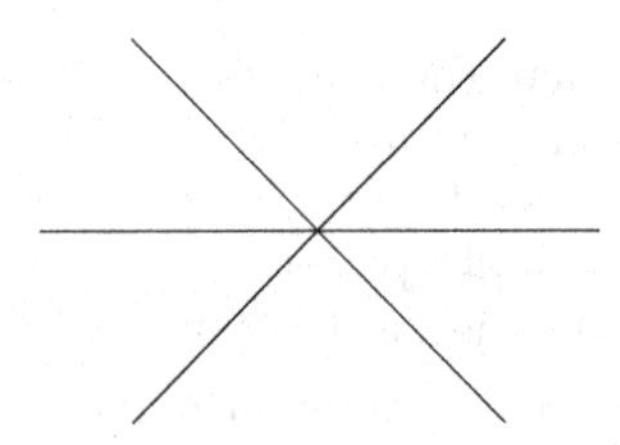

For irreducible curves the estimate of the number of singularities can be sharpened. To do this we need a few preparations. We begin with a fundamental concept.

Definition 7.12. A set L of curves of degree d in $\mathbb{P}^2(K)$ is called a *linear system of degree* d if there is a subspace V of the K-vector space of all homogeneous polynomials of degree d in $K[X_0, X_1, X_2]$ such that L consists of all curves F with $F \in V \setminus \{0\}$. If $\dim V =: \delta + 1$, then we set $\dim L := \delta$.

That the dimension of L has been chosen to be 1 less than the dimension of V can be explained by the fact that two polynomials that differ by a non-zero constant factor define the same curve. In fact, $L = \mathbb{P}^2(V)$ is the projective space associated with V.

Remark 7.13. For $P_1, \dots, P_s \in \mathbb{P}^2(K)$, let L be the linear system of all curves of degree d with $\{P_1, \dots, P_s\} \subset \operatorname{Supp}(F)$. Then

$$\dim L \geq \binom{d+2}{2} - s - 1.$$

Proof. We have $\dim_K K[X_0, X_1, X_2]_d = \binom{d+2}{2}$. A homogeneous polynomial F of degree d has therefore $\binom{d+2}{2}$ coefficients in K. The requirement that $\{P_1, \dots, P_s\} \subset \operatorname{Supp}(F)$ gives rise to s linear conditions on the coefficients. At least $\binom{d+2}{2} - s$ linearly independent polynomials satisfy these conditions.

Through 2 points there passes a line, through 5 points a quadric, through 9 points a cubic, etc.

d	1	2	3	4	$\cdots$
$\binom{d+2}{d} - 1$	2	5	9	14	$\cdots$

We will now consider, in addition to the points $P_1, \dots, P_s \in \mathbb{P}^2(K)$ on the curve F of degree d, given integers $m_1, \dots, m_s \geq 1$ for which

$$m_{P_i}(F) \geq m_i \qquad (i = 1, \dots, s).$$

Let $F = \sum_{\nu_0+\nu_1+\nu_2=d} a_{\nu_0\nu_1\nu_2} X_0^{\nu_0} X_1^{\nu_1} X_2^{\nu_2}$ and let $P = \langle 1, a, b \rangle$. Then $m_P(F) \geq m$ if and only if the leading form of the polynomial

$$F(1, X+a, Y+b) = \sum_{\nu_0+\nu_1+\nu_2=d} a_{\nu_0\nu_1\nu_2} (X+a)^{\nu_1} (Y+b)^{\nu_2} = \sum b_{\mu_1\mu_2} X^{\mu_1} Y^{\mu_2}$$

(with respect to the standard grading) has degree $\geq m$; i.e., we must have $b_{\mu_1\mu_2} = 0$ for $\mu_1 + \mu_2 < m$. The $b_{\mu_1\mu_2}$ are linear combinations of the $a_{\nu_0\nu_1\nu_2}$ with coefficients in K. The requirement $m_P(F) \geq m$ gives rise to $\binom{m+1}{2}$ linear conditions on the coefficients of F. The next theorem now follows.

Theorem 7.14. *The curves F of degree d with $m_{P_i}(F) \geq m_i$ $(i = 1, \dots, s)$ form a linear system L with*

$$\dim L \geq \binom{d+2}{d} - \sum_{i=1}^{s} \binom{m_i+1}{2} - 1.$$

We now come to the announced sharpening of 7.10 for irreducible curves.

Theorem 7.15. *Let F be an irreducible curve of degree d and let* $\mathrm{Sing}(F) = \{P_1, \dots, P_s\}$. *Then*

$$\binom{d-1}{2} \geq \sum_{i=1}^{s} \binom{m_{P_i}(F)}{2}.$$

Proof. By 7.9 we certainly have $\binom{d+1}{2} > \binom{d}{2} \geq \sum_{i=1}^{s} \binom{m_{P_i}(F)}{2}$. By 7.14 the linear system of all curves G of degree $d-1$ with $m_{P_i}(G) \geq m_{P_i}(F) - 1$ $(i = 1, \dots, s)$ is nonempty, and one can even find a curve G that additionally passes through

$$\binom{d+1}{2} - \sum_{i=1}^{s} \binom{m_{P_i}(F)}{2} - 1$$

prescribed simple points of F.

Since F is irreducible and $\deg G = \deg F - 1$, the curves F and G have no common component. Thus it follows from Bézout's theorem and 7.4 that

$$\begin{aligned} d(d-1) &\geq \sum_{i=1}^{s} m_{P_i}(F)(m_{P_i}(F) - 1) + \binom{d+1}{2} - \sum_{i=1}^{s} \binom{m_{P_i}(F)}{2} - 1 \\ &= \sum_{i=1}^{s} \binom{m_{P_i}(F)}{2} + \binom{d+1}{2} - 1, \end{aligned}$$

and we get

$$\binom{d-1}{2} \geq \sum_{i=1}^{s} \binom{m_{P_i}(F)}{2}.$$

Corollary 7.16. *An irreducible curve of degree d has at most $\binom{d-1}{2}$ singularities.*

For example, an irreducible cubic has at most one singularity with multiplicity 2. At that singular point, it can have two distinct tangents (folium of Descartes) or a double tangent (Neil's semicubical parabola). An irreducible curve of degree 4 (a quartic) can have up to 3 singularities of multiplicity 2, or one of multiplicity 3.

The following theorem classifies the singular cubics.

Theorem 7.17. *In case* $\mathrm{Char}\, K \neq 3$ *every irreducible singular cubic curve in* $\mathbb{P}^2(K)$ *is given, in a suitable coordinate system, by one of the following two equations:*

$$\begin{aligned} &(a)\ X_0X_1X_2 + X_1^3 + X_2^3 = 0, \\ &(b)\ X_0X_2^2 - X_1^3 \qquad\quad\; = 0. \end{aligned}$$

If $\mathrm{Char}\, K = 3$, *the curve is given by one of the equations (a), (b) or by the equation*

$$(c)\ X_0X_2^2 - X_1^3 - X_1^2X_2 = 0.$$

Proof. Let F be an irreducible cubic curve with a singularity P. We can assume $P = (0,0)$ without loss of generality.

(a) If F has two distinct tangents at P, then without loss of generality we can take these to be the X-axis and the Y-axis. The affine polynomial associated with F then has the form

$$XY + aX^3 + bX^2Y + cXY^2 + dY^3 \qquad (a, b, c, d \in K).$$

Here $a \neq 0$ and $d \neq 0$, for otherwise, F would be reducible. Set $a = \alpha^3$, $d = \delta^3$ $(\alpha, \delta \in K)$, so then the polynomial has the form

$$XY(1 + bX + cY) + (\alpha X)^3 + (\delta Y)^3,$$

and the substitution $\alpha X \mapsto X$, $\delta Y \mapsto Y$ gives

$$X \cdot Y \cdot \left(\frac{1}{\alpha\delta} + \frac{b}{\alpha^2\delta}X + \frac{c}{\alpha\delta^2}Y\right) + X^3 + Y^3.$$

Homogenizing this polynomial, and relabeling the expression in the parentheses X_0, we get

$$X_0X_1X_2 + X_1^3 + X_2^3 = 0.$$

(b) If F has a double tangent at P, then the associated affine polynomial, in a suitable coordinate system, has the form

$$Y^2 + aX^3 + bX^2Y + cXY^2 + dY^3 \qquad (a, b, c, d \in K,\ a \neq 0).$$

If $\operatorname{Char} K \neq 3$, then by the substitution $X \mapsto X - \frac{b}{3a}Y$, we can assume that $b = 0$. Now write the polynomial in the form

$$Y^2 \cdot (1 + cX + dY) + aX^3 + bX^2Y$$

and proceed as in (a) to get the equations (b) and (c).

We will deal with nonsingular cubics (elliptic curves) in detail in Chapter 10. The next theorem gives sufficient conditions under which the intersection scheme of two curves is a subscheme of a further curve, and so complements Noether's fundamental theorem 5.14 and the Cayley–Bacharach theorem 5.17.

Theorem 7.18. *Let F_1, F_2, and G be curves in $\mathbb{P}^2(K)$. Suppose one of the following conditions is satisfied for a point $P \in \mathcal{V}_+(F_1) \cap \mathcal{V}_+(F_2)$:*

(a) *P is a simple point of F_1 and $\mu_P(F_1, G) \geq \mu_P(F_1, F_2)$.*
(b) *F_1 and F_2 have no tangents in common at P and*

$$m_P(G) \geq m_P(F_1) + m_P(F_2) - 1.$$

Then

$$\dim_K \mathcal{O}_{F_1 \cap F_2 \cap G, P} = \mu_P(F_1, F_2).$$

Proof. Under the assumption (a), the ring $\mathcal{O}_{F_1,P}$ is a discrete valuation ring. Let $I = (\gamma)$ and $J = (\eta)$ be the ideals in $\mathcal{O}_{F_1,P}$ associated with F_2 and G. Then by 7.2,

$$\nu_P(\eta) = \mu_P(F_1, G) \geq \mu_P(F_1, F_2) = \nu_P(\gamma),$$

and therefore $J \subset I$. It follows that

$$\dim_K \mathcal{O}_{F_1 \cap F_2 \cap G,P} = \dim_K \mathcal{O}_{F_1,P}/I + J = \dim_K \mathcal{O}_{F_1,P}/I = \mu_P(F_1, F_2).$$

Under the assumption (b), we let $P = (0,0)$ without loss of generality, and let f_1, f_2, g be the corresponding affine curves. Denote the maximal ideal of the local ring $\mathcal{O}'_P$ by $\mathfrak{M}$, and let Lf_1, Lf_2, Lg be the leading forms of f_1, f_2, g with respect to the $\mathfrak{M}$-adic filtration of $\mathcal{O}'_P$. We have $\operatorname{gr}_{\mathfrak{M}} \mathcal{O}'_P \cong K[X,Y]$. Since F_1 and F_2 have no tangents in common at P, it follows that Lf_1 and Lf_2 are relatively prime polynomials in $K[X,Y]$. Then by B.12,

$$\operatorname{gr}_{\mathfrak{M}} \mathcal{O}'_P/(f_1, f_2) \cong K[X,Y]/(Lf_1, Lf_2).$$

If $\deg Lf_1 = m_P(F_1) =: m$, $\deg Lf_2 = m_P(F_2) =: n$, then the homogeneous component of the largest degree in $K[X,Y]/(Lf_1, Lf_2)$ (the socle, cf. 3.14) is of degree $m+n-2$ according to A.12(b). Since, by assumption, $\deg Lg \geq m+n-1$, we must have $Lg \in (Lf_1, Lf_2)$, and therefore $g \in (f_1, f_2)\mathcal{O}'_P + \mathfrak{M}^{m+n}$. By induction we get $g \in \bigcap_{i=m+n}^{\infty} (f_1, f_2)\mathcal{O}'_P + \mathfrak{M}^i$, and by the Krull intersection theorem E.8 it follows that $g \in (f_1, f_2)\mathcal{O}'_P$. But then $\dim_K \mathcal{O}_{F_1 \cap F_2 \cap G,P} = \dim_K \mathcal{O}'_P/(f_1, f_2, g) = \dim_K \mathcal{O}'_P/(f_1, f_2) = \mu_P(F_1, F_2)$.

Corollary 7.19. *Let F_1 and F_2 be curves with no component in common, and let G be a further curve. For each $P \in \mathcal{V}_+(F_1) \cap \mathcal{V}_+(F_2)$ suppose one of the following conditions is satisfied:*

(a) *F_1 is smooth at P and $\mu_P(F_1, G) \geq \mu_P(F_1, F_2)$.*
(b) *F_1 and F_2 have no tangent line in common at P and*

$$m_P(G) \geq m_P(F_1) + m_P(F_2) - 1.$$

Then $F_1 \cap F_2$ is a subscheme of G.

Corollary 7.19, like the Cayley–Bacharach theorem, gives sufficient conditions for the conclusion that $F_1 \cap F_2$ is a subscheme of G. Theorem 7.18 itself can sometimes be used to give an easy proof that condition (4) of the Cayley–Bacharach theorem holds.

We illustrate an application of 7.18 and 7.19 by means of Pascal's theorem.

Examples 7.20.

(a) Let F_1 and F_2 each be the union of 3 distinct lines. Suppose F_1 and F_2 intersect as in the following figure.

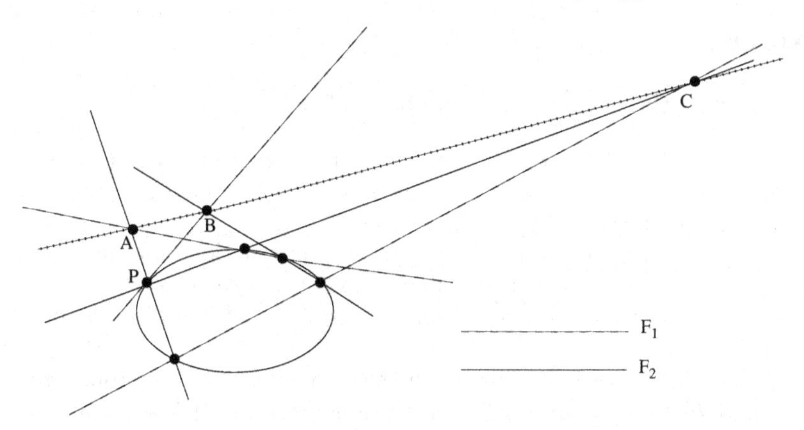

Here P is a point of multiplicity 2 of F_2, and a simple point of F_1. Besides this point, $\mathcal{V}_+(F_1) \cap \mathcal{V}_+(F_2)$ contains 7 other points of intersection multiplicity 1.

A quadric Q contains the 5 distinct points of intersection other than A, B, C, and is tangent to F_1 at P. Then $\dim_K \mathcal{O}_{F_1 \cap F_2 \cap Q, P} = \mu_P(F_1, F_2) = 2$ by 7.18(a). Let G be the union of Q with the line through A and B. Then the hypotheses of the Cayley–Bacharach theorem 5.17 are satisfied for F_1, F_2, G, and it follows that $C \in \mathrm{Supp}(G)$; i.e., A, B, C lie on a line.

This "degenerate case" of Pascal's theorem can be used to construct, using only a straightedge, a tangent line to a quadric Q through a given point P.

(b) The situation can even "degenerate" more strongly:

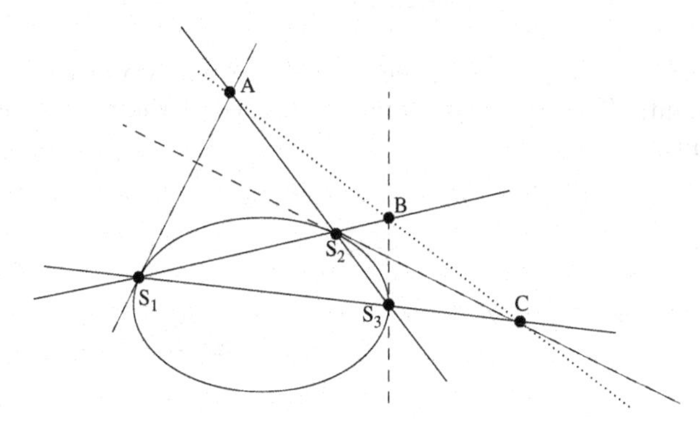

The quadric Q contains the double points S_1, S_2, and S_3 of F_2 and is tangent to F_1 at these points. Then A, B, C lie on a line.

Exercises

1. Calculate the intersection multiplicity of the curves

$$Y^2 - X^3 = 0, \qquad Y^p - X^2 = 0 \qquad (3 \leq p \leq q \neq 3 \cdot \tfrac{p}{2})$$

at the origin (use 5.9 and 7.4).
2. Carry out the above-mentioned tangent construction on a quadric.
3. How can one construct the line segment between two points with a straightedge that is shorter than the distance between the points? (Pappus's theorem).
4. Deduce the following theorems in the geometry of circles from 7.18:
   (a) Miguel's theorem: In $\mathbb{R}^2$ let a triangle $\{A, B, C\}$ be given, let A' be on the line through B and C, let B' be on the line through A and C, and let C' be a point on the line through A and B. Assume that $A', B', C' \notin \{A, B, C\}$. Consider the circles through A, B', C', through A', B, C', and through A', B', C. These circles intersect in a point.

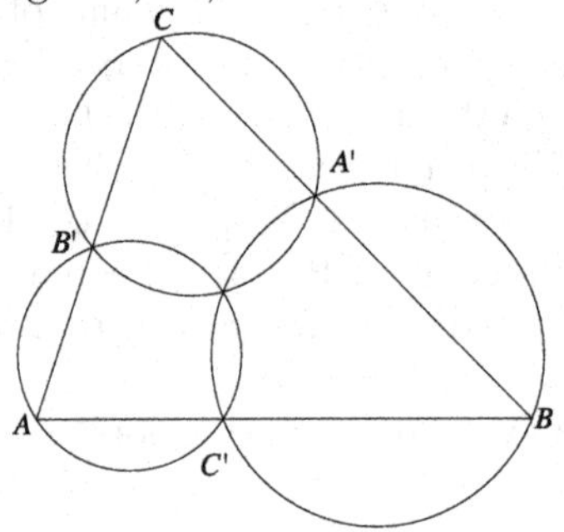

   (b) In $\mathbb{R}^2$, let 3 circles be given such that each two of them intersect in 2 points. Then the common secant lines of the circles intersect in a point.

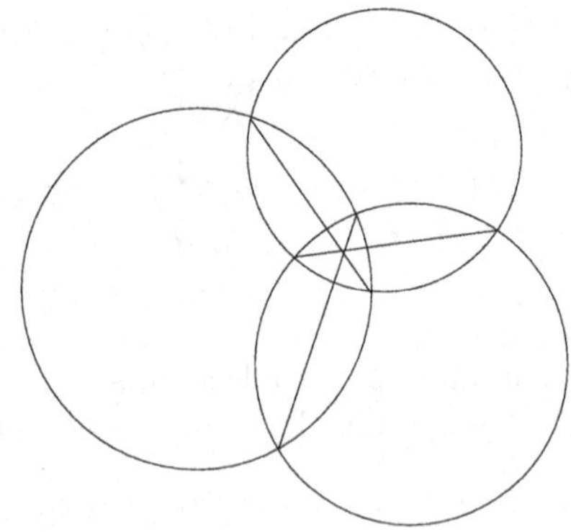

5. Give a proof using elementary geometry for 7.20(b) in the case that the quadric is a circle.

8

Rational Maps. Parametric Representations of Curves

Rational maps of the projective plane are given by homogeneous polynomials of the same degree. Above all, we are interested in the characterization of birational equivalence by rational maps. It will also be shown that a curve is rational precisely when it has a "parametric representation." This chapter depends on Chapter 4, but it also uses parts of Chapter 6.

Definition 8.1. For relatively prime homogeneous polynomials $\Phi_0, \Phi_1, \Phi_2 \in K[X_0, X_1, X_2]$ of the same degree, denote by $\Phi = \langle \Phi_0, \Phi_1, \Phi_2 \rangle$ the mapping

$$\Phi : \mathbb{P}^2(K) \setminus \bigcap_{i=0}^{2} \mathcal{V}_+(\Phi_i) \to \mathbb{P}^2(K)$$

given by $\Phi(\langle x_0, x_1, x_2 \rangle) = \langle \Phi_0(x_0, x_1, x_2), \Phi_1(x_0, x_1, x_2), \Phi_2(x_0, x_1, x_2) \rangle$. It is called the *rational map given by* Φ_0, Φ_1, Φ_2. We call $\operatorname{Def}(\Phi) := \mathbb{P}^2(K) \setminus \cap_{i=0}^{2} \mathcal{V}_+(\Phi_i)$ the *domain of definition* and $\cap_{i=0}^{2} \mathcal{V}_+(\Phi_i)$ the set of *indeterminate points* of Φ.

It is clear that Φ has only finitely many indeterminate points, since the Φ_i are relatively prime.

Examples 8.2.

(a) *Coordinate transformations of* $\mathbb{P}^2(K)$ are rational maps. If c is a coordinate transformation and Φ is a rational map, then $c \circ \Phi$ and $\Phi \circ c$ are also rational maps. In order to prove the rationality of a mapping, one can choose a "suitable" coordinate system.

(b) The *central projection* from a point onto a line. Let G be a line in $\mathbb{P}^2(K)$ and $P \in \mathbb{P}^2(K) \setminus G$. For each line G' through P, we associate to $G' \setminus \{P\}$ the point P' of intersection of G and G'.

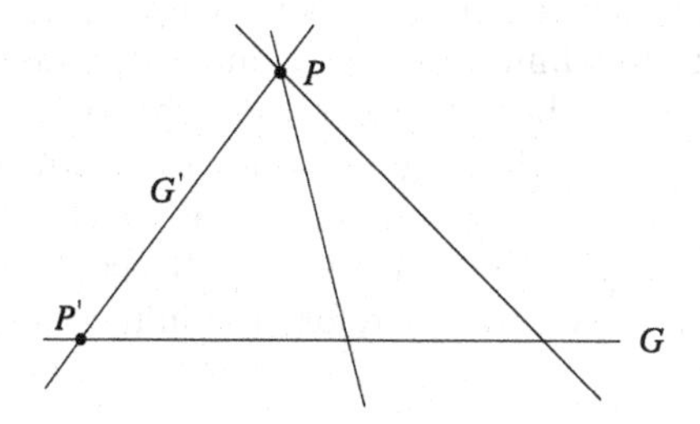

The map $\mathbb{P}^2(K) \setminus \{P\} \to G \subset \mathbb{P}^2(K)$ given here is rational: In a suitable coordinate system $P = \langle 0, 0, 1 \rangle$ and G is the line $X_2 = 0$. The central projection from P onto G will then be given by $\langle x_0, x_1, x_2 \rangle \mapsto \langle x_0, x_1, 0 \rangle$; i.e., it is $\Phi = \langle X_0, X_1, 0 \rangle$. The point P is called the *center* of the projection.

(c) *Quadratic transformations* (Cremona transformations). These are the maps $\Phi = \langle \Phi_0, \Phi_1, \Phi_2 \rangle$ with relatively prime homogeneous polynomials Φ_i of degree 2. We will investigate the specific map

$$\Phi = \langle X_1 X_2, X_0 X_2, X_0 X_1 \rangle$$

in some detail. Its indeterminate points are

$$P_0 = \langle 1, 0, 0 \rangle, \quad P_1 = \langle 0, 1, 0 \rangle, \quad P_2 = \langle 0, 0, 1 \rangle.$$

The lines $X_i = 0$ are mapped by Φ onto P_i $(i = 0, 1, 2)$.

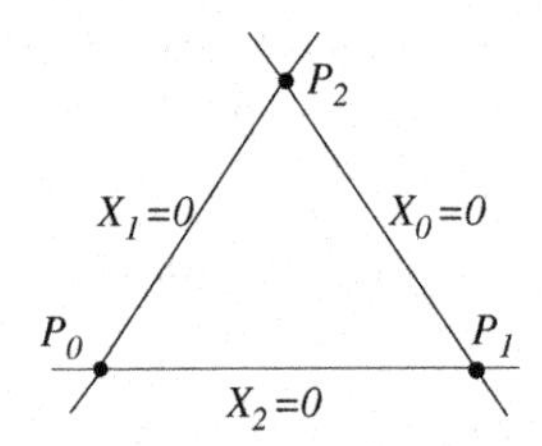

For points not in $\mathcal{V}_+(X_0 X_1 X_2)$, the map Φ^2 is given by

$$\langle x_0, x_1, x_2 \rangle \mapsto \langle x_0^2 x_1 x_2, x_0 x_1^2 x_2, x_0 x_1 x_2^2 \rangle,$$

and so is equal to the identity. The function Φ therefore maps $\mathbb{P}^2 \setminus \mathcal{V}_+(X_0 X_1 X_2)$ bijectively onto itself.

In a moment we shall use the following notation: If $A \in K[X_0, X_1, X_2]$ is a homogeneous polynomial and $\Phi = \langle \Phi_0, \Phi_1, \Phi_2 \rangle$ is a rational map, then we set

$$A^\Phi := A(\Phi_0, \Phi_1, \Phi_2).$$

This is also a homogeneous polynomial, and we have

(1) $$A^\Phi(x_0, x_1, x_2) = A(\Phi(x_0, x_1, x_2)) \quad \text{for} \quad \langle x_0, x_1, x_2 \rangle \in \mathrm{Def}(\Phi).$$

We will now show that two reduced curves in $\mathbb{P}^2(K)$ are birationally equivalent if and only if after finitely many points are removed from the curves, the remaining points can be mapped bijectively to one another by a rational mapping of the plane. Here again we identify reduced curves with their support $\mathcal{V}_+(F)$. As in Chapter 4 we call a subset $F^* \subset F$ dense if it contains infinitely many points from each irreducible component of F. The dense subsets of irreducible curves are therefore the infinite subsets.

Theorem 8.3. *For two reduced curves F and G in $\mathbb{P}^2(K)$, the following are equivalent:*

(a) *F and G are birationally equivalent.*
(b) *There are cofinite subsets $F^* \subset F$, $G^* \subset G$ and rational maps Φ, Ψ of $\mathbb{P}^2(K)$ with $F^* \subset \mathrm{Def}(\Phi)$, $G^* \subset \mathrm{Def}(\Psi)$, such that $\Phi|_{F^*}$ and $\Psi|_{G^*}$ are bijective inverses of each other.*
(c) *There are dense subsets $F^* \subset F$, $G^* \subset G$ and rational maps Φ, Ψ such that the conditions given in* (b) *are satisfied.*

Proof. (c) $\Rightarrow$ (a). If $A \in K[X_0, X_1, X_2]$ is a homogeneous polynomial with no components in common with G, then also F and A^Φ have no components in common, because otherwise by (1), A would vanish at infinitely many points of $G^* = \Phi(F^*)$, and therefore must have a component in common with G (3.10). This shows that there is a K-homomorphism

$$\mathcal{O}_G \longrightarrow \mathcal{O}_F \qquad \left(\frac{A}{B} \mapsto \frac{A^\Phi}{B^\Phi}\right)$$

defined by Φ. This map sends I_G to I_F, for if A is divisible by G, then by (1) A^Φ vanishes on F^*, hence is divisible by F, because F is reduced. Passing to the residue class rings, we get a K-homomorphism

$$\Phi^* : \mathcal{R}(G) \longrightarrow \mathcal{R}(F).$$

Similarly, Ψ induces a K-homomorphism

$$\Psi^* : \mathcal{R}(F) \longrightarrow \mathcal{R}(G).$$

For $\frac{A}{B} \in \mathcal{O}_G$ the rational functions $\frac{A}{B}$ and $\frac{(A^\Phi)^\Psi}{(B^\Phi)^\Psi}$ agree on the dense subset $G^* \subset G$. By 4.3 these fractions give the same functions of $\mathcal{R}(G)$. This shows that Φ^* and Ψ^* are inverse K-isomorphisms, i.e., that F and G are birationally equivalent.

(a) $\Rightarrow$ (b). Conversely, suppose two inverse K-isomorphisms

$$\alpha : \mathcal{R}(G) \to \mathcal{R}(F) \qquad \text{and} \qquad \beta : \mathcal{R}(F) \to \mathcal{R}(G)$$

are given. Suppose further, without loss of generality, that X_0 is not a component of F or G. Then $K\left[\frac{X_1}{X_0}, \frac{X_2}{X_0}\right] \subset \mathcal{O}_F \cap \mathcal{O}_G$. Denote by $\tilde{\alpha}$ the composite of the K-homomorphisms

$$K\left[\frac{X_1}{X_0}, \frac{X_2}{X_0}\right] \hookrightarrow \mathcal{O}_G \xrightarrow{\text{can.}} \mathcal{R}(G) \xrightarrow{\alpha} \mathcal{R}(F).$$

Then $\ker(\tilde{\alpha}) = I_G \cap K\left[\frac{X_1}{X_0}, \frac{X_2}{X_0}\right] = \left(G\left(1, \frac{X_1}{X_0}, \frac{X_2}{X_0}\right)\right)$. Write

$$\tilde{\alpha}\left(\frac{X_i}{X_0}\right) = \frac{\Phi_i}{\Phi_0} + I_F \quad \text{with} \quad \frac{\Phi_i}{\Phi_0} \in \mathcal{O}_F \quad (i = 1, 2).$$

By getting a least common denominator we can make sure that both fractions have the same denominator and that $\gcd(\Phi_0, \Phi_1, \Phi_2) = 1$. Since $\frac{\Phi_i}{\Phi_0} \in \mathcal{O}_F$ $(i = 1, 2)$, we also have $\gcd(\Phi_0, F) = 1$.

For the rational map $\Phi := (\Phi_0, \Phi_1, \Phi_2)$, because $G(1, \frac{\Phi_1}{\Phi_0}, \frac{\Phi_2}{\Phi_0}) \in I_F$, it follows that $G^{\Phi}(P) = G(\Phi(P)) = 0$ for all $P \in F$ with $\Phi_0(P) \neq 0$, hence on a cofinite subset of F and therefore on all of F. But it could be that P is an indeterminate point of Φ. In any case, $\Phi(P) \in G$ for all $P \in F$ that are not indeterminate points of Φ.

If $\tilde{\beta}$ and $\Psi = (\Psi_0, \Psi_1, \Psi_2)$ are defined similarly, then also $\Psi(Q) \in F$ for all $Q \in G$ that are not indeterminate points of Ψ. And it follows from $\tilde{\beta}(\frac{X_i}{X_0}) = \frac{\Psi_i}{\Psi_0} + I_G$ that

$$\tilde{\beta}\left(\frac{\Phi_i}{\Phi_0}\right) = \frac{\Phi_i^{\Psi}}{\Phi_0^{\Psi}} + I_G \qquad (i = 1, 2).$$

In particular, $\frac{\Phi_i^{\Psi}}{\Phi_0^{\Psi}} \in \mathcal{O}_G$ and hence $\gcd(\Phi_0^{\Psi}, G) = 1$.

Let G^* be the set of all points of G that are not indeterminate points of Ψ and that are not mapped under Ψ to indeterminate points of Φ. Also, the second set of exceptions is finite, because $\Phi_0(\Psi(P)) = \Phi_0^{\Psi}(P) = 0$ for only finitely many $P \in G$, since $\gcd(\Phi_0^{\Psi}, G) = 1$. The set $F^* \subset F$ is defined similarly. Then $F^* \subset F$ and $G^* \subset G$ are cofinite subsets.

Because $\beta \circ \alpha = \mathrm{id}$, the rational functions $\frac{\Phi_i^{\Psi}}{\Phi_0^{\Psi}}$ and $\frac{X_i}{X_0}$ agree $(i = 1, 2)$. We have then $\Phi(\Psi(P)) = P$ for all $P \in G^*$, and it follows that $\Phi|_{F^*}$ and $\Psi|_{G^*}$ are bijections that are inverse to each other.

Since (b) $\Rightarrow$ (c) is trivial, the theorem is proved.

An irreducible curve F in $\mathbb{P}^2(K)$ is therefore rational if and only if it is birationally equivalent to the line $X_2 = 0$. We may identify $\mathbb{P}^1(K)$ with this line by means of the map

$$\mathbb{P}^1(K) \to \mathbb{P}^2(K) \qquad (\langle u, v \rangle \mapsto \langle u, v, 0 \rangle).$$

If F is rational, then according to Theorem 8.3 there are homogeneous polynomials Φ_0, Φ_1, $\Phi_2 \in K[U, V]$ of the same degree such that, except for a finite number of choices of point P, every point of F can be represented as

$$P = \langle \Phi_0(u, v), \Phi_1(u, v), \Phi_2(u, v) \rangle$$

for a uniquely determined $\langle u, v \rangle \in \mathbb{P}^1(K)$. Without loss of generality, one can take the Φ_0, Φ_1, Φ_2 to be relatively prime, because the representation does not change at all if one cancels the greatest common divisor. Since $F(\Phi_0, \Phi_1, \Phi_2)$ vanishes for infinitely many points $\langle u, v \rangle \in \mathbb{P}^1(K)$, this polynomial in $K[U, V]$ is therefore the zero polynomial, and thus we have a mapping

$$(2) \qquad \Phi : \mathbb{P}^1(K) \to F \qquad (\langle u, v \rangle \mapsto \langle \Phi_0(u, v), \Phi_1(u, v), \Phi_2(u, v) \rangle)$$

that is bijective on cofinite subsets of $\mathbb{P}^1(K)$ and F.

Suppose conversely that arbitrary homogeneous polynomials $\Phi_0, \Phi_1, \Phi_2 \in K[U,V]$ of the same degree are given, suppose they are relatively prime, and suppose

$$\Phi : \mathbb{P}^1(K) \to \mathbb{P}^2(K)$$

is given as in (2). We will show that the image of Φ is an irreducible curve in $\mathbb{P}^2(K)$, but first we will show only that the image is contained in a uniquely determined irreducible curve.

Let $\phi_0, \phi_1, \phi_2 \in K[T]$ be the dehomogenizations of Φ_0, Φ_1, Φ_2 with respect to U, i.e., $\phi_i(T) = \Phi_i(1,T)$ $(i = 0,1,2)$. Without loss of generality we can assume that the rational functions

$$\frac{\phi_1}{\phi_0}, \frac{\phi_2}{\phi_0} \in K(T)$$

are not both constant, for otherwise dehomogenize with respect to V. It is then clear that the image of Φ contains infinitely many points (in affine coordinates) $(\frac{\phi_1(t)}{\phi_0(t)}, \frac{\phi_2(t)}{\phi_0(t)})$ $(t \in K,\ \phi_0(t) \neq 0)$. The K-homomorphism

$$\phi^* : K[X,Y] \to K(T) \qquad (\ X \mapsto \tfrac{\phi_1}{\phi_0}, \quad Y \mapsto \tfrac{\phi_2}{\phi_0}\ )$$

is not injective, since it is well known that any two elements in $K(T)$ are algebraically dependent over K, as is in fact easy to show. In addition, $\ker(\phi^*)$ can not be a maximal ideal, because the $\frac{\phi_i}{\phi_0}$ are not both constant. Therefore, $\ker(\phi^*) = (f)$ for some irreducible polynomial $f \in K[X,Y]$. Let $F \in K[X_0, X_1, X_2]$ be its homogenization. From $f(\frac{\phi_1}{\phi_0}, \frac{\phi_2}{\phi_0}) = 0$ it follows that

$$F(\Phi_0, \Phi_1, \Phi_2) = 0$$

and hence $\operatorname{im}\Phi \subset F$. Since the image of Φ contains infinitely many points, there can be only one irreducible curve F of this kind.

Definition 8.4. We say that F is given by the *parametric representation*

$$X_i = \Phi_i(U,V) \qquad (i = 0,1,2).$$

An arbitrary curve F *has a parametric representation* if it can be given by a parametric representation.

Certainly a line $F : a_0X_0 + a_1X_1 + a_2X_2 = 0$ $(a_i \in K)$ has a parametric representation $\Phi : \mathbb{P}^1(K) \to F$, which is moreover bijective. If, say, $a_2 \neq 0$, then it can be given by $X_0 = U$, $X_1 = V$, $X_2 = -\frac{1}{a_2}(a_0U + a_1V)$.

Theorem 8.5. *An irreducible curve F in $\mathbb{P}^2(K)$ is rational if and only if it has a parametric representation.*

Proof. By 8.3 it has already been shown that a rational curve has a parametric representation. To prove the converse, we may regard the curve as given by a parametric representation $X_i = \Phi_i(U, V)$ $(i = 0, 1, 2)$ as above. There is an injective K-homomorphism of $K[f] = K[X, Y]/(f)$ into $K(T)$ induced by ϕ^*, and hence an injection of $\mathcal{R}(F) = Q(K[f])$ into $K(T)$.

The theorem of Lüroth from field theory (which we assume is known here), implies that every field extension of K contained in $K(T)$ is generated by one element. Therefore $\mathcal{R}(F) = K(T')$ for some $T' \in K(T) \setminus K$, and hence F is a rational curve.

Using valuation-theoretic arguments from Chapter 6 we show

Theorem 8.6. *Suppose a curve F is given by a parametric representation $X_i = \Phi_i(U, V)$, for $i = 0, 1, 2$. Then the mapping*

$$\Phi : \mathbb{P}^1(K) \to F, \qquad \langle u, v \rangle \mapsto \langle \Phi_0(u, v), \Phi_1(u, v), \Phi_2(u, v) \rangle,$$

is surjective; i.e., the parametric representation "hits" every point of F.

The proof requires a few more preparations. The field $\mathcal{R}(\mathbb{P}^1)$ of rational functions on $\mathbb{P}^1$ is the set of all quotients $\frac{a}{b}$ where $a, b \in K[U, V]$ are homogeneous polynomials of the same degree and $b \neq 0$. Such a quotient will, as usual, be considered as a function defined on that part of $\mathbb{P}^1$ where the denominator does not vanish. It is clear that

$$\mathcal{R}(\mathbb{P}^1) = K\left(\frac{V}{U}\right) = K(T) \qquad \text{with } T := \frac{V}{U}.$$

Given a curve F as in the theorem, we identify the embedding $\mathcal{R}(F) \hookrightarrow K(T)$ constructed above with the mapping

$$\mathcal{R}(F) \hookrightarrow \mathcal{R}(\mathbb{P}^1)$$

that assigns to each rational function $r \in \mathcal{R}(F)$ the composition $r \circ \Phi \in \mathcal{R}(\mathbb{P}^1)$. This is independent of the choice of coordinates in $\mathbb{P}^2(K)$ and $\mathbb{P}^1(K)$.

The discrete valuation rings of $K(T)/K$ correspond one-to-one with the points $\langle u, v \rangle \in \mathbb{P}^1(K)$, and are the rings $R_{\langle u,v \rangle}$ of all rational functions on $\mathcal{R}(\mathbb{P}^1)$ that are defined at $\langle u, v \rangle$. Their description as subrings of $K(T)$ is as follows: If $u \neq 0$, then $R_{\langle u,v \rangle} = K[T]_{(uT-v)}$, the localization of $K[T]$ at the prime ideal $(uT - v)$. If $u = 0$, then $R_{\langle u,v \rangle} = K[T^{-1}]_{(T^{-1})}$. Since $\mathbb{P}^1(K)$ can be identified with an arbitrary line $F \subset \mathbb{P}^2(K)$, these statements follow from 6.13; see also Chapter 6, Exercise 1.

We denote the maximal ideal of $R_{\langle u,v \rangle}$ by $\mathfrak{m}_{\langle u,v \rangle}$. The proof of 8.6 is based on a valuation-theoretic description of the parametric representation Φ.

Lemma 8.7. *To each $\langle u, v\rangle \in \mathbb{P}^1(K)$ there exists a unique $P \in F$ with*

$$\mathcal{O}_{F,P} \subset \mathcal{R}_{\langle u,v\rangle}, \qquad \mathfrak{m}_{F,P} = \mathfrak{m}_{\langle u,v\rangle} \cap \mathcal{O}_{F,P}\,.$$

Here

$$P = \Phi(\langle u, v\rangle)$$

is the point corresponding to the "parameter" $\langle u, v\rangle$.

Proof. To show that P is uniquely determined by $\langle u, v\rangle$, we appeal to the corresponding uniqueness theorem in 6.12(a).

To prove the existence of P, by a suitable choice of coordinates, we can assume that $u \neq 0$. With $t := \frac{v}{u}$ and $\phi_i(T) := \Phi_i(1, T)$ $(i = 0, 1, 2)$, we can further assume that $\phi_0(t) \neq 0$. Then by the inclusion $\mathcal{R}(F) \hookrightarrow \mathcal{R}(\mathbb{P}^1) = K(T)$, the affine coordinate ring $K[f] = K[X, Y]/(f) = K[x, y]$ is identified with $K[\frac{\phi_1}{\phi_0}, \frac{\phi_2}{\phi_0}] \subset \mathcal{R}_{\langle u,v\rangle}$. The composition of $K[x, y] \to \mathcal{R}_{\langle u,v\rangle}$ with the canonical epimorphism $\mathcal{R}_{\langle u,v\rangle} \to K$ maps x to $a := \frac{\phi_1}{\phi_0}$ and y to $b := \frac{\phi_2}{\phi_0}$. The kernel of this map is $\mathfrak{m}_P := (x - a, y - b)$, and hence

$$\mathcal{O}_{F,P} \subset K[x, y]_{\mathfrak{m}_P} \subset \mathcal{R}_{\langle u,v\rangle}, \qquad \mathfrak{m}_{F,P} = \mathfrak{m}_{\langle u,v\rangle} \cap \mathcal{O}_{F,P}\,.$$

PROOF OF 8.6:
By 6.12(b), for each $P \in F$ there is a discrete valuation ring R' of $\mathcal{R}(F)/K$ with maximal ideal $\mathfrak{m}'$ such that

$$\mathcal{O}_{F,P} \subset R', \qquad \mathfrak{m}_{F,P} = \mathfrak{m}' \cap \mathcal{O}_{F,P}\,.$$

We have $\mathcal{R}(F) = K(T') \subset K(T)$ for some $T' \in K(T) \setminus K$. Therefore it is enough to show that there is a discrete valuation ring R of $K(T)/K$ with maximal ideal $\mathfrak{m}$ such that

$$R' \subset R, \qquad \mathfrak{m}' = \mathfrak{m} \cap R'.$$

Without loss of generality we can assume that $R' = K[T']_{(T')}$. Write in short form $T' = \frac{f}{g}$ $(f, g \in K[T],\ g \neq 0)$. If f is not a constant, then there is an element $a \in K$ with $f(a) = 0$, $g(a) \neq 0$. In this case, $R' \subset K[T]_{(T-a)}$ and $\mathfrak{m}' = (T - a)K[T]_{(T-a)} \cap R'$. If f is constant, then g is not constant. In this case we have $R' \subset K[T^{-1}]_{(T^{-1})}$ and $\mathfrak{m}' = (T^{-1})K[T^{-1}]_{(T^{-1})} \cap R'$.

Parametric representations of curves in the affine plane are given by two rational functions α, $\beta \in K(T)$ that are not both constant. The kernel of the K-homomorphism

$$K[X, Y] \to K(T) \qquad (X \mapsto \alpha,\ Y \mapsto \beta)$$

is a principal ideal (f) generated by an irreducible polynomial f. Since $f(\alpha, \beta) = 0$, we see that the curve $f = 0$ contains all points $(\alpha(t), \beta(t))$ for which $t \in K$ is not a pole of α and β. In this way f is uniquely determined. We say that f is given by the (rational) parametrization

$$X = \alpha(T), \qquad Y = \beta(T).$$

In contrast to the projective case, given a parameter value t there does not necessarily correspond a point on the curve, and also not all points on the curve are necessarily given by the parametric representation. For an example, see Chapter 1, Exercise 2. However, this is the case if the curve has a *polynomial* parametric representation (Exercise 1 below). It is clear that an affine irreducible curve has a parametric representation if and only if its projective closure has one.

Exercises

1. Suppose a curve f in $\mathbb{A}^2(K)$ is given by a "polynomial" parametric representation

$$X = \alpha(T), \quad Y = \beta(T) \qquad\qquad (\alpha, \beta \in K[T]).$$

   Show that the mapping $K \to \mathcal{V}(f)$ $(t \to (\alpha(t), \beta(t)))$ is surjective.
2. Show that all irreducible singular cubics are rational. (Theorem 7.17).
3. The reader may already be familiar with the *epicycloid* and *hypocycloid.* For $r, \rho, a \in \mathbb{R}_+$ and a variable $t \in \mathbb{R}$, the epicycloid is a plane (in general transcendental) curve with a parametric representation

$$x = (r+\rho)\cos t - a\cos(\tfrac{r+\rho}{\rho}t),$$

$$y = (r+\rho)\sin t - a\sin(\tfrac{r+\rho}{\rho}t).$$

   The hypocycloid is given by

$$x = (r-\rho)\cos t + a\cos(\tfrac{r-\rho}{\rho}t),$$

$$y = (r-\rho)\sin t - a\sin(\tfrac{r-\rho}{\rho}t).$$

   Show that if $\frac{r}{\rho}$ is a rational number, then these are rational algebraic curves (and only then). Which of the curves sketched in Chapter 1 are of this form?
4. Describe the image of the quadric $X_0^2 + X_1^2 + X_2^2 = 0$ under the quadratic transformation $\phi = \langle X_1X_2, X_0X_2, X_0X_1\rangle$.
5. Determine the divisor class group of $K(T)/K$.

9

Polars and Hessians of Algebraic Curves

The study of the tangents to an algebraic curve is continued in this chapter. We are concerned with the question of how many tangents of an algebraic curve can pass through a given point of the plane. We also investigate the "flex tangents," the tangent lines at inflection points.

For a point $P \in \mathbb{P}^2(K)$ and a line G with $P \notin G$, let

$$\pi_P : \mathbb{P}^2(K) \setminus \{P\} \to G$$

be the central projection from P onto G (8.2(b)).

If F is a projective algebraic curve of degree d with $P \notin \mathcal{V}_+(F)$, then π_P induces a mapping

$$\pi_P : \mathcal{V}_+(F) \to G.$$

This is surjective, since each line G' through P intersects the curve F, and for each $Q \in G$, the set $\pi_P^{-1}(Q)$ consists of d points P', when these are counted with their multiplicity $\mu_{P'}(F, G')$, where $G' := g(Q, P)$. One says that $\pi_P : \mathcal{V}_+(F) \to G$ is a "d-fold covering."

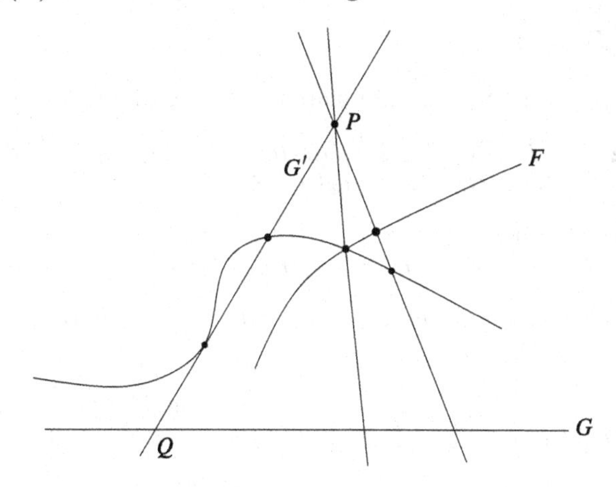

If $\pi_P^{-1}(Q)$ contains fewer than d distinct points, then $\mu_{P'}(F, G') > 1$ for at least one $P' \in \pi_P^{-1}(Q)$; i.e., the "projection line" G' must be tangent to F at P', or P' is a singularity of F. The question is, for how many points $Q \in G$ does this case occur?

For $P = \langle x_0, x_1, x_2 \rangle$ we consider the homogeneous polynomial

$$D_P := x_0 \frac{\partial F}{\partial X_0} + x_1 \frac{\partial F}{\partial X_1} + x_2 \frac{\partial F}{\partial X_2}$$

of degree $\deg D_P = \deg F - 1$. If F is irreducible with $\deg F > 1$ and the characteristic of K is either 0 or $> \deg F$, then D_P is nonzero for all $P \in \mathbb{P}^2(K)$, for otherwise (in a suitable coordinate system), F would be dependent on only two variables and so would be reducible.

Definition 9.1. If D_P is nonzero for a point $P \in \mathbb{P}^2(K)$, then we call the curve associated with D_P the *polar of F with respect to the pole P.*

The polar does not depend on the choice of the projective coordinate system: Let $A \in GL(3, K)$ be the matrix of a projective coordinate transformation and let $(y_0, y_1, y_2) = (x_0, x_1, x_2) \cdot A$, so that

$$F^A(Y_0, Y_1, Y_2) = F((Y_0, Y_1, Y_2) \cdot A^{-1})$$

with indeterminates Y_0, Y_1, Y_2. Then by the chain rule (using the shorthand $YA^{-1} = (Y_0, Y_1, Y_2) \cdot A^{-1}$):

$$\left(\frac{\partial F^A}{\partial Y_0}, \frac{\partial F^A}{\partial Y_1}, \frac{\partial F^A}{\partial Y_2} \right)^t = A^{-1} \cdot \left(\frac{\partial F}{\partial X_0}(YA^{-1}), \frac{\partial F}{\partial X_1}(YA^{-1}), \frac{\partial F}{\partial X_2}(YA^{-1}) \right)^t,$$

and therefore

$$\begin{aligned} \sum_{i=0}^{2} y_i \frac{\partial F^A}{\partial Y_i} &= (x_0, x_1, x_2) A A^{-1} \left(\frac{\partial F}{\partial X_0}(YA^{-1}), \frac{\partial F}{\partial X_1}(YA^{-1}), \frac{\partial F}{\partial X_2}(YA^{-1}) \right)^t \\ &= D_P(YA^{-1}). \end{aligned}$$

The geometrical significance of polars is given by the following theorem:

Theorem 9.2. *For $P \in \mathbb{P}^2(K)$ suppose the polar D_P of F is defined (i.e., $D_P \neq 0$). Then $\mathcal{V}_+(D_P) \cap \mathcal{V}_+(F)$ consists of:*

(a) *the singularities of F,*
(b) *the points of contact of all the tangent lines to F that pass through P.*

If $\deg F$ does not divide the characteristic of K, then we have $P \in \mathcal{V}_+(D_P)$ if and only if $P \in \mathcal{V}_+(F)$.

Proof. Let $Q = \langle y_0, y_1, y_2 \rangle \in \mathcal{V}_+(F)$. If Q is a singularity of F, then $Q \in \mathcal{V}_+(D_P)$ by the Jacobian Criterion 6.8. On the other hand, if Q is a regular point of F, then

$$X_0 \frac{\partial F}{\partial y_0} + X_1 \frac{\partial F}{\partial y_1} + X_2 \frac{\partial F}{\partial y_2} = 0$$

is the equation of the tangent to F at Q. This contains the point P if and only if $Q \in \mathcal{V}_+(D_P)$.

The Euler relation

$$X_0\frac{\partial F}{\partial X_0} + X_1\frac{\partial F}{\partial X_1} + X_2\frac{\partial F}{\partial X_2} = (\deg F)\cdot F$$

gives the last statement of the theorem.

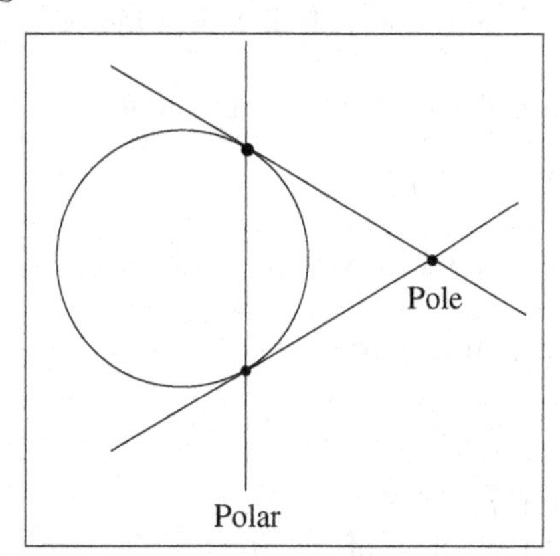

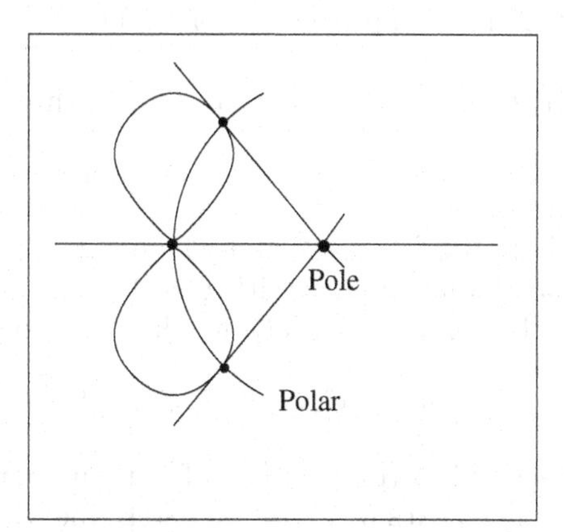

Corollary 9.3. *Let F be a smooth curve of degree $d > 1$ and suppose* $\operatorname{Char} K = 0$ *or* $\operatorname{Char} K > d$. *Then for each point $P \in \mathbb{P}^2(K)$ there are at most $d(d-1)$ tangents to F passing through P.*

Proof. F and D_P are relatively prime. By Bézout the set $\mathcal{V}_+(D_P) \cap \mathcal{V}_+(F)$ has at most $d(d-1)$ points.

If $P \notin \mathcal{V}_+(F)$ and $\pi_P : \mathcal{V}_+(F) \to G$ is the central projection onto the line G, then there are at most $d(d-1)$ points $Q \in G$ for which $\pi_P^{-1}(Q)$ contains fewer than d distinct points. The corollary no longer remains valid if the condition on the characteristic is violated. There can even be infinitely many tangents to a smooth curve through P (Exercise 1 below).

In the rest of this section we study inflection points (or flexes) and tangent lines at inflection points (flex tangents).

Definition 9.4. A point $P \in \mathbb{P}^2(K)$ is called a *flex* or an *inflection point* of F if

(a) P is a simple point of F, and
(b) if G is the tangent to F at P, then $\mu_P(F, G) > 2$.

A tangent at an inflection point is called a *flex tangent*.

The definition allows for G to be a component of F. A flex where the tangent is not a component of F is called a *proper flex*.

For example, all points of a line are (improper) flexes. A curve F of degree 2 has no proper flexes, for if a line G is not a component of F, then by Bézout $\sum_P \mu_P(F, G) = 2$ and therefore $\mu_P(F, G) \leq 2$ for all $P \in \mathcal{V}_+(F)$. It is clear that the concept of a flex is independent of the coordinate system.

In the following we assume that P is a regular point of F, where $\deg F \geq 3$. Let G be the tangent line to F at P. In order to derive conditions under which the point P is a flex of F we will assume that $P = (0,0)$ and that G is given by the affine equation $Y = 0$. Also let f be the affine curve associated with F. Two cases are possible:
(I) P is an improper flex of F. This is the case if and only if Y is a factor of F.
(II) G is not a component of F. In this case f can be written in the form

$$f = X^{\mu} \cdot \phi(X) + Y \cdot \psi(X, Y),$$

where $\mu \in \mathbb{N}$, $\mu \geq 2$ and where ϕ is a polynomial in X alone with $\phi(0) \neq 0$ and ψ a polynomial with $\psi(0,0) \neq 0$.

In the local ring $\mathcal{O}'_P$ we then have that ϕ and ψ are units. Also

$$\mu_P(F, G) = \dim_K \mathcal{O}'_P / (f, Y) = \dim_K K[X]_{(X)}(X^{\mu}) = \mu.$$

Then P is a (proper) flex of F if and only if $\mu > 2$.

The flexes of F can be determined with the help of the *Hessian determinant*

$$H_F := \det \left(\frac{\partial^2 F}{\partial X_i \partial X_j} \right)_{i,j=0,1,2} .$$

We have $\deg H_F = 3 \cdot (\deg F - 2)$. We will see that we can have $H_F = 0$. If, however $H_F \neq 0$, then one calls the curve in $\mathbb{P}^2(K)$ corresponding to H_F the *Hessian curve* (*or Hessian*) *of* F. This is independent of the choice of coordinates: If $F^A(Y_0, Y_1, Y_2)$ is given as in 9.1, an easy calculation using the chain rule shows that

$$H_{F^A}(Y_0, Y_1, Y_2) = (\det A)^2 \cdot H_F((Y_0, Y_1, Y_2) \cdot A^{-1}).$$

In the following we write $F_{X_i} := \frac{\partial F}{\partial X_i}$, $F_{X_i X_j} := \frac{\partial^2 F}{\partial X_i \partial X_j}$.

Lemma 9.5. *We always have*

$$X_0^2 \cdot H_F = \begin{vmatrix} n(n-1)F & (n-1)F_{X_1} & (n-1)F_{X_2} \\ (n-1)F_{X_1} & F_{X_1X_1} & F_{X_1X_2} \\ (n-1)F_{X_2} & F_{X_1X_2} & F_{X_2X_2} \end{vmatrix} .$$

Proof. Multiply the first row of the Hessian determinant by X_0 and add to that X_1 times the second row and X_2 times the third row. Using the Euler formula

$$(n-1)F_{X_i} = \sum_{j=0}^{2} F_{X_i X_j} \cdot X_j \qquad (i = 0, 1, 2)$$

yields the determinant

$$X_0 \cdot \begin{vmatrix} (n-1)F_{X_0} & (n-1)F_{X_1} & (n-1)F_{X_2} \\ F_{X_0X_1} & F_{X_1X_1} & F_{X_1X_2} \\ F_{X_0X_2} & F_{X_1X_2} & F_{X_2X_2} \end{vmatrix}.$$

Now do the analogous operations on the columns of this determinant and use $nF = \sum_{i=0}^{2} F_{X_i} \cdot X_i$ to get $X_0^2 H_F$ in the form given in the lemma.

Corollary 9.6. *For every singular point P of F we have $H_F(P) = 0$. Also, $H_F = 0$ if the characteristic of K divides $n-1$.*

Next we want to prove the following theorem.

Theorem 9.7. *Let F be a reduced curve of degree $n \geq 3$. Let p be the characteristic of K, and assume that either $p = 0$ or $p > n$. Then*

(a) *$H_F \equiv 0 \pmod F$ if and only if F is a union of lines.*
(b) *If $H_F \not\equiv 0 \pmod F$, then the intersection of F with its Hessian curve consists of the singular points of F and the flexes of F.*
(c) *For every regular point P of F whose tangent line G at P is not a component of F we have*

$$\mu_P(F, G) = \mu_P(F, H_F) + 2.$$

Proof. Let P be a regular point of F and let G be the tangent to F at P. To determine whether P is a flex point of F and whether $H_F(P) = 0$ we can assume that $P = (0,0)$ and that G is given by $Y = 0$. Let f be the affine curve corresponding to F.

By 9.5, $H_F(P)$ is the value of the determinant

$$\begin{aligned} \Delta := &\begin{vmatrix} n(n-1)f & (n-1)f_X & (n-1)f_Y \\ (n-1)f_X & f_{XX} & f_{XY} \\ (n-1)f_Y & f_{XY} & f_{YY} \end{vmatrix} \\ &= n(n-1)f(f_{XX}f_{YY} - f_{XY}^2) - (n-1)^2(f_X^2 f_{YY} + f_Y^2 f_{XX} - 2f_X f_Y f_{XY}) \end{aligned}$$

at the point $(0,0)$. If P is an improper flex of F (case I above), then Y is a divisor of f. In this case f, f_X, and f_{XX} all vanish at the point P and therefore $H_F(P) = 0$.

Case (II) above still remains, and now f will be written as it was there. A calculation with partial derivatives gives

$$\begin{aligned} f_X &= \mu X^{\mu-1}\phi + X^\mu \phi' + Y \cdot \psi_X, \\ f_{XX} &= \mu(\mu-1)X^{\mu-2}\phi + 2\mu X^{\mu-1}\phi' + X^\mu\phi'' + Y \cdot \psi_{XX}, \\ f_Y &= \psi + Y \cdot \psi_Y, \\ f_{YY} &= 2\psi_Y + Y \cdot \psi_{YY}, \\ f_{XY} &= \psi_X + Y \cdot \psi_{XY}. \end{aligned}$$

We consider now the value of the image of Δ in $\mathcal{O}_{F,P} \cong K[X,Y]_{(X,Y)}/(f)$ under the associated valuation. Since ϕ and ψ are units, the congruence

$Y \cdot \psi(X,Y) \equiv -X^\mu \phi(X) \bmod (f)$ shows that the maximal ideal of $\mathcal{O}_{F,P}$ is generated by the image of X. Also, the image of Y in $\mathcal{O}_{F,P}$ has the value μ. We see now that in the above expression for Δ the image of $f_Y^2 f_{XX}$ has value $\mu - 2$, while the images of the remaining terms have higher values. Therefore the image of Δ has value $\mu - 2$.

In particular, this shows that $\Delta \not\equiv 0 \bmod (f)$. Hence $H_F \not\equiv 0 \bmod (F)$, provided there exists a regular point P of F whose tangent is not a component of F. This is the case precisely when F is not a union of lines.

On the other hand, if F is such a union, then one can directly calculate that $H_F \equiv 0 \bmod (F)$: Without loss of generality suppose that $F = X_0 G$, where G is a product of linear homogeneous polynomials. Then 9.5 shows that

$$X_0^2 \cdot H_F \equiv \left(\frac{n-1}{n-2}\right)^2 X_0^5 \cdot H_G \bmod (F).$$

By induction we can assume that $H_G \equiv 0 \bmod (G)$, and therefore $H_F \equiv 0 \bmod (F)$.

This proves part (a) of the theorem. The formula

$$\mu_P(F, H_F) = \dim_K K[X,Y]_{(X,Y)}/(f, \Delta) = \mu - 2$$

shows that statement (c) of the theorem is also correct. Furthermore, the formula says that $P \in \mathcal{V}_+(F) \cap \mathcal{V}_+(H_F)$ precisely when $\mu > 2$, i.e., when P is a flex of F. Together with 9.6 this gives the assertion (b) of the theorem.

Example 9.8. If one drops the assumption about the characteristic, then the theorem is no longer true in general. Suppose $\operatorname{Char} K = 3$ and $F := X_0^2 X_2 - X_1^3$. This curve is irreducible, its singularity $\langle 0,0,1 \rangle$ is the only point at infinity, and one sees easily that $H_F = 0$.

The regular points of F are the points at finite distance. They satisfy the equation $Y = X^3$. Obviously $(0,0)$ is a flex of F. For an arbitrary point (a,b) of F at finite distance,

$$Y - X^3 = Y - X^3 - (b - a^3) = (Y - b) - (X - a)^3$$

and therefore (a,b) is also a flex of F.

In contrast to this we have the following.

Corollary 9.9. *Under the assumptions of 9.7 let F be irreducible and let s be the number of singularities of F. Then F has at most $3n(n-2) - s$ flexes.*

Proof. By 9.7(a) we have $H_F \not\equiv 0 \pmod F$ and by 9.7(c) we have $\mu_P(F, H_F) < \infty$ for every regular point of F. In particular, F is not a divisor of H_F. According to Bézout,

$$\sum_P \mu_P(F, H_F) = \deg F \cdot \deg H_F = 3n(n-2).$$

Since s terms in the sum come from the singularities of F, the statement follows.

Corollary 9.10. *Under the assumptions of 9.7 let F be smooth. Then F has at least one flex. If $\{P_1, \ldots, P_r\}$ is the set of all flexes of F, and $\{G_1, \ldots, G_r\}$ the set of corresponding flex tangents, then*

$$\sum_{i=1}^{r} (\mu_{P_i}(F, G_i) - 2) = 3n(n-2).$$

Exercises

1. Let F be a curve in $\mathbb{P}^2(K)$. A point $P \in \mathbb{P}^2(K)$ is called *strange* for F if there are infinitely many tangent lines to F through P.
   (a) Show that if there is a strange point for F, then $\operatorname{Char} K > 0$.
   (b) Give an example of a smooth curve with a strange point.
2. Determine the flexes of the curves in 1.2.
3. Let F be the irreducible quadric $X_0^2 + X_1^2 + X_2^2$ ($\operatorname{Char} K \neq 2$). For $P \in \mathbb{P}^2(K)$, let D_P denote the polar of P with respect to F. Show that the map $P \mapsto D_P$ gives a bijection of $\mathbb{P}^2(K)$ onto the set of all lines of $\mathbb{P}^2(K)$. What is the image of a line in $\mathbb{P}^2(K)$ under this map? Give an elementary geometric description of the mapping $P \mapsto D_P$.

10

Elliptic Curves

Next to the quadrics these are the most studied curves. They are the object of an extensive and deep theory with many connections to analysis and arithmetic (Husemöller [Hus], Lang [L], Silverman [S₁], [S₂]). On the role of elliptic curves in cryptography, see Koblitz [K] and Washington [W]. After choosing a point O, an elliptic curve may be given a group structure using a geometric construction. We first concern ourselves with this construction. Finally, we classify elliptic curves up to a coordinate transformation. This chapter contains only the rudiments of the algebraic theory of elliptic curves.

Definition 10.1. An *elliptic curve* in $\mathbb{P}^2(K)$ is a smooth curve of degree 3.

Theorem 10.2. *Suppose* $\operatorname{Char} K \neq 2$ *or* 3. *Every elliptic curve has exactly* 9 *flexes.*

Proof. Let $P_1, \dots, P_r$ be the flexes of an elliptic curve E and let $G_1, \dots, G_r$ be the corresponding flex tangents. Since $\deg E = 3$, we have $\mu_{P_i}(E, G_i) = 3$ for $i = 1, \dots, r$, and from the formula in 9.10 it follows that $r = 9$.

Example 10.3. The Fermat curve $X_0^3 + X_1^3 + X_2^3 = 0$ in $\mathbb{P}^2(\mathbb{C})$ is elliptic. The corresponding Hessian curve is given by $X_0X_1X_2 = 0$. The flexes of the Fermat curve are

$$\langle 1, \xi, 0 \rangle, \quad \langle 1, 0, \xi \rangle, \quad \langle 0, 1, \xi \rangle,$$

where ξ runs through the set of solutions of $X^3 + 1 = 0$.

Now let E be an elliptic curve and let G be a line in $\mathbb{P}^2(K)$. By Bézout G intersects the curve E in three points P, Q, R, where two or even all three of these may coincide. The case $P = Q$ occurs exactly when G is tangent to E at P, and $P = Q = R$ when G is a flex tangent at P. In any case, we write $G \cap E = \{P, Q, R\}$, where a point is repeated according to its intersection multiplicity of G with E.

Now let $O \in E$ be an arbitrarily chosen point. For A, $B \in E$, denote by $g(A, B)$ the line through A and B when $A \neq B$, and the tangent line to E at A when $A = B$. In addition, let

$$g(A, B) \cap E = \{A, B, R\}$$

and

$$g(O, R) \cap E = \{O, R, S\}.$$

There is then a well-defined operation

$$E \times E \to E, \qquad (A, B) \mapsto S,$$

which we call the *addition on* E (with respect to O), and which we write as the sum $S = A + B$.

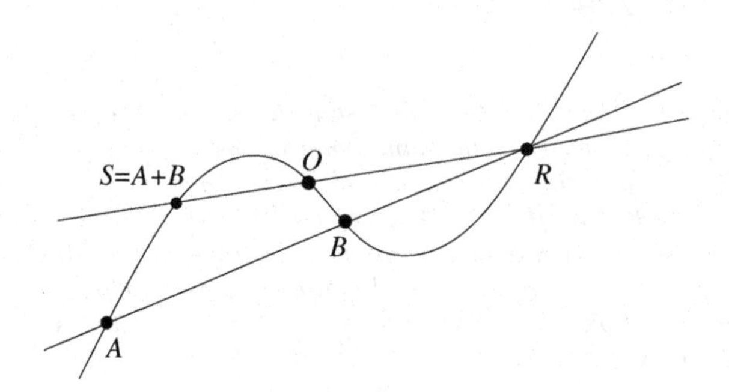

Theorem 10.4. $(E, +)$ *is an abelian group with identity element* O.

Proof. (a) By definition, $O + B = B$ for all $B \in E$. The commutativity of the addition is likewise obvious.

(b) Existence of inverses: Let $A \in E$ be given and suppose $g(O, O) \cap E = \{O, O, R\}$. Suppose further that $g(A, R) \cap E = \{A, R, B\}$. Then $A + B = O$ by definition of addition, and therefore $B = -A$.

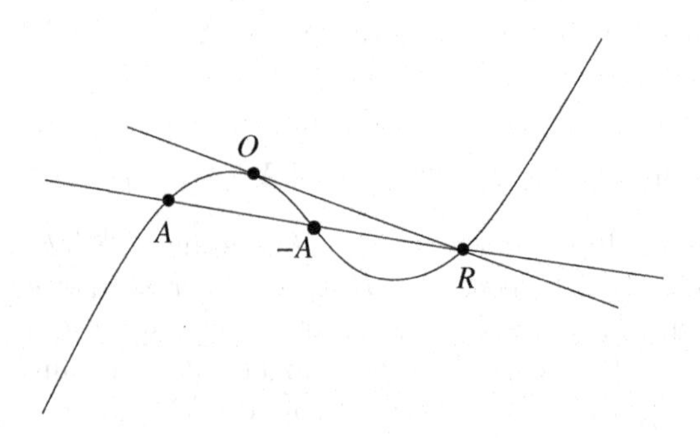

(c) The verification of associativity is somewhat more complicated. It uses a special case of the Cayley–Bacharach theorem. Let $A, B, C \in E$ be given, where of course two or all three of the points can coincide.

We define certain lines and intersection points with E one after another according to the following sketch and description:

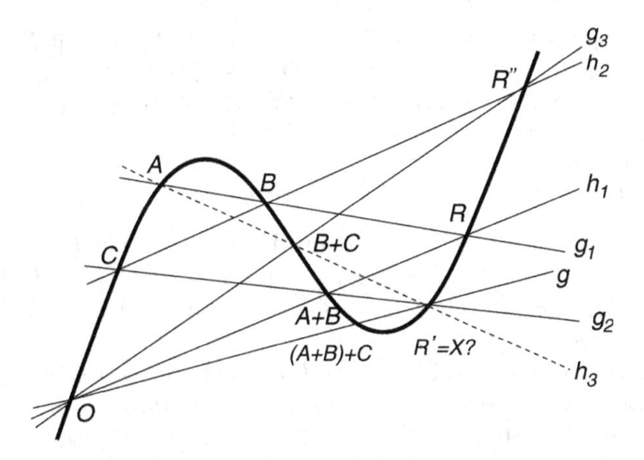

$$\begin{aligned}
g_1 &:= g(A,B), & g_1 \cap E &= \{A,B,R\},\\
h_1 &:= g(O,R), & h_1 \cap E &= \{O,R,A+B\},\\
g_2 &:= g(C,A+B), & g_2 \cap E &= \{C,A+B,R'\},\\
g &:= g(O,R'), & g \cap E &= \{O,R',(A+B)+C\},\\
h_2 &:= g(B,C), & h_2 \cap E &= \{B,C,R''\},\\
g_3 &:= g(O,R''), & g_3 \cap E &= \{O,R'',B+C\},\\
h_3 &:= g(A,B+C), & h_3 \cap E &= \{A,B+C,X\}.
\end{aligned}$$

We will show that $X = R'$, and then it follows that

$$(A+B)+C = A+(B+C).$$

Consider the cubic curves

$$\Gamma_1 := g_1 + g_2 + g_3, \quad \Gamma_2 := h_1 + h_2 + h_3.$$

We then have

$$\mathcal{V}_+(E) \cap \mathcal{V}_+(\Gamma_1) = \{O,A,B,C,R,R',R'',A+B,B+C\},$$

$$\mathcal{V}_+(E) \cap \mathcal{V}_+(\Gamma_2) = \{O,A,B,C,R,X,R'',A+B,B+C\},$$

where some of these points may also coincide, and then we count them according to their intersection multiplicities. We have to show equality of the intersection cycles

$$E * \Gamma_1 = E * \Gamma_2.$$

To do that we will use the Cayley–Bacharach theorem (5.17) and Theorem 7.18.

Let $S := \{O,A,B,C,R,R'',A+B,B+C\}$. Then for all $P \in S$ we have

$$\mu_P(E,\Gamma_1) = \mu_P(E,\Gamma_2),$$

and by 7.18 this number is also equal to $\dim_K \mathcal{O}_{\Gamma_1\cap\Gamma_2\cap E,P}$. Furthermore, $\sum_{P\in S} \dim_K \mathcal{O}_{\Gamma_1\cap\Gamma_2\cap E,P} \geq 8$, and therefore by Cayley–Bacharach, $E \cap \Gamma_1$ is a subscheme of Γ_2; i.e., $\dim_K \mathcal{O}_{\Gamma_1\cap\Gamma_2\cap E,P} = \mu_P(E,\Gamma_1)$ for all $P \in \mathcal{V}_+(E) \cap \mathcal{V}_+(\Gamma_1)$. In particular, $R' \in \mathcal{V}_+(E) \cap \mathcal{V}_+(\Gamma_2)$. Similarly, $\dim_K \mathcal{O}_{\Gamma_1\cap\Gamma_2\cap E,P} = \mu_P(E,\Gamma_2)$ for all $P \in \mathcal{V}_+(E)\cap\mathcal{V}_+(\Gamma_2)$, and in particular, $X \in \mathcal{V}_+(E)\cap\mathcal{V}_+(\Gamma_1)$. Therefore $E * \Gamma_1 = E * \Gamma_2$.

Remarks 10.5.

(a) Suppose the elliptic curve E is defined over a subfield $K_0 \subset K$, and let $E(K_0)$ be the set of K_0-rational points of E. If $O \in E(K_0)$, then for $A, B \in E(K_0)$, we also have $A + B \in E(K_0)$ and $-A \in E(K_0)$ and hence $(E(K_0), +)$ is a subgroup of $(E, +)$.

In particular, for an elliptic curve E defined over $\mathbb{Q}$ the set of $\mathbb{Q}$-rational points of E is a subgroup of $(E, +)$. By a deep *theorem of Mordell–Weil* this group is finitely generated (cf. Silverman [S1], Chapter VIII).

(b) Under the assumptions of 10.4, let $O^* \in E$ be another point and let $+^*$ be the addition defined by means of O^* on E. If c is a given coordinate transformation of $\mathbb{P}^2(K)$ with $c(E) = E$ and $c(O) = O^*$, then c induces a group isomorphism of $(E, +)$ and $(E, +^*)$. This is clear, because the construction of the sum is compatible with coordinate transformations.

However, one cannot map any point of E to any other by a coordinate transformation with $c(E) = E$; e.g., a flex point cannot be mapped to any point that is not a flex point. Nevertheless, $(E, +)$ and $(E, +^*)$ are always isomorphic groups (Exercise 3).

Theorem 10.6. *Let E be an elliptic curve. For $O \in E$ suppose $g(O,O)\cap E = \{O, O, T\}$. Then for points $A, B, C \in E$ the following are equivalent:*

(a) *There is a line g such that $g \cap E = \{A, B, C\}$.*
(b) *In $(E, +)$ we have $A + B + C = T$.*

Proof. Suppose $g(A, B) = \{A, B, R\}$. Then $T = A + B + R$ by definition of addition. Now we have $T = A + B + C$ precisely when $R = C$, i.e., precisely when $g(A, B) \cap E = \{A, B, C\}$.

Corollary 10.7. *If O is a flex of E, then $A+B+C$ is the intersection cycle of E with a line g if and only if $A + B + C = O$ in $(E, +)$.*

Corollary 10.8. *Suppose O is a flex of E. Then $P \in E$ is a flex if and only if $3P = O$. The set of flexes forms a subgroup of $(E, +)$ isomorphic to $\mathbb{Z}_3 \times \mathbb{Z}_3$.*

Proof. A point P is a flex of E if and only if $3P$ is the intersection cycle of E with a line. By 10.7 this is equivalent to $3P = O$ in $(E, +)$.

Each flex of E is a torsion point of $(E, +)$ of order 3. Since there are 9 flexes, these form a group isomorphic to $\mathbb{Z}_3 \times \mathbb{Z}_3$.

Notice that with regard to the 9 flexes, every two of them are collinear with a third, a situation that cannot be illustrated in $\mathbb{R}^2$. Remember that a curve over $\mathbb{C}$ may be thought of as a real surface, and a complex line as a real plane.

We come now to the classification of elliptic curves in the case that $\operatorname{Char} K \neq 2$ or 3 (see Husemöller [Hus] for $\operatorname{Char} K = 2$ and 3). If P is a flex of an elliptic curve E, then the coordinate system can be chosen so that $P = \langle 0, 0, 1\rangle$ and $X_0 = 0$ is the flex tangent to E at P. In such a coordinate system E has the equation

$$\text{(1)} \qquad a_0 X_2^3 + a_1 X_2^2 + a_2 X_2 + a_3 = 0,$$

where $a_i \in K[X_0, X_1]$ are homogeneous of degree i ($i = 0, \dots, 3$). Because $P = \langle 0, 0, 1\rangle \in E$, we must have $a_0 = 0$. Dehomogenizing with respect to X_2, we get an affine equation

$$a_1(X, Y) + a_2(X, Y) + a_3(X, Y) = 0.$$

Because $X_0 = 0$ is the tangent to E at P, we must have $a_1 = cX$ for some $c \in K^*$. However, $\mu_P(E, X_0) = 3$, and therefore X must be a divisor of a_2. We can take $c = 1$ without loss of generality. The equation (1) then has the form

$$X_0 X_2^2 + X_0(\alpha X_0 + \beta X_1)X_2 + a_3(X_0, X_1) = 0 \qquad (\alpha, \beta \in K).$$

Using the substitution

$$X_2 \mapsto X_2 - \frac{1}{2}(\alpha X_0 + \beta X_1), \quad X_1 \mapsto X_1, \quad X_0 \mapsto X_0,$$

we get the equation

$$X_0 X_2^2 + a_3(X_0, X_1) = 0 \qquad (\deg a_3 = 3).$$

Since X_0 is not a divisor of a_3, using another substitution this equation can eventually be written in the form

$$\text{(2)} \quad X_0 X_2^2 - (X_1 - aX_0)(X_1 - bX_0)(X_1 - cX_0) = 0 \qquad (a, b, c \in K).$$

Here a, b, c are distinct, for if $a = b$, say, then $\langle 1, a, 0\rangle$ is a singularity of E, as one sees immediately by taking partial derivatives.

The polar D_P of E with respect to $P = \langle 0, 0, 1\rangle$ is given by $\frac{\partial E}{\partial X_2} = 2X_0X_2 = 0$. It consists of the line at infinity $X_0 = 0$ and the affine X-axis. In addition to the point P, the polar intersects the curve E at the points with affine coordinates

$$(a, 0), \ (b, 0), \ (c, 0).$$

By 9.2 these are the points of contact of the tangents to E passing through P and different from the flex tangent. They are assigned to the flex P in a way independent of the coordinates. We have therefore shown:

Theorem 10.9. *If P is a flex of an elliptic curve E, then except for the flex tangent there are exactly three further tangents to E containing the point P. The points of contact of these tangents are collinear.*

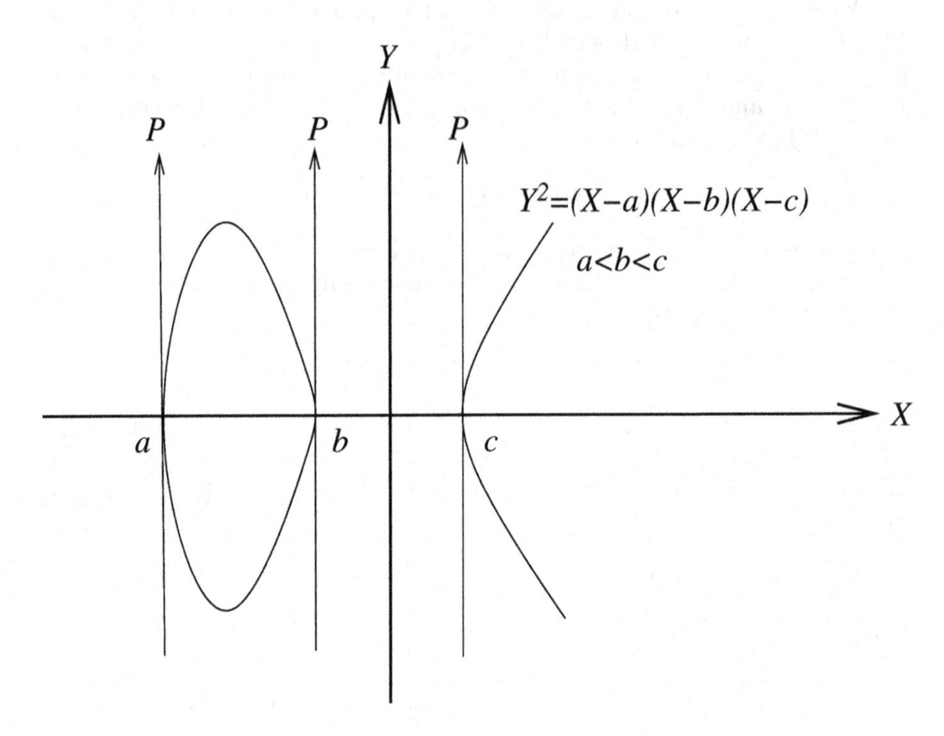

Using the substitution

$$X_0' = (b-a)X_0, \quad X_1' = X_1 - aX_0, \quad X_2' = (\sqrt{b-a})^{-1} \cdot X_2,$$

equation (2) becomes

$$(3) \qquad X_0'X_2'^2 - X_1'(X_1' - X_0')(X_1' - \lambda X_0') = 0 \quad \text{with} \quad \lambda := \frac{c-a}{b-a} \neq 0, 1.$$

We have now shown

Theorem 10.10. *Every elliptic curve is given in a suitable coordinate system by*

$$E_\lambda : Y_0Y_2^2 - Y_1(Y_1 - Y_0)(Y_1 - \lambda Y_0) = 0$$

with $\lambda \in K \setminus \{0, 1\}$.

The question arises when E_λ is projectively equivalent to $E_{\bar{\lambda}}$ with $\bar{\lambda} \in K \setminus \{0, 1\}$. We observe next:

Lemma 10.11. *If A and B are two flex points of E, then there is a coordinate transformation c of $\mathbb{P}^2(K)$ with $c(E) = E$ and $c(A) = B$.*

Proof. We may assume that $A \neq B$. The points A and B are collinear with a third flex P. As above, we choose this point to be the point at infinity $P = \langle 0, 0, 1\rangle$, and write the equation of E in the form (3). The corresponding affine equation is then

$$Y^2 - X(X-1)(X-\lambda) = 0. \tag{4}$$

The line $g(A, B)$ is parallel to the Y-axis, since it contains P. Under the substitution $Y \mapsto -Y$, equation (4) remains invariant and the line $g(A, B)$ is mapped to itself. Also, since flex points are mapped to flex points and $A \neq B$, the points A and B must necessarily be switched by the substitution.

The point $P = \langle 0, 0, 1\rangle$ is a flex of E_λ and $E_{\bar{\lambda}}$. If there exists a coordinate transformation c with $c(E_\lambda) = E_{\bar{\lambda}}$, then $c(P) = P$. The points $(0,0)$, $(1,0)$, and $(\lambda, 0)$ of the X-axis are the points of contact of the tangents to E_λ through P. Similarly for the points $(0,0)$, $(1,0)$, $(\bar{\lambda}, 0)$ and the curve $E_{\bar{\lambda}}$. Therefore c must fix the X-axis and map $\{(0,0),(1,0),(\bar{\lambda},0)\}$ to $\{(0,0),(1,0),(\lambda,0)\}$.

On the X-axis, c is given by the substitution $X \mapsto \gamma(X)$ with $\gamma(X) = aX + b$ ($a \in K^*$, $b \in K$), where $\gamma(\{0,1,\lambda\}) = \{0,1,\bar{\lambda}\}$, and every such transformation leads to an equation (4) for $E_{\bar{\lambda}}$. One can show easily that such a γ exists if and only if $\bar{\lambda}$ belongs to the set

$$M_\lambda := \{\lambda, \lambda^{-1}, 1-\lambda, (1-\lambda)^{-1}, \lambda(\lambda-1)^{-1}, (\lambda-1)\lambda^{-1}\}.$$

We have therefore shown:

Theorem 10.12. *E_λ is mapped by a coordinate transformation to $E_{\bar{\lambda}}$ if and only if $\bar{\lambda} \in M_\lambda$.*

The function j given by

$$j(\lambda) = 2^8 \frac{(\lambda^2 - \lambda + 1)^3}{\lambda^2(\lambda-1)^2} \qquad (\lambda \neq 0, 1)$$

is invariant under the substitutions $\lambda \mapsto \lambda$, $\lambda \mapsto \lambda^{-1}$, $\lambda \mapsto 1-\lambda$, etc. It is therefore an invariant of the class of curves "projectively equivalent" to E_λ, i.e., curves that can be mapped onto E_λ by a coordinate transformation. We set $j(E) = j(\lambda)$ for all elliptic curves E in the class of E_λ. (The number 2^8 is a "normalization factor," but we will not go into this any further here.)

Definition 10.13. $j(E)$ is called the *j-invariant* of the elliptic curve E.

Theorem 10.14. *For each $a \in K$ there is one and, up to projective equivalence, only one elliptic curve E with $j(E) = a$.*

Proof. For each $a \in K$ the degree 6 equation

$$2^8(\lambda^2 - \lambda + 1)^3 - a\lambda^2(\lambda-1)^2 = 0 \tag{5}$$

has a solution $\lambda_0 \neq 0, 1$, and all elements of M_{λ_0} are also solutions. We have $j(E_{\lambda_0}) = a$, and each λ_0' with $j(E_{\lambda_0'}) = a$ solves (5).

If M_{λ_0} consists of 6 different values, then these are all the solutions of the corresponding equation and we are finished. It is easy to check that M_{λ_0} contains fewer than 6 elements only in the following cases:

$$M_{-1} = M_{\frac{1}{2}} = M_2 = \left\{-1, \frac{1}{2}, 2\right\}$$

$$M_\rho = M_{\rho^{-1}} = \{\rho, \rho^{-1}\}, \quad \text{where } \rho := \frac{1}{2} + \frac{1}{2}\sqrt{-3}.$$

In the first case, $a = 2^6 \cdot 3^3$, and in the second case $a = 0$. In every case M_{λ_0} is the set of all solutions of the corresponding equation (5).

We have now solved the *classification problem for elliptic curves* in the following sense: There is a bijective map j from the set of projective equivalence classes of elliptic curves onto the field K.

In the complex numbers, elliptic curves can be parametrized by elliptic functions. This explains their name. Let $\Omega = \mathbb{Z}\omega_1 \oplus \mathbb{Z}\omega_2$ be a "lattice," i.e., $\omega_1, \omega_2 \in \mathbb{C}$ are linearly independent over $\mathbb{R}$. The Weierstraß $\wp$-function of the lattice Ω is well known to solve the differential equation

$$\wp'^2 - 4(\wp - e_1)(\wp - e_2)(\wp - e_3) = 0$$

with $e_1 := \wp(\frac{\omega_1}{2})$, $e_2 := \wp(\frac{\omega_2}{2})$, $e_3 := \wp(\frac{\omega_1+\omega_2}{2})$. The points $(\wp(z), \wp'(z))$ for $z \notin \Omega$ thus lie on the affine curve with equation

$$E_\Omega : Y^2 - 4(X - e_1)(X - e_2)(X - e_3) = 0.$$

One can show that if one assigns the $z \in \Omega$ to the points at infinity of E_λ, then one gets a bijection between $\mathbb{C}/\Omega$ and $\hat{E}_\Omega$, the projective completion of E_Ω. The group structure of $\mathbb{C}/\Omega$ corresponds to the group structure of the elliptic curve $\hat{E}_\Omega$ with the point at infinity O.

From well-known theorems about elliptic functions it follows using Theorem 10.12 that every elliptic curve over $\mathbb{C}$ is projectively equivalent to a curve $\hat{E}_\Omega$ for a suitably chosen lattice Ω in $\mathbb{C}$: The numbers e_1, e_2, e_3 can be assumed to be arbitrary distinct a, b, c in $\mathbb{C}$ when Ω is chosen properly.

Exercises

1. Let F be an irreducible singular cubic curve. When counted with multiplicity, every line intersects F in 3 points. By analogy with elliptic curves, one can then try to construct a group structure on F. Consider $F \setminus \operatorname{Sing} F$ and try to carry out the construction. What can you conclude?
2. Let K be a field of characteristic 3 and let $E \subset \mathbb{A}^2(K)$ be an elliptic curve. Show that no point $P \in \mathbb{A}^2(K)$ is a strange point for E (cf. Chapter 9, Exercise 1).

3. Under the assumptions of 10.5(b) let $g(O, O^*) \cap E = \{O, O^*, T\}$. For each $P \in E$, let $\alpha(P) \in E$ be defined by $g(P, T) \cap E = \{P, T, \alpha(P)\}$. Show that $\alpha : E \to E$ is an isomorphism from $(E, +)$ onto $(E, +^*)$.
4. Let O be the identity element for addition on an elliptic curve E, where here the addition will be denoted by $\tilde{+}$. The addition on the divisor group $\operatorname{Div}(E)$ will be denoted by $+$. Let $\operatorname{Div}^0(E)$ be the group of divisors of degree 0, $\mathcal{H}(E)$ the group of principal divisors, and $\operatorname{Cl}^0(E) := \operatorname{Div}^0(E)/\mathcal{H}(E)$ the group of divisor classes of degree 0. Show that:
   (a) If $P, Q \in E$, then $(P \tilde{+} Q) + O - P - Q \in \mathcal{H}(E)$.
   (b) The mapping

$$(E, +) \to \operatorname{Cl}^0(E) \qquad (P \mapsto (P - O) + \mathcal{H}(E))$$

   is an isomorphism of groups.
5. Show that an elliptic curve defined over $\mathbb{R}$ has exactly 3 real flex points.

11

Residue Calculus

We assign "residues" to the intersection points of two affine algebraic curves. The residues depend on a further curve (or more precisely, on a differential form $\omega = h\,dXdY$). They generalize the intersection multiplicity of two curves in a certain sense, and they contain more precise information about the intersection behavior. The elementary and purely algebraic construction of the residue that we present here is based on Appendix H and goes back to Scheja and Storch [SS₁], [SS₂]. Their work is also the basis of residue theory in higher-dimensional affine spaces, which can be developed in a similar fashion as here. What we are talking about is sometimes called Grothendieck residue theory. It was originally introduced in [H], Chapter 3, §9, in great generality. For different approaches, see also [Li₁] and [Li₂]. Chapters 11 and 12 will not be used in Chapter 13 and later. The reader may go directly from here to the Riemann–Roch theorem.

Let F and G be two algebraic curves in $\mathbb{P}^2(K)$ with $\deg F =: p > 0$ and $\deg G =: q > 0$, and with no common components. We assume that the coordinate system has been so chosen so that F and G have no points in common on the line at infinity $X_0 = 0$. Let f and g denote the dehomogenizations of F and G with respect to X_0, so that by 3.8 the projective coordinate ring

$$S := K[X_0, X_1, X_2]/(F, G)$$

of $F \cap G$ is the Rees algebra of the affine coordinate ring

$$A := K[X, Y]/(f, g)$$

of $f \cap g$ with respect to the degree filtration $\mathcal{F}$, and the associated graded algebra of A with respect to $\mathcal{F}$ is of the form

$$B := \operatorname{gr}_{\mathcal{F}} A = K[X, Y]/(Gf, Gg),$$

where Gf and Gg are the degree forms of f and g respectively. By 3.9, A/K and B/K are finite-dimensional algebras, and $S/K[X_0]$ has a finite basis consisting of homogeneous elements. Furthermore, $A \cong S/(X_0 - 1)$ and $B \cong S/X_0S$. The canonical modules (cf. Appendix H) $\omega_{S/K[X_0]}$ and $\omega_{B/K}$ are graded. The following connection between the canonical modules and the canonical traces results from H.5 and H.6:

Theorem 11.1. *There is a canonical isomorphism of graded B-modules*

$$\omega_{B/K} \cong \omega_{S/K[X_0]}/X_0\omega_{S/K[X_0]}$$

and a canonical isomorphism of A-modules

$$\omega_{A/K} \cong \omega_{S/K[X_0]}/(X_0 - 1)\omega_{S/K[X_0]}.$$

Here the canonical trace $\sigma_{S/K[X_0]}$ *corresponds in* $\omega_{B/K}$ *(respectively* $\omega_{A/K}$*) to the canonical trace* $\sigma_{B/K}$ *(respectively* $\sigma_{A/K}$*).*

By 3.14 the socle $\mathfrak{S}(B)$ of B is a K-vector space of dimension 1, and $\mathfrak{S}(B) = B_{p+q-2}$ is the homogeneous component of B of degree $p + q - 2$. Hence H.18 gives us the following result:

Theorem 11.2. *The algebras* $S/K[X_0]$ *and* B/K *have homogeneous traces of degree* $-(p + q - 2)$. *In particular, there are isomorphisms of graded modules*

$$\omega_{S/K[X_0]} \cong S, \qquad \omega_{B/K} \cong B.$$

The algebra A/K *also has a trace:*

$$\omega_{A/K} \cong A.$$

Now let F and G be two arbitrary curves in $\mathbb{P}^2(K)$. Suppose that at a point $P \in \mathcal{V}_+(F) \cap \mathcal{V}_+(G)$, the curves F and G do not have a component in common. Then $\mathcal{O}_{F\cap G,P}$ is a finite-dimensional algebra over K (5.3).

Corollary 11.3. $\mathcal{O}_{F\cap G,P}/K$ *has a trace.*

Proof. By 5.2 one can assume that F and G have no components in common at all. Since $\mathcal{O}_{F\cap G,P}$ is independent of the coordinates, one can further assume that F and G do not intersect on the line at infinity. Then we are in the situation as above. Since $\mathcal{O}_{F\cap G,P}$ is a direct factor of A and A/K has a trace (11.2), by H.10 this is also the case for $\mathcal{O}_{F\cap G,P}/K$.

Let f and g be two arbitrary affine curves with no common components that may now have common points at infinity. Then $A := K[X,Y]/(f,g)$ is always a finite-dimensional K-algebra and is the direct product of the local rings at the points of $\mathcal{V}(f) \cap \mathcal{V}(g)$. Since these each have a trace over K by 11.3, we get by H.10, via 11.2, also in this somewhat more general situation, the following corollary.

Corollary 11.4. A/K *has a trace.*

Under the assumptions of 11.3, we set $\mathcal{O} := \mathcal{O}_{F\cap G,P}$. Let $\mathfrak{m}$ denote the maximal ideal of $\mathcal{O}$, let $R := \mathcal{R}_\mathfrak{m}\,\mathcal{O} = \bigoplus_{k\in\mathbb{N}} \mathfrak{m}^k T^{-k} \oplus \bigoplus_{k=1}^{\infty} \mathcal{O}T^k$ be the Rees algebra, and let $G := \operatorname{gr}_\mathfrak{m}\mathcal{O}$ be the associated graded algebra of $\mathcal{O}$ with respect to the $\mathfrak{m}$-adic filtration. If P is a point at finite distance and $\mathfrak{M}$ is the maximal ideal of $K[X,Y]$ corresponding to P, then $\mathcal{O} \cong \mathcal{O}'_P/(f,g)$ with $\mathcal{O}'_P = K[X,Y]_\mathfrak{M}$. Let m be the multiplicity of F at P, and let n be the multiplicity of G at P. Then $\operatorname{ord}_\mathfrak{M} f = -m$ and $\operatorname{ord}_\mathfrak{M} g = -n$.

The algebra $\mathrm{gr}_{\mathfrak{m}} \mathcal{O}$ is a finite-dimensional K-algebra, and $\mathcal{R}_{\mathfrak{m}} \mathcal{O}$ has a basis as a $K[T]$-module consisting of homogeneous elements. As in 11.1, there are canonical isomorphisms

$$(1) \qquad \begin{aligned} \omega_{\mathcal{O}/K} &\cong \omega_{R/K[T]}/(T-1)\omega_{R/K[T]}, \\ \omega_{G/K} &\cong \omega_{R/K[T]}/T\omega_{R/K[T]}, \end{aligned}$$

and the corresponding statements about the canonical traces are valid.

Theorem 11.5. *Suppose F and G do not have a common tangent at P. Then $\mathcal{R}_{\mathfrak{m}} \mathcal{O}/K[T]$ and $\mathrm{gr}_{\mathfrak{m}} \mathcal{O}/K$ have homogeneous traces of degree $m+n-2$.*

Proof. Because of the assumption on the tangents, the $\mathfrak{M}$-leading forms Lf and Lg are relatively prime polynomials in $K[X,Y]$ (cf. 6.3(b)). By B.12, $G = \mathrm{gr}_{\mathfrak{m}} \mathcal{O} \cong K[X,Y]/(Lf, Lg)$, and the socle $\mathfrak{S}(G)$ is a K-vector space of dimension 1; hence $\mathfrak{S}(G) = G_{-(m+n-2)}$. The rest of the proof proceeds analogously to that of 11.2.

Remark. In general, $\mathrm{gr}_{\mathfrak{m}} \mathcal{O}/K$ need not have a trace (Exercise 1).

In the following it is important to specify certain traces. Some preparations are necessary. In the situation as at the beginning of this chapter we consider the enveloping algebra (cf. Appendix G)

$$S^e := S \otimes_{K[X_0]} S = S \otimes_{K[X_0]} K[X_0, X_1, X_2]/(F,G) = S[X_1, X_2]/(F,G)$$

of $S/K[X_0]$. The grading of S can be extended to the polynomial algebra $S[X_1, X_2]$, where the indeterminates X_1, X_2 have degree 1. Then the residue class algebra S^e is also positively graded.

Let x_1, x_2 be the images of X_1, X_2 in S; the image of X_0 in S will be denoted by X_0 again. We then have

$$S[X_1, X_2] = S[X_1 - x_1, X_2 - x_2] \quad \text{and} \quad (F,G)S[X_1, X_2] \subset (X_1 - x_1, X_2 - x_2).$$

Here the $X_i - x_i$ $(i = 1, 2)$ can be considered as variables over S of degree 1. We write

$$(2) \qquad \begin{aligned} F &= a_{11}(X_1 - x_1) + a_{12}(X_2 - x_2), \\ G &= a_{21}(X_1 - x_1) + a_{22}(X_2 - x_2), \end{aligned}$$

with homogeneous $a_{ij} \in S[X_1, X_2]$ $(i, j = 1, 2)$ and set $\Delta := \det(a_{ij})$. This determinant is homogeneous of degree $p + q - 2$. Similarly, let

$$A^e := A \otimes_K A = A \otimes_K K[X,Y]/(f,g) = A[X,Y]/(f,g)$$

and

$$B^e := B \otimes_K B = B \otimes_K K[X,Y]/(Gf, Gg) = B[X,Y]/(Gf, Gg)$$

be the enveloping algebras of A/K and B/K. The epimorphisms $S \to A$ (modulo $(X_0 - 1)$) and $S \to B$ (modulo X_0) induce epimorphisms

$$\varepsilon : S[X_1, X_2] \to A[X, Y] \quad (X_1 \mapsto X, \quad X_2 \mapsto Y)$$

and

$$\delta : S[X_1, X_2] \to B[X, Y] \quad (X_1 \mapsto X, \quad X_2 \mapsto Y).$$

Here $\varepsilon(F) = f$, $\varepsilon(G) = g$, $\delta(F) = Gf$, $\delta(G) = Gg$, so that by ε the induced epimorphism

$$S[X_1, X_2]/(F, G) \to A[X, Y]/(f, g)$$

can be identified with the canonical epimorphism $S^e \to A^e$ and

$$S[X_1, X_2]/(F, G) \to B[X, Y]/(Gf, Gg)$$

with $S^e \to B^e$.

We denote the images of X, Y in A by x, y and the images of X, Y in B by ξ, η. Applying ε to the system of equations (2), we get in $A[X, Y]$ the system of equations

$$\text{(3)} \qquad \begin{aligned} f &= \alpha_{11}(X_1 - x_1) + \alpha_{12}(X_2 - x_2), \\ g &= \alpha_{21}(X_1 - x_1) + \alpha_{22}(X_2 - x_2) \quad (\alpha_{ij} \in A[X, Y]). \end{aligned}$$

Similarly, the application of δ gives a system of equations in $B[X, Y]$,

$$\text{(4)} \qquad \begin{aligned} Gf &= \overline{a_{11}}(X_1 - x_1) + \overline{a_{12}}(X_2 - x_2), \\ Gg &= \overline{a_{21}}(X_1 - x_1) + \overline{a_{22}}(X_2 - x_2), \end{aligned}$$

with homogeneous $\overline{a_{ij}} \in B[X, Y]$. Let $\Delta^{F,G}_{x_1,x_2}$ be the image of $\det(a_{ij})$ in $S^e = S[X_1, X_2]/(F, G)$. We consider also systems (3) and (4), which do not necessarily arise as specializations of (2) by means of ε respectively δ, and define $\Delta^{f,g}_{x,y} \in A^e$ and $\Delta^{Gf,Gg}_{\xi,\eta} \in B^e$ similarly to the way we defined $\Delta^{F,G}_{x_1,x_2}$.

Theorem 11.6. *$\Delta^{F,G}_{x_1,x_2}$ is independent of the special choice of the coefficients a_{ij} in equation (2). Similarly for $\Delta^{f,g}_{x,y}$ and $\Delta^{Gf,Gg}_{\xi,\eta}$. Moreover, $\Delta^{F,G}_{x_1,x_2}$ is mapped by the canonical epimorphism $S^e \to A^e$ to $\Delta^{f,g}_{x,y}$ and by $S^e \to B^e$ to $\Delta^{Gf,Gg}_{\xi,\eta}$.*

Proof. By assumption, F and G are relatively prime in $K[X_0, X_1, X_2]$; i.e., F is not a zerodivisor mod G and G is not a zerodivisor mod F. The same holds also in $S[X_1, X_2]$ by G.4(b). By I.4, $\Delta^{F,G}_{x_1,x_2}$ does not depend on the choice of the coefficients a_{ij} in (2). The proof is similar for $\Delta^{f,g}_{x,y}$ and $\Delta^{Gf,Gg}_{\xi,\eta}$. The remaining statements follow because (3) and (4), as we have shown, can be considered to be specializations of (2).

Now let I^S be defined to be the kernel of the map $S^e \to S$ $(a \otimes b \mapsto ab)$, and similarly for I^A and I^B.

Theorem 11.7. $\operatorname{Ann}_{A^e}(I^A) = A \cdot \Delta_{x,y}^{f,g}$.

Proof. We apply I.5 with $R = A[X,Y]$, $a_1 = X - x$, $a_2 = Y - y$, $b_1 = f$, and $b_2 = g$. Before we start, by using a linear transformation of the variables X, Y, we make sure that g is a monic polynomial in Y. Then g is also monic as a polynomial in $Y - y$. The transformation of the variables has no effect on the statement of the theorem, since $\Delta_{x,y}^{f,g}$ will only be multiplied by the determinant of the transformation.

As a monic polynomial in $Y - y$, g is not a zerodivisor of $A[X,Y]/(X - x)$. Then also $X - x$ is not a zerodivisor of $A[X,Y]/(g)$. Since $Y - y$ is not a zerodivisor of $A[X,Y]/(X - x) \cong A[Y]$, the conditions of I.5 are satisfied. In $A^e = A[X,Y]/(f,g)$ we identify I^A with the ideal $(X - x, Y - y)/(f,g)$. Therefore, by I.5 we deduce the desired equality $\operatorname{Ann}_{A^e}(I^A) = (\Delta_{x,y}^{f,g})$.

According to H.20, the element $\Delta_{x,y}^{f,g}$ corresponds to a trace of A/K that we will denote by $\tau_{f,g}^{x,y}$. Whatever we say about A/K will also hold in particular for B/K. Therefore,

$$\operatorname{Ann}_{B^e}(I^B) = B \cdot \Delta_{\xi,\eta}^{Gf,Gg}$$

and a trace $\tau_{Gf,Gg}^{\xi,\eta}$ of B/K is specified by $\Delta_{\xi,\eta}^{Gf,Gg}$. Since $\Delta_{\xi,\eta}^{Gf,Gg}$ is homogeneous of degree $p + q - 2$, we have

$$\deg \tau_{Gf,Gg}^{\xi,\eta} = -(p + q - 2).$$

Finally, it follows from H.23 that

$$\operatorname{Ann}_{S^e}(I^S) = S \cdot \Delta_{x_1,x_2}^{F,G}.$$

The trace of $S/K[X_0]$ determined by $\Delta_{x_1,x_2}^{F,G}$ will be denoted by $\tau_{F,G}^{x_1,x_2}$. For this trace we also have

$$\deg \tau_{F,G}^{x_1,x_2} = -(p + q - 2).$$

Theorem 11.8. *$\tau_{f,g}^{x,y}$ is the image of $\tau_{F,G}^{x_1,x_2}$ under the canonical epimorphism $\omega_{S/K[X_0]} \to \omega_{A/K}$, and $\tau_{Gf,Gg}^{\xi,\eta}$ is the image of $\tau_{F,G}^{x_1,x_2}$ under the canonical epimorphism $\omega_{S/K[X_0]} \to \omega_{B/K}$.*

This follows from H.21.

Theorem 11.9. *The following formulas are valid:*

$$\sigma_{S/K[X_0]} = \frac{\partial(F,G)}{\partial(x_1,x_2)} \cdot \tau_{F,G}^{x_1,x_2}, \qquad \sigma_{A/K} = \frac{\partial(f,g)}{\partial(x,y)} \cdot \tau_{f,g}^{x,y},$$

and

$$\sigma_{B/K} = \frac{\partial(Gf,Gg)}{\partial(\xi,\eta)} \cdot \tau_{Gf,Gg}^{\xi,\eta}.$$

Proof. Let $\{s_1, \ldots, s_m\}$ be a basis of $S/K[X_0]$ and let $\{s_1', \ldots, s_m'\}$ be the dual basis to this basis with respect to $\tau_{F,G}^{x_1,x_2}$. Then by H.9,

$$\sigma_{S/K[X_0]} = \left(\sum_{i=1}^{m} s_i' s_i \right) \cdot \tau_{F,G}^{x_1,x_2} = \mu \left(\sum_{i=1}^{m} s_i' \otimes s_i \right) \cdot \tau_{F,G}^{x_1,x_2},$$

where $\mu : S^e \to S$ is the canonical surjection. By H.20(a) and the definition of $\tau_{F,G}^{x_1,x_2}$, we have $\Delta_{x_1,x_2}^{F,G} = \sum_{i=1}^{m} s_i' \otimes s_i$ and therefore

$$\sigma_{S/K[X_0]} = \mu(\Delta_{x_1,x_2}^{F,G}) \cdot \tau_{F,G}^{x_1,x_2}.$$

But (2) shows that the $\mu(a_{ij})$ are precisely the partial derivatives

$$\frac{\partial F}{\partial x_i} = \frac{\partial F}{\partial X_i}(x_1, x_2) \quad \text{and} \quad \frac{\partial G}{\partial x_j} = \frac{\partial G}{\partial X_j}(x_1, x_2).$$

Therefore $\mu(\Delta_{x_1,x_2}^{F,g})$ is the corresponding Jacobian determinant $\frac{\partial(F,G)}{\partial(x_1,x_2)}$. The proofs of the remaining formulas are similar.

The theorem shows in particular that the standard traces are traces if and only if the corresponding Jacobian determinants are units of S, A, and B respectively.

We will now describe the action of the trace $\tau_{Gf,Gg}^{\xi,\eta} : B \to K$ more precisely. In $K[X,Y]$ we have a system of equations

$$(5) \qquad \begin{aligned} Gf &= c_{11}X + c_{12}Y, \\ Gg &= c_{21}X + c_{22}Y, \end{aligned}$$

with homogeneous $c_{ij} \in K[X,Y]$. By 3.14 the image $d_{\xi,\eta}^{Gf,Gg}$ of $\det(c_{ij})$ in B is a generator of the socle $\mathfrak{S}(B) = B_{p+q-2}$ of B. Recall that by 3.12(b),

$$d_{\xi,\eta}^{Gf,Gg} = \frac{1}{pq} \cdot \frac{\partial(Gf, Gg)}{\partial(\xi, \eta)}$$

when the characteristic of K does not divide pq.

Theorem 11.10. *Let $\rho := p + q - 2$. Then*

$$\begin{aligned} \tau_{Gf,Gg}^{\xi,\eta}(B_k) &= \{0\} \textit{ for } k < \rho, \\ \tau_{Gf,Gg}^{\xi,\eta}(d_{\xi,\eta}^{Gf,Gg}) &= 1. \end{aligned}$$

Proof. The first formula holds because the trace is homogeneous of degree $-\rho$. To prove the second, we consider the following relations, which come from (5) and the corresponding equations in B,

$$\begin{aligned} Gf &= c_{11}'(X - \xi) + c_{12}'(Y - \eta) + (c_{11} - c_{11}')X + (c_{12} - c_{12}')Y, \\ Gg &= c_{21}'(X - \xi) + c_{22}'(Y - \eta) + (c_{21} - c_{21}')X + (c_{22} - c_{22}')Y, \end{aligned}$$

where the c'_{ij} are the images of the c_{ij} in B. The polynomials $c_{ij} - c'_{ij}$ vanish at the points (ξ, η), and are therefore linear combinations of $X - \xi$ and $Y - \eta$. We deduce that there are equations

$$\begin{aligned} Gf &= \widetilde{c_{11}}(X - \xi) + \widetilde{c_{12}}(Y - \eta), \\ Gg &= \widetilde{c_{21}}(X - \xi) + \widetilde{c_{22}}(Y - \eta), \end{aligned}$$

with homogeneous $\widetilde{c_{ij}} \in B[X, Y]$, where $\widetilde{c_{ij}} \equiv c'_{ij} \bmod (X, Y)$. Therefore

$$\det(\widetilde{c_{ij}}) \equiv d_{\xi,\eta}^{Gf,Gg} \bmod (X, Y)B[X, Y]. \tag{6}$$

Now let $\{1, b_1, \ldots, b_{pq-1}\}$ be a homogeneous basis for B/K with $1 \le \deg b_1 \le \deg b_2 \le \cdots \le \deg b_{pq-1} = \rho$. Then $\{1 \otimes 1, 1 \otimes b_1, \ldots, 1 \otimes b_{pq-1}\}$ is a basis for B^e/B, where B^e is the B-algebra with respect to $B \to B^e$ $(a \mapsto a \otimes 1)$ (G.4). By (6) there is an equation

$$\Delta_{\xi,\eta}^{Gf,Gg} = d_{\xi,\eta}^{Gf,Gg} \otimes 1 + \sum_{i=1}^{pq-1} b'_i \otimes b_i$$

with homogeneous elements b'_i of degree $< \rho$ $(i = 1, \ldots, pq - 1)$. Since $\tau_{Gf,Gg}^{\xi,\eta}$ is by definition the trace of $\Delta_{\xi,\eta}^{Gf,Gg}$, by H.20,

$$\tau_{Gf,Gg}^{\xi,\eta}\left(d_{\xi,\eta}^{Gf,Gg}\right) \cdot 1 + \sum_{i=1}^{pq-1} \tau_{Gf,Gg}^{\xi,\eta}(b'_i) \cdot b_i = 1.$$

The coefficients of the b_i vanish because of degree considerations, and it follows that

$$\tau_{Gf,Gg}^{\xi,\eta}\left(d_{\xi,\eta}^{Gf,Gg}\right) = 1.$$

The construction of the trace $\tau_{f,g}^{x,y} : A \to K$ makes no use of the fact that f and g have no points at infinity in common. Therefore $\tau_{f,g}^{x,y}$ is defined whenever f and g have positive degree and are relatively prime. We consider now the somewhat more general situation where $P \in \mathcal{V}(f) \cap \mathcal{V}(g)$, where we require only that f and g have no components with zero P in common. Let $\mathcal{O} = \mathcal{O}'_P/(f, g)\mathcal{O}'_P$. Furthermore, let $\mathfrak{M}$ be the maximal ideal in $K[X, Y]$ belonging to P and let $M := K[X, Y] \setminus \mathfrak{M}$. Then

$$\begin{aligned} \mathcal{O}^e &:= \mathcal{O} \otimes_K \mathcal{O} = \mathcal{O} \otimes_K \mathcal{O}'_P/(f, g)\mathcal{O}'_P \\ &= \mathcal{O} \otimes_K (K[X, Y]_{\mathfrak{M}}/(f, g)) = \mathcal{O}[X, Y]_M/(f, g). \end{aligned}$$

If x and y are the images of X and Y in $\mathcal{O}$, then in $\mathcal{O}[X, Y]_M$ there are equations

$$\begin{aligned} f &= \alpha_{11}(X - x) + \alpha_{12}(Y - y), \\ g &= \alpha_{21}(X - x) + \alpha_{22}(Y - y), \end{aligned} \tag{7}$$

with $\alpha_{ij} \in \mathcal{O}[X,Y]_M$. It is easy to see that I.4 can also be applied here: The image $(\Delta^{f,g}_{x,y})_P$ of $\det(\alpha_{ij})$ in $\mathcal{O}^e$ is independent of the choice of coefficients α_{ij} in (7). Also,

$$\text{(8)} \qquad \operatorname{Ann}_{\mathcal{O}^e}(I^{\mathcal{O}}) = \mathcal{O} \cdot (\Delta^{f,g}_{x,y})_P,$$

where $I^{\mathcal{O}}$ is the kernel of $\mathcal{O}^e \to \mathcal{O}$. By omitting factors in f and g that are units in $\mathcal{O}$, we can assume that f and g have no components in common. Then (7) can be considered as a system (3) to be read in $\mathcal{O}[X,Y]_M$. We then have that the image of $\Delta^{f,g}_{x,y}$ under the canonical homomorphism $A^e \to \mathcal{O}^e$ is $(\Delta^{f,g}_{x,y})_P$, and from $\operatorname{Ann}_{A^e}(I^A) = A \cdot \Delta^{f,g}_{x,y}$, equation (8) follows using G.9.

We denote the trace of $\mathcal{O}/K$ corresponding to $\Delta^{f,g}_{x,y}$ by $(\tau^{x,y}_{f,g})_P$, and ask how this trace is connected with $\tau^{x,y}_{f,g}$.

Theorem 11.11. *Suppose that f and g have no components in common and that $P \in \mathcal{V}(f) \cap \mathcal{V}(g)$. Then $(\tau^{x,y}_{f,g})_P$ is the restriction of $\tau^{x,y}_{f,g}$ to the direct factor $\mathcal{O}$ of A. In particular, for all $a \in A$,*

$$\tau^{x,y}_{f,g}(a) = \sum_{P \in \mathcal{V}(f) \cap \mathcal{V}(g)} (\tau^{x,y}_{f,g})_P(a_P),$$

where a_P denotes the image of a in the localization at P.

Proof. Write $A = \mathcal{O}_1 \times \cdots \times \mathcal{O}_h$, where the $\mathcal{O}_i$ are the localizations of A at its maximal ideals. Then by G.6(f),

$$A^e = A \otimes_K A = \prod_{i,j=1}^{h} \mathcal{O}_i \otimes_K \mathcal{O}_j\,.$$

Under the canonical mapping $A^e \to A$, if $i \neq j$, $\mathcal{O}_i \otimes_K \mathcal{O}_j$ is mapped to $\{0\}$, since $\mathcal{O}_i$ and $\mathcal{O}_j$ mutually annihilate each other in A. On the other hand, $\mathcal{O}_i \otimes_K \mathcal{O}_i$ is mapped as usual onto $\mathcal{O}_i$. With the usual notation the following formulas are valid:

$$I^A = I^{\mathcal{O}_1} \times \cdots \times I^{\mathcal{O}_h} \times \prod_{i \neq j} \mathcal{O}_i \otimes \mathcal{O}_j$$

and

$$\operatorname{Ann}_{A^e}(I^A) = \operatorname{Ann}_{\mathcal{O}_1^e}(I^{\mathcal{O}_1}) \times \cdots \times \operatorname{Ann}_{\mathcal{O}_h^e}(I^{\mathcal{O}_h}) \times \{0\}.$$

In particular, $\operatorname{Ann}_{\mathcal{O}^e}(I^{\mathcal{O}})$ is the image of $\operatorname{Ann}_{A^e}(I^A)$ under the projection of A^e onto the factor of $\mathcal{O}^e$ corresponding to P.

There is a commutative diagram

$$\begin{array}{ccc} \omega_{A/K} & \xrightarrow{\sim} & \operatorname{Hom}_A(\operatorname{Ann}_{A^e}(I^A), A) \\ \downarrow & & \downarrow \\ \omega_{\mathcal{O}/K} & \xrightarrow{\sim} & \operatorname{Hom}_{\mathcal{O}}(\operatorname{Ann}_{\mathcal{O}^e}(I^{\mathcal{O}}), \mathcal{O}) \end{array}$$

in which the horizontal isomorphisms come from H.19 and the vertical arrows are given by projection onto the appropriate direct factors. Since the linear form of $\operatorname{Hom}_A(\operatorname{Ann}_{A^e}(I^A), A)$ specified by $\Delta^{f,g}_{x,y} \mapsto 1$ is mapped to $(\Delta^{f,g}_{x,y})_P \mapsto 1$, it follows that $(\tau^{x,y}_{f,g})_P$ is the restriction of $\tau^{x,y}_{f,g}$ on $\mathcal{O}$. The last statement of the theorem follows similarly as in the proof of the formula in H.3.

As in 11.5 we additionally assume that f and g have no common tangents. Then $G := \operatorname{gr}_{\mathfrak{m}} \mathcal{O} = K[X,Y]/(Lf, Lg)$, where Lf and Lg are the $\mathfrak{M}$-leading forms of f respectively g, and these are relatively prime. If ξ and η represent the residue classes of X and Y in G, then the trace

$$\tau^{\xi,\eta}_{Lf,Lg} : G \longrightarrow K$$

is defined. Observe that here the variables X, Y are of degree -1 and that G consists of only homogeneous components of degrees ≤ 0.

Completely analogous to 11.10 we have the following:

Theorem 11.12. *Let $m := m_P(f)$, $n := m_P(g)$, and $\rho := m + n - 2$. Then the trace $\tau^{\xi,\eta}_{Lf,Lg}$ is homogeneous of degree ρ. In particular,*

$$\tau^{\xi,\eta}_{Lf,Lg}(G_k) = \{0\} \quad \textit{for} \quad k = -\rho + 1, \ldots, 0.$$

Using notation analogous to that in 11.10 we have furthermore

$$\tau^{\xi,\eta}_{Lf,Lg}(d^{Lf,Lg}_{\xi,\eta}) = 1.$$

This completely describes $\tau^{\xi,\eta}_{Lf,Lg}$. By 3.12(b) we have

$$d^{Lf,Lg}_{\xi,\eta} = \frac{1}{mn} \cdot \frac{\partial(Lf, Lg)}{\partial(\xi, \eta)}$$

if the characteristic of K does not divide mn.

Now the question arises as to how $\tau^{\xi,\eta}_{Lf,Lg}$ is related to $(\tau^{\xi,\eta}_{f,g})_P$. The relationship follows once again correspondingly as in 11.8 using the Rees algebra $R := \mathcal{R}_{\mathfrak{m}} \mathcal{O}$ of $\mathcal{O}/K$.

The Rees algebra

$$Q := \mathcal{R}_{\mathfrak{M}} K[X,Y]_{\mathfrak{M}} = \bigoplus_{k \in \mathbb{N}} \mathfrak{M}^k K[X,Y]_{\mathfrak{M}} \cdot T^{-k} \oplus \bigoplus_{k=1}^{\infty} K[X,Y]_{\mathfrak{M}} T^k$$

of $\mathcal{O}'_P$ with respect to its maximal ideal can be identified with $K[T, X^*, Y^*]_M$ according to C.14, where $M := K[X,Y] \setminus \mathfrak{M}$. Then in the polynomial ring $K[T, X^*, Y^*]$ we have the degree relations $\deg T = 1$, $\deg X^* = \deg Y^* = -1$, and the polynomial algebra $K[X,Y]$ is embedded in $K[T, X^*, Y^*]$ by means of $X = TX^*$, $Y = TY^*$.

If $f = f_m + \cdots + f_p$, $g = g_n + \cdots + g_q$ are the decompositions of f and g into homogeneous polynomials, then

$$f^* = f_m(X^*, Y^*) + T f_{m+1}(X^*, Y^*) + \cdots + T^{p-m} f_p(X^*, Y^*),$$
$$g^* = g_n(X^*, Y^*) + T g_{n+1}(X^*, Y^*) + \cdots + T^{q-n} g_q(X^*, Y^*).$$

Since $R = Q/(f^*, g^*)$, we have

$$R^e = R \otimes_{K[T]} (Q/(f^*, g^*)) = (R \otimes_{K[T]} Q)/(1 \otimes f^*, 1 \otimes g^*) = R[X^*, Y^*]_M/(f^*, g^*).$$

Let x^* and y^* be the images of X^* and Y^* in R. Then there is a system of equations

$$\begin{aligned} f^* &= a_{11}(X^* - x^*) + a_{12}(Y^* - y^*), \\ g^* &= a_{21}(X^* - x^*) + a_{22}(Y^* - y^*), \end{aligned} \tag{9}$$

with $a_{ij} \in R[X^*, Y^*]_M$, and we define $\Delta_{x^*,y^*}^{f^*,g^*}$ as the image of $\det(a_{ij})$ in R^e. We check that this image does not depend on the special choice of the a_{ij}.

Since Lf and Lg are relatively prime in $K[X,Y] = \operatorname{gr}_{\mathfrak{M}} \mathcal{O}'_P$, it follows that g^* is a nonzerodivisor on $Q/(f^*)$, and f^* is a nonzerodivisor on $Q/(g^*)$ (B.12). Then, however, g^* is also a nonzerodivisor on $R \otimes_{K[T]} Q/(f^*) = R[X^*, Y^*]_M/(f^*)$, and f^* is a nonzerodivisor on $R[X^*, Y^*]_M/(g^*)$ (G.4(b)). Hence the hypotheses of I.4 are satisfied and the claim follows.

In formula (9), if one specializes the variable T to 0, then one gets a system of equations in $G[X^*, Y^*]$, as in the construction of $\tau_{\xi,\eta}^{Lf,Lg}$. Therefore, $\Delta_{\xi,\eta}^{Lf,Lg}$ is the image of $\Delta_{x^*,y^*}^{f^*,g^*}$ under the epimorphism $R^e \to G^e$. By H.23 we have

$$\operatorname{Ann}_{R^e}(I^R) = R \cdot \Delta_{x^*,y^*}^{f^*,g^*},$$

and hence a trace $\tau_{f^*,g^*}^{x^*,y^*} : R \to K[T]$ is defined. As was the case for the standard trace (cf. formula (1)), we also have here the following:

Theorem 11.13. *Under the canonical epimorphism* $\omega_{\mathcal{R}_{\mathfrak{M}} \mathcal{O}/K[T]} \to \omega_{\mathcal{O}/K}$ *the trace* $\tau_{f^*,g^*}^{x^*,y^*}$ *is mapped onto* $(\tau_{f^*,g^*}^{x^*,y^*})_P$*, and under the canonical epimorphism* $\omega_{\mathcal{R}_{\mathfrak{M}} \mathcal{O}/K[T]} \to \omega_{\operatorname{gr}_{\mathfrak{m}} \mathcal{O}/K}$ *onto* $\tau_{Lf,Lg}^{\xi,\eta}$.

The proof is analogous to that of 11.8.

Now again let $A = K[X,Y]/(f,g) = K[x,y]$ with relatively prime polynomials f and g. As the notation suggests, the traces $\tau_{f,g}^{x,y}$ and $(\tau_{f,g}^{x,y})_P$ depend on the generators (coordinates) x, y of the algebra A/K and on the ordered pair of relations $\{f, g\}$. The following lemma is simple.

Lemma 11.14. *Let* $\tilde{X}$, $\tilde{Y} \in K[X,Y]$ *be given and suppose*

$$X = \gamma_{11}\tilde{X} + \gamma_{12}\tilde{Y},$$
$$Y = \gamma_{21}\tilde{X} + \gamma_{22}\tilde{Y},$$

with $\gamma_{ij} \in K$, $\det(\gamma_{ij}) \neq 0$. *For* $h \in K[X,Y]$, *let* $\tilde{h}$ *be defined by* $\tilde{h}(\tilde{X}, \tilde{Y}) = h(\gamma_{11}\tilde{X} + \gamma_{12}\tilde{Y}, \gamma_{21}\tilde{X} + \gamma_{22}\tilde{Y})$. *Let* $\tilde{x}$ *and* $\tilde{y}$ *denote the residue classes of* $\tilde{X}$ *and* $\tilde{Y}$ *in* A. *Then*

$$\tau_{f,g}^{x,y} = \left(\det(\gamma_{ij}) \cdot \tau_{\tilde{f},\tilde{g}}^{\tilde{x},\tilde{y}}\right) \circ c,$$

where $c : A \to A$ denotes the K-automorphism induced by $h \mapsto \tilde{h}$. Similarly for the local traces.

We now introduce a new notation and a new name for the trace.

Definition 11.15. For $h \in K[X,Y]$ with image $\bar{h}$ in A, we call

$$\int \begin{bmatrix} \omega \\ f, g \end{bmatrix} := \tau_{f,g}^{x,y}(\bar{h})$$

the *integral* of $\omega := h\, dX dY$ with respect to f, g. For $h \in \mathcal{O}'_P$ with image $\bar{h}$ in $\mathcal{O}$, we call

$$\operatorname{Res}_P \begin{bmatrix} \omega \\ f, g \end{bmatrix} := (\tau_{f,g}^{x,y})_P(\bar{h})$$

the *residue* of $\omega = h\, dX dY$ with respect to f, g at the point P. We set $\operatorname{Res}_P \begin{bmatrix} \omega \\ f, g \end{bmatrix} = 0$ if $P \notin \mathcal{V}(f) \cap \mathcal{V}(g)$.

Without going further into differential forms, we understand by $\omega = h\, dX dY$ a symbol that is changed under a coordinate transformation $X = \gamma_{11}\tilde{X} + \gamma_{12}\tilde{Y}$, $Y = \gamma_{21}\tilde{X} + \gamma_{22}\tilde{Y}$ ($\gamma_{ij} \in K$) by the factor $\det(\gamma_{ij})$: $dX dY = \det(\gamma_{ij})\, d\tilde{X} d\tilde{Y}$. By 11.14, $\int$ and Res_P are independent of the choice of coordinates. The residue is defined even if f and g have no components in common at P. The symbol $\operatorname{Res}_P \begin{bmatrix} \omega \\ f,\ g \end{bmatrix}$ is sometimes called the *Grothendieck residue symbol.*

In the following, the basic properties of the integral and the residue will be described. Obviously,

$$\int \begin{bmatrix} \omega \\ f, g \end{bmatrix} = 0 \qquad \text{if } h \in (f,g)K[X,Y], \tag{10}$$

$$\operatorname{Res}_P \begin{bmatrix} \omega \\ f, g \end{bmatrix} = 0 \qquad \text{if } h \in (f,g)K[X,Y]_{\mathfrak{M}}. \tag{11}$$

Since the traces are K-linear maps, the integral and residue are also K-linear functions of ω; i.e., for $\kappa_1, \kappa_2 \in K$ and differential forms ω_1, ω_2,

$$\int \begin{bmatrix} \kappa_1\omega_1 + \kappa_2\omega_2 \\ f, g \end{bmatrix} = \kappa_1 \int \begin{bmatrix} \omega_1 \\ f, g \end{bmatrix} + \kappa_2 \int \begin{bmatrix} \omega_2 \\ f, g \end{bmatrix}, \tag{12}$$

and similarly for the residues.

Furthermore, by the formula in 11.11 with $\omega = h\, dX dY$ and $h \in K[X,Y]$,

$$\int \begin{bmatrix} \omega \\ f,g \end{bmatrix} = \sum_P \operatorname{Res}_P \begin{bmatrix} \omega \\ f,g \end{bmatrix}, \tag{13}$$

the sum being extended over all $P \in \mathbb{A}^2(K)$.

We now turn to the question, How do the integral and the residue depend on the polynomials f and g?

Let $P \in \mathcal{V}(f) \cap \mathcal{V}(g)$, let $\phi, \psi \in K[X,Y]$, and also let f, g be two polynomials that have no irreducible factor with zero P in common. Suppose $(\phi,\psi)\,\mathcal{O}'_P \subset (f,g)\,\mathcal{O}'_P$. Then, for $\mathcal{O}' := \mathcal{O}'_P/(\phi,\psi)\,\mathcal{O}'_P$, there is a canonical epimorphism $\varepsilon : \mathcal{O}' \to \mathcal{O}$ with kernel $(f,g)\,\mathcal{O}'_P/(\phi,\psi)\,\mathcal{O}'_P$, and a canonical injection of $\omega_{\mathcal{O}/K} = \operatorname{Hom}_K(\mathcal{O},K)$ into $\omega_{\mathcal{O}'/K} = \operatorname{Hom}_K(\mathcal{O}',K)$, where each $\ell \in \operatorname{Hom}_K(\mathcal{O},K)$ is mapped to the composition $\ell \circ \varepsilon$.

If we write, as we have so often,

$$\begin{aligned} \phi &= c_{11} f + c_{12} g, \\ \psi &= c_{21} f + c_{22} g, \end{aligned} \tag{14}$$

with $c_{ij} \in \mathcal{O}'_P$, then the image $(d^{\phi,\psi}_{f,g})_P$ of $\det(c_{ij})$ in $\mathcal{O}'$ is independent of the special choice of coefficients c_{ij} in (14). If f is a nonzerodivisor mod $(\psi\,\mathcal{O}'_P)$, then I.5 can be applied, and it follows that $(d^{\phi,\psi}_{f,g})_P$ generates the annihilator of $(f,g)\,\mathcal{O}'_P/(\phi,\psi)\,\mathcal{O}'_P$. If f is a zerodivisor mod $(\psi\,\mathcal{O}'_P)$, replace ψ by $\phi+\psi$. Then f is a nonzerodivisor mod $(\phi+\psi)\,\mathcal{O}'_P$ and $\det(c_{ij})$ is unchanged. In any case, $(d^{\phi,\psi}_{f,g})_P$ generates the above annihilator. Multiplication by $(d^{\phi,\psi}_{f,g})_P$ in $\mathcal{O}'_P$ induces therefore an $\mathcal{O}'$-linear map $\mathcal{O} \to \mathcal{O}'$, which we also denote by $(d^{\phi,\psi}_{f,g})_P$. If $(\phi,\psi)\,\mathcal{O}'_P = (f,g)\,\mathcal{O}'_P$, then of course $\mathcal{O}' = \mathcal{O}$ and $(d^{\phi,\psi}_{f,g})_P$ is a unit of $\mathcal{O}$.

Theorem 11.16 (Chain Rule). *Let x', y' be the images of X, Y in $\mathcal{O}'$. Under the canonical injection $\omega_{\mathcal{O}/K} \to \omega_{\mathcal{O}'/K}$ the trace $(\tau^{x,y}_{f,g})_P$ is mapped to $(d^{\phi,\psi}_{f,g})_P \cdot (\tau^{x',y'}_{\phi,\psi})_P$. In other words, there is a commutative diagram*

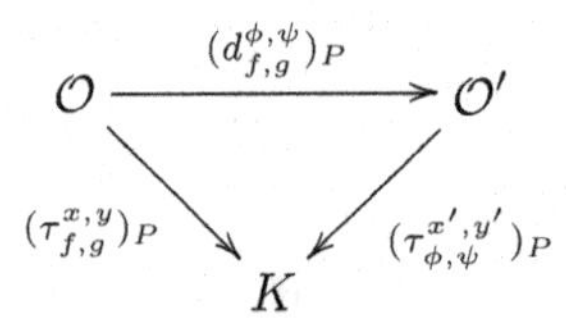

Proof. Consider in $\mathcal{O}'[X,Y]_M$ a system of equations

$$\begin{aligned} \phi &= a'_{11}(X-x') + a'_{12}(Y-y'), \\ \psi &= a'_{21}(X-x') + a'_{22}(Y-y'). \end{aligned}$$

Using the canonical epimorphism $\mathcal{O}'[X,Y]_M \to \mathcal{O}[X,Y]_M$, this is mapped over to a system

$$\begin{aligned} \phi &= a_{11}(X-x) + a_{12}(Y-y), \\ \psi &= a_{21}(X-x) + a_{22}(Y-y). \end{aligned} \tag{15}$$

On the other hand, we can write in $\mathcal{O}[X,Y]_M$,

$$\begin{aligned} f &= b_{11}(X-x) + b_{12}(Y-y), \\ g &= b_{21}(X-x) + b_{22}(Y-y), \end{aligned}$$

substitute in (14), and get a system similar to (15). By I.4, we have in $\mathcal{O} \otimes_K \mathcal{O}' = \mathcal{O}[X,Y]_M/(\phi,\psi)$ an equation

$$(\varepsilon \otimes 1)((\Delta_{x',y'}^{\phi,\psi})_P) = (1 \otimes (d_{f,g}^{\phi,\psi})_P) \cdot \Delta,$$

where Δ is mapped onto $(\Delta_{x,y}^{f,g})_P$ under the map $\mathrm{id}_{\mathcal{O}} \otimes \varepsilon : \mathcal{O} \otimes_K \mathcal{O}' \to \mathcal{O} \otimes_K \mathcal{O}$. If we choose for Δ a representation $\Delta = \sum a_i \otimes b_i'$ ($a_i \in \mathcal{O}$, $b_i' \in \mathcal{O}'$), then $(\Delta_{x,y}^{f,g})_P = \sum a_i \otimes \varepsilon(b_i')$.

Consider now the canonical commutative diagram

$$\begin{array}{ccc} \mathcal{O}' \otimes_K \mathcal{O}' & \xrightarrow[\phi']{\sim} & \mathrm{Hom}_K(\omega_{\mathcal{O}'/K}, \mathcal{O}') \\ {\scriptstyle \varepsilon\otimes\mathrm{id}}\downarrow & & \downarrow \\ \mathcal{O} \otimes_K \mathcal{O}' & \xrightarrow[\phi]{\sim} & \mathrm{Hom}_K(\omega_{\mathcal{O}/K}, \mathcal{O}') \end{array}$$

in which ϕ' is defined as in H.19; and similarly, by $\phi(\sum a_i \otimes b_i')$, each $\ell \in \omega_{\mathcal{O}/K}$ is mapped to $\sum \ell(a_i) b_i'$. We will show that

$$\phi'\left(\left(\Delta_{x',y'}^{\phi,\psi}\right)_P\right)\left(\left(\tau_{f,g}^{x,y}\right)_P\right) = \left(d_{f,g}^{\phi,\psi}\right)_P,$$

and then by definition of $\left(\tau_{\phi,\psi}^{x',y'}\right)_P$ the desired equation

$$\left(\tau_{f,g}^{x,y}\right)_P = \left(d_{f,g}^{\phi,\psi}\right)_P \cdot \left(\tau_{\phi,\psi}^{x',y'}\right)_P$$

follows.

But in fact,

$$\begin{aligned} \phi'((\Delta_{x',y'}^{\phi,\psi})_P)((\tau_{f,g}^{x,y})_P) &= \phi((\varepsilon \otimes \mathrm{id})((\Delta_{x',y'}^{\phi,\psi})_P)((\tau_{f,g}^{x,y})_P) \\ &= \phi(1 \otimes (d_{f,g}^{\phi,\psi})_P)(\sum a_i \otimes b_i'))((\tau_{f,g}^{x,y})_P) \\ &= \sum (\tau_{f,g}^{x,y})_P(a_i) \cdot (d_{f,g}^{\phi,\psi})_P \cdot b_i' \\ &= \sum (\tau_{f,g}^{x,y})_P(a_i) \cdot (d_{f,g}^{\phi,\psi})_P \cdot \varepsilon(b_i') = (d_{f,g}^{\phi,\psi})_P \end{aligned}$$

where we have used $\left(d_{f,g}^{\phi,\psi}\right)_P \cdot b_i' = \left(d_{f,g}^{\phi,\psi}\right)_P \cdot \varepsilon(b_i')$ and $\sum \left(\tau_{f,g}^{x,y}\right)_P (a_i)\varepsilon(b_i') = 1$ (H.20). The theorem has therefore been proved.

From this we get immediately

Theorem 11.17 (Transformation Formula for Residues). *Under the above assumptions we have for every $h \in K[X,Y]$,*

$$\mathrm{Res}_P \begin{bmatrix} h\,dX\,dY \\ f, g \end{bmatrix} = \mathrm{Res}_P \begin{bmatrix} \det(c_{ij})h\,dX\,dY \\ \phi, \psi \end{bmatrix}.$$

Some special cases of this are

(16) $$\operatorname{Res}_P \begin{bmatrix} h\,dX dY \\ f - ag, g \end{bmatrix} = \operatorname{Res}_P \begin{bmatrix} h\,dX dY \\ f, g \end{bmatrix} \quad \text{for every } a \in K[X,Y], \text{ and}$$

(17) $$\operatorname{Res}_P \begin{bmatrix} hf_2\,dX dY \\ f_1 f_2, g \end{bmatrix} = \operatorname{Res}_P \begin{bmatrix} h\,dX dY \\ f_1, g \end{bmatrix} \quad \text{(cancellation rule)}$$

if $f_1 f_2$ and g have no component with zero P in common, and

(18) $$\operatorname{Res}_P \begin{bmatrix} h\,dX dY \\ g, f \end{bmatrix} = -\operatorname{Res}_P \begin{bmatrix} h\,dX dY \\ f, g \end{bmatrix}.$$

Theorem 11.18 (Transformation Formula for Integrals). *Let $f, g \in K[X,Y]$ as well as $\phi, \psi \in K[X,Y]$ be relatively prime polynomials with $(\phi, \psi) \subset (f, g)$. Consider a system of equations (14) with coefficients $c_{ij} \in K[X,Y]$. Then for every $h \in K[X,Y]$ we have*

$$\int \begin{bmatrix} h\,dX dY \\ f, g \end{bmatrix} = \int \begin{bmatrix} \det(c_{ij}) h\,dX dY \\ \phi, \psi \end{bmatrix}.$$

The proof of this last formula is similar to that of 11.16. One can also appeal to 11.17 and (12), if one considers the following: If $P \notin \mathcal{V}(f) \cap \mathcal{V}(g)$, $P \in \mathcal{V}(\phi) \cap \mathcal{V}(\psi)$, then $\det(c_{ij}) \in (\phi, \psi)\,\mathcal{O}'_P$ and therefore

$$\operatorname{Res}_P \begin{bmatrix} \det(c_{ij}) h\,dX dY \\ \phi, \psi \end{bmatrix} = 0.$$

We come now to the main theorem of this chapter. Many classical theorems about algebraic curves can be derived from it, as we will show in Chapter 12. We will assume that Gf and Gg are relatively prime. Further, let $d_{\xi,\eta}^{Gf,Gg}$ be as in 11.10 and $\rho := p + q - 2$. Observe that every $h \in K[X,Y]$ can be represented modulo (f, g) by a polynomial of degree $\leq \rho$. Therefore it is possible to calculate the integral by the following theorem.

Theorem 11.19 (Residue Theorem). *Let O denote the origin of $\mathbb{A}^2(K)$. For $h \in K[X,Y]$ let $\overline{Gh}$ be the residue class of Gh in $K[X,Y]/(Gf, Gg)$. If $\deg h = \rho$, then there is a unique $\kappa \in K$ with $\overline{Gh} = \kappa \cdot d_{\xi,\eta}^{Gf,Gg}$. With this notation we have*

$$\int \begin{bmatrix} h\,dX dY \\ f, g \end{bmatrix} = \operatorname{Res}_O \begin{bmatrix} Gh\,dX dY \\ Gf, Gg \end{bmatrix} = \begin{cases} \kappa & \deg h = \rho, \\ 0 & \deg h < \rho. \end{cases}$$

Proof. Consider the homogenization F, G, and H of f, g, respectively h in $K[X_0, X_1, X_2]$. The hypotheses of 11.8 are satisfied, and therefore it follows, if $\overline{H}$ denotes the residue class of H in $S = K[X_0, X_1, X_2]/(F, G)$, that

$$\int \begin{bmatrix} h\,dX dY \\ f, g \end{bmatrix} = \tau_{F,G}^{x_1,x_2}(\overline{H})|_{X_0=1},$$
$$\mathrm{Res}_O \begin{bmatrix} Gh\,dX dY \\ Gf, Gg \end{bmatrix} = \tau_{F,G}^{x_1,x_2}(\overline{H})|_{X_0=0}.$$

If $\deg h < \rho$, and hence also $\deg \overline{H} < \rho$, then $\tau_{F,G}^{x_1,x_2}(\overline{H}) \in K[X_0]$ has negative degree. Therefore $\tau_{F,G}^{x_1,x_2}(\overline{H}) = 0$, and consequently,

$$\int \begin{bmatrix} h\,dX dY \\ f, g \end{bmatrix} = \mathrm{Res}_O \begin{bmatrix} Gh\,dX dY \\ Gf, Gg \end{bmatrix} = 0.$$

On the other hand, if $\deg h = \rho$, then $\tau_{F,G}^{x_1,x_2}(\overline{H})$ has degree 0, and therefore is an element of K, and it follows that

$$\int \begin{bmatrix} h\,dX dY \\ f, g \end{bmatrix} = \tau_{F,G}^{x_1,x_2}(\overline{H}) = \mathrm{Res}_O \begin{bmatrix} Gh\,dX dY \\ Gf, Gg \end{bmatrix}.$$

By 11.10 this residue coincides with κ.

There is an analogous theorem for the calculation of residues, in which, however, one must assume that f and g have no tangent line at $P \in \mathcal{V}(f) \cap \mathcal{V}(g)$ in common. We will use the notation in 11.12, in particular $m = m_P(f)$, $n = m_P(g)$. Further, let $\overline{Lh}$ be the residue class of the leading form Lh in $G = K[X,Y]/(Lf, Lg)$.

Theorem 11.20. *Let $h \in K[X,Y]_{\mathfrak{M}}$. In case $\mathrm{ord}_{\mathfrak{M}}\, h = -(m+n-2)$, there is a unique $\kappa \in K$ with $\overline{Lh} = \kappa \cdot d_{\xi,\eta}^{Lf,Lg}$. If $\rho = m+n-2$, we have*

$$\mathrm{Res}_P \begin{bmatrix} h\,dX dY \\ f, g \end{bmatrix} = \mathrm{Res}_O \begin{bmatrix} Lh\,dX dY \\ Lf, Lg \end{bmatrix} = \begin{cases} \kappa & \mathrm{ord}_{\mathfrak{M}}\, h = -\rho, \\ 0 & \mathrm{ord}_{\mathfrak{M}}\, h < -\rho. \end{cases}$$

Proof. Let $\mathcal{O}$ be the local ring of $f \cap g$ at P and $\mathfrak{m}$ its maximal ideal. In $G = \mathrm{gr}_{\mathfrak{m}}\, \mathcal{O}$ we have $G_k = \{0\}$ for $k < -\rho$, i.e., $\mathfrak{m}^{\rho+1} = \mathfrak{m}^{\rho+2}$, and therefore $\mathfrak{m}^{\rho+1} = \{0\}$ by Nakayama. In $K[X,Y]_{\mathfrak{M}}$ this means that $\mathfrak{M}^{\rho+1}K[X,Y]_{\mathfrak{M}} = (f,g)K[X,Y]_{\mathfrak{M}}$.

If $\mathrm{ord}_{\mathfrak{M}}\, h < -\rho$, then it follows that $h \in (f,g)K[X,Y]_{\mathfrak{M}}$ and hence by (11),

$$\mathrm{Res}_P \begin{bmatrix} h\,dX dY \\ f, g \end{bmatrix} = \mathrm{Res}_O \begin{bmatrix} Lh\,dX dY \\ Lf, Lg \end{bmatrix} = 0.$$

If, on the other hand, $\mathrm{ord}_{\mathfrak{M}}\, h = -\rho$, one can finish using 11.13 and 11.12 as in the proof of 11.19.

Example 11.21. Let $P = (0,0)$, and so $\mathfrak{M} = (X,Y)$. If $\mu := \mu_P(f,g) = \dim_K \mathcal{O}$ is the intersection multiplicity of f and g at P, then $\mathfrak{m}^\mu = (0)$ and therefore X^μ, $Y^\mu \in (f,g)\,\mathcal{O}'_P$. Set

$$X^\mu = c_{11} f + c_{12} g,$$
$$Y^\mu = c_{21} f + c_{22} g,$$

with $c_{ij} \in \mathcal{O}'_P$. Then by 11.17 for $h \in K[X, Y]_{\mathfrak{M}}$,

$$\operatorname{Res}_P \begin{bmatrix} h\,dX dY \\ f, g \end{bmatrix} = \operatorname{Res}_P \begin{bmatrix} h \det(c_{ij})\, dX dY \\ X^\mu, Y^\mu \end{bmatrix}.$$

Now write

$$h \det(c_{ij}) = \sum_{0 \le \alpha, \beta < \mu} a_{\alpha\beta} X^\alpha Y^\beta + R$$

with $a_{\alpha\beta} \in K$ and a "remainder" $R \in (X^\mu, Y^\mu)\, \mathcal{O}'_P$. Then by (12) and (11),

$$\operatorname{Res}_P \begin{bmatrix} h\,dX dY \\ f, g \end{bmatrix} = \sum_{0 \le \alpha, \beta < \mu} a_{\alpha\beta} \operatorname{Res}_P \begin{bmatrix} X^\alpha Y^\beta\, dX dY \\ X^\mu, Y^\mu \end{bmatrix} = a_{\mu-1, \mu-1},$$

where for the last equation we used Theorems 11.19 and 11.17. The formula makes clear the analogy to residues of a function of a complex variable (see also Exercise 3).

Exercises

1. Give an example of the following situation: f and g are algebraic curves that have no components in common at $P \in \mathcal{V}(f) \cap \mathcal{V}(g)$. Furthermore, let $\mathcal{O}$ be the local ring of $f \cap g$ at P and $G = \operatorname{gr}_{\mathfrak{m}} \mathcal{O}$ the associated graded ring of $\mathcal{O}$ with respect to its maximal ideal $\mathfrak{m}$. The algebra G/K has no trace.
2. Denote by ϑ the Noether different (G.10). Show that
   (a) If, under the hypotheses of H.7, the algebra S/R has a trace, then $\vartheta(S/R)$ is a principal ideal.
   (b) Under the assumptions of 11.1 we have

$$\vartheta(S/K[X_0]) = \left(\frac{\partial(F, G)}{\partial(x_1, x_2)} \right)$$

   and under the assumptions of 11.4 we have

$$\vartheta(A/K) = \left(\frac{\partial(f, g)}{\partial(x, y)} \right).$$

3. Let f be an affine algebraic curve with $P := (0,0) \in \mathcal{V}(f)$ and suppose Y is not a divisor of f. For $h \in K[X, Y]$ let the "Laurent series" of $\frac{h(X,0)}{f(X,0)}$ be given by $\sum_{i \ge \mu} a_i X^i$, with some $\mu \in \mathbb{Z}$. Show that

$$\operatorname{Res}_P \begin{bmatrix} h\,dX dY \\ f, Y \end{bmatrix} = a_{-1}.$$

4. Calculate

$$\int \begin{bmatrix} X^6 \, dX dY \\ X^2Y^2 - 1, X^3 + Y^3 - 1 \end{bmatrix}$$

and

$$\mathrm{Res}_P \begin{bmatrix} (X + X^2 - Y^3) \, dX dY \\ XY, X^2 - Y^2 + X^3 \end{bmatrix} \quad \text{if } P := (0, 0).$$

12

Applications of Residue Theory to Curves

The formulas and theorems in Chapter 11 on residues in the affine plane allow uniform proofs and generalizations of classical theorems about intersection theory of plane curves. Maybe B. Segre [Se] was the first who proceeded in a way similar to ours, but he used another concept of residue, the residue of differentials on a smooth curve. See also Griffiths–Harris [GH], Chapter V. The theorems presented here have far-reaching higher-dimensional generalizations ([Hü], [HK], [Ku3], [Ku4], [KW]). In his thesis [Q] Gerhard Quarg has discovered further global geometric applications of algebraic residue theory. [Ku4] contains an outline of part of this thesis.

Suppose we are given two curves f and g in $\mathbb{A}^2(K)$ with no common components, with $\deg f =: p$, $\deg g =: q$, and let $A := K[X,Y]/(f,g) = K[x,y]$. For the differential form $\omega = \frac{\partial(f,g)}{\partial(X,Y)}\, dXdY$ we also write $\omega = df dg$.

Formulas 12.1. We have

$$\int \begin{bmatrix} df dg \\ f, g \end{bmatrix} = (\dim_K A) \cdot 1_K$$

and

$$\operatorname{Res}_P \begin{bmatrix} df dg \\ f, g \end{bmatrix} = \mu_P(f,g) \cdot 1_K,$$

where $\mu_P(f,g)$ is the intersection multiplicity of f and g at the point P.

Proof. By 11.9,

$$\int \begin{bmatrix} df dg \\ f, g \end{bmatrix} = \tau_{f,g}^{x,y}\left(\frac{\partial(f,g)}{\partial(x,y)}\right) = \sigma_{A/K}(1) = (\dim_K A) \cdot 1_K.$$

Denote by $\mathcal{O}$ the local ring of P on $f \cap g$. We have furthermore

$$\operatorname{Res}_P \begin{bmatrix} df dg \\ f, g \end{bmatrix} = (\tau_{f,g}^{x,y})_P\left(\frac{\partial(f,g)}{\partial(x,y)}\right) = \left(\frac{\partial(f,g)}{\partial(x,y)} \cdot (\tau_{f,g}^{x,y})_P\right)(1).$$

Since $(\tau_{f,g}^{x,y})_P$ is the restriction of $\tau_{f,g}^{x,y}$, and $\sigma_{\mathcal{O}/K}$ is the restriction of $\sigma_{A/K} = \frac{\partial(f,g)}{\partial(x,y)} \cdot \tau_{f,g}^{x,y}$ on $\mathcal{O}$ (H.3), it follows that

$$\operatorname{Res}_P \begin{bmatrix} df dg \\ f, g \end{bmatrix} = \sigma_{\mathcal{O}/K}(1) = (\dim_K \mathcal{O}) \cdot 1_K = \mu_P(f,g) \cdot 1_K.$$

Formula (13) in Chapter 11 yields

$$\dim_K A \equiv \sum_P \mu_P(f,g) \pmod{\chi},$$

where χ is the characteristic of K. This is of course no surprise, because the Chinese remainder theorem entered the theory of Chapter 11 several times. If f and g have no points at infinity in common, then this is Bézout's theorem up to congruence mod χ.

If f and g intersect transversally at P, then one has a residue formula, which is analogous to that about a pole of order 1 in functions of a complex variable. We set $J := \frac{\partial(f,g)}{\partial(X,Y)}$ and denote the maximal ideal of P in $K[X,Y]$ by $\mathfrak{M}$. Let $\mathcal{O}'_P := K[X,Y]_{\mathfrak{M}}$.

Formula 12.2. If f and g intersect transversally at P, then $J(P) \neq 0$ and for each $h \in K[X,Y]_{\mathfrak{M}}$ we have

$$\operatorname{Res}_P \begin{bmatrix} h\,dX dY \\ f, g \end{bmatrix} = \frac{h(P)}{J(P)}.$$

Proof. If f and g intersect transversally at P, then $J(P) \neq 0$ by 7.7. Furthermore, $\mathcal{O} := \mathcal{O}'_P/(f,g)\,\mathcal{O}'_P \cong K$ and therefore $\sigma_{\mathcal{O}/K} = \mathrm{id}_K$. By the formula $\sigma_{\mathcal{O}/K} = J(P) \cdot (\tau^{x,y}_{f,g})_P$ we get

$$\operatorname{Res}_P \begin{bmatrix} h\,dX dY \\ f, g \end{bmatrix} = (\tau^{x,y}_{f,g})_P(h(P)) = \frac{1}{J(P)} \cdot \sigma_{\mathcal{O}/K}(h(P)) = \frac{h(P)}{J(P)}.$$

We will now use the residue theorem 11.19 for the first time. From this theorem, we get the following immediately from 12.2.

Theorem 12.3 (Residue Theorem for Transversal Intersections). *Suppose f and g have no points at infinity in common and intersect transversally at all points of intersection. For $h \in K[X,Y]$ denote by $\overline{Gh}$ the residue class of Gh in $K[X,Y]/(Gf,Gg)$ and suppose (with the notation of 11.10)*

$$\overline{Gh} = \kappa \cdot d^{Gf,Gg}_{\xi,\eta} \qquad (\kappa \in K)$$

in case $\deg h = \rho = p + q - 2$. *Then*

(a) *If* $\deg h < \rho$, *then*

$$\sum_{P \in \mathcal{V}(f) \cap \mathcal{V}(g)} \frac{h(P)}{J(P)} = 0 \qquad \text{(*Formula of Jacobi* [J], *1835*).}$$

(b) *If* $\deg h = \rho$, *then*

$$\sum_{P \in \mathcal{V}(f) \cap \mathcal{V}(g)} \frac{h(P)}{J(P)} = \kappa.$$

The right side of the formula in (b) depends only on the degree forms Gf, Gg, and Gh. Therefore the left side does not change if other curves with the same degree forms are substituted for the given curves.

The formula of Jacobi contains the following special case of the Cayley–Bacharach theorem: Under the assumptions of the theorem, suppose that a curve h of degree $< \rho$ passes through $pq-1$ points of $\mathcal{V}(f) \cap \mathcal{V}(g)$. Then it goes through all the pq intersection points.

Applications of this theorem, for example Pascal's theorem, have already been discussed in 5.16. One can also consider this result as follows: Suppose the intersection points $P_\nu = (a_\nu, b_\nu)$ of f and g for $\nu = 1, \ldots, pq-1$ have already been calculated, and the last intersection point $P = (x, y)$ is still unknown. Suppose $p + q \geq 4$; hence $\rho \geq 2$. By the equations in 12.3(a)

$$\frac{1}{J(P)} + \sum_{i=1}^{pq-1} \frac{1}{J(P_i)} = 0,$$

$$\frac{x}{J(P)} + \sum_{i=1}^{pq-1} \frac{a_i}{J(P_i)} = 0, \text{ and}$$

$$\frac{y}{J(P)} + \sum_{i=1}^{pq-1} \frac{b_i}{J(P_i)} = 0,$$

one can successively determine $J(P)$, x, and y. If $\rho > 2$, then there are more equations to consider, and knowing a few intersection points can sometimes be sufficient to determine the rest. But it is difficult to decide in general whether two curves f and g intersect transversally at all points of intersection.

To two curves f and g that have no components in common at $P \in \mathcal{V}(f) \cap \mathcal{V}(g)$ we assign the invariant

$$a_P(f,g) := \operatorname{Res}_P \begin{bmatrix} (f_X g_X + f_Y g_Y)\, dX dY \\ f, g \end{bmatrix},$$

where $f_X = \frac{\partial f}{\partial X}$, etc. By 11.17, if f or g is multiplied by a nonzero constant, then $a_P(f,g)$ does not change. However, $a_P(f,g)$ is not independent of the coordinates, since f_X, f_Y, etc. are not. But we have

Lemma 12.4. *$a_P(f,g)$ is invariant under orthogonal coordinate transformations. By this we mean transformations of the form*

$$(X,Y) \mapsto (X,Y) \cdot A + (b_1, b_2),$$

where $(b_1, b_2) \in K^2$ and $A \in SO(2,K)$; i.e., A is a 2×2 matrix such that $A \cdot A^t = I$ the identity matrix, and $\det A = 1$.

Proof. Let $(X,Y) = (X',Y')A + (b_1, b_2)$, as given in the lemma. Then by the chain rule,

$$(f_{X'}, f_{Y'}) = (f_X, f_Y) \cdot A^t,$$

and therefore

$$f_{X'}g_{X'} + f_{Y'}g_{Y'} = (f_X, f_Y) \cdot A^t \cdot A \cdot (g_X, g_Y)^t = f_X g_X + f_Y g_Y.$$

Because $\det A = 1$, we have $dXdY = dX'dY'$, and the residue defining $a_P(f,g)$ remains invariant under the transformation.

Definition 12.5. We call $a_P(f,g)$ the *angle* between f and g at the point P.

We will see to what extent this designation is justifiable in the following. We set $a_P(f,g) = 0$ in case $P \notin \mathcal{V}(f) \cap \mathcal{V}(g)$.

Example 12.6. Let $f = aX + bY$, $g = cX + dY$ be two different lines through $P = (0,0)$; hence $ad - bc \neq 0$. Then we have

$$a_P(f,g) = \frac{ac + bd}{ad - bc}.$$

In the reals, if $v_1 := (a,b)$, $v_2 := (c,d)$, then

$$ac + bd = |v_1| \cdot |v_2| \cdot \cos\phi,$$

$$ad - bc = |v_1| \cdot |v_2| \cdot \sin\phi,$$

where ϕ is the oriented angle between v_1 and v_2.

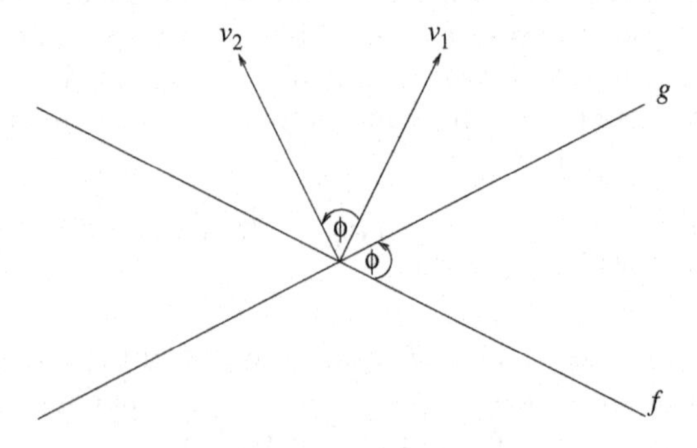

It follows that

$$a_P(f,g) = \cot\phi.$$

The intersection angle is additive, as was the intersection multiplicity (5.8).

Lemma 12.7. *Let $f = f_1 \cdots f_r$ and $g = g_1 \cdots g_s$ be factorizations of f and g. Then for all $P \in \mathbb{A}^2(K)$,*

$$a_P(f,g) = \sum_{\substack{i=1,\dots,r \\ j=1,\dots,s}} a_P(f_i, g_j).$$

Proof. It is enough to treat the case $f = f_1 \cdot f_2$, $g = g_1$, and we can assume that $P \in \mathcal{V}(f) \cap \mathcal{V}(g)$. Then $a_P(f, g) =$

$$\operatorname{Res}_P \begin{bmatrix} f_1(f_{2X}g_X + f_{2Y}g_Y)\,dXdY \\ f_1 f_2, g \end{bmatrix} + \operatorname{Res}_P \begin{bmatrix} f_2(f_{1X}g_X + f_{2Y}g_Y)\,dXdY \\ f_1 f_2, g \end{bmatrix}.$$

By the cancellation rule (17) from Chapter 11, the first residue equals $a_P(f_2, g)$ and the second equals $a_P(f_1, g)$.

If two curves have no tangents in common at an intersection point, then the intersection angle is given by the angle between the two tangents according to the following theorem:

Theorem 12.8. *Suppose $t_1, \dots, t_m$ are the tangents to f at P, and $t'_1, \dots, t'_n$ are the tangents to g at P, counted with their multiplicities (so $m = m_P(f)$, $n = m_P(g)$). If $t_i \neq t'_j$ for all i and j, then*

$$a_P(f, g) = \sum_{\substack{i=1,\dots,m \\ j=1,\dots,n}} a_P(t_i, t_j).$$

Proof. Without loss of generality, we can assume that $P = O$ is the origin. By 12.7 we have

$$\sum_{i,j} a_O(t_i, t'_j) = a_O\left(\prod_i t_i, \prod_j t'_j\right) = a_O(Lf, Lg).$$

However,

$$(Lf)_X \cdot (Lg)_X + (Lf)_Y \cdot (Lg)_Y = L(f_X g_X + f_Y g_Y)$$

or

$$(Lf)_X \cdot (Lg)_X + (Lf)_Y \cdot (Lg)_Y = 0.$$

In the second case, $\operatorname{ord}_{\mathfrak{M}}(f_X g_X + f_Y g_Y) < -(m + n - 2)$ if $\mathfrak{M} := (X, Y)$. In any case the statement follows from 11.20.

Suppose f and g are real curves and all their tangents at O are real. Then by 12.8, $a_O(f, g)$ is the sum of all the cotangents of oriented angles between the tangents of f and those of g, assuming that f and g have no tangents in common at O.

The *asymptotes* of a curve f of degree p are the lines $a_i X - b_i Y = 0$, where $\langle 0, b_i, a_i \rangle$ $(i = 1, \dots, p)$ are the points at infinity of f. These are the lines through O in the "direction of the points at infinity" of f. The $a_i X - b_i Y$ are also the linear factors of Gf. Asymptotes will be counted with multiplicity, according to how many times the $a_i X - b_i Y$ appear in Gf.

The following theorem concerns the sum of all the intersection angles of two curves.

Theorem 12.9 (Humbert's Theorem [Hu]). *Suppose f and g do not intersect on the line at infinity. Let $\ell_1, \dots, \ell_p$ be the asymptotes of f and let $\ell'_1, \dots, \ell'_q$ be those of g. Then*

$$\sum_{P \in \mathcal{V}(f) \cap \mathcal{V}(g)} a_P(f,g) = \sum_{\substack{i=1,\dots,p \\ j=1,\dots,q}} a_O(\ell_i, \ell'_j).$$

Proof. By Chapter 11 (13) the left side of the equation equals

$$\int \begin{bmatrix} (f_X g_X + f_Y g_Y)\, dX dY \\ f, g \end{bmatrix},$$

and by 12.8 the right side equals

$$a_O(Gf, Gg) = \operatorname{Res}_O \begin{bmatrix} ((Gf)_X (Gg)_X + (Gf)_Y (Gg)_Y)\, dX dY \\ Gf, Gg \end{bmatrix}.$$

And $(Gf)_X(Gg)_X + (Gf)_Y(Gg)_Y = G(f_X g_X + f_Y g_Y)$ or $(Gf)_X(Gg)_X + (Gf)_Y(Gg)_Y = 0$. In the first case, $h := f_X g_X + f_Y g_Y$ has degree $\rho = p+q-2$, and the desired formula follows from 11.19. In the second case, $\deg h < \rho$ and both sides vanish.

Over the reals, Humbert's theorem has the following interpretation: Let f and g be real curves that intersect in $p \cdot q$ different real points P_i $(i = 1, \dots, pq)$ and let ϕ_i be the oriented angle between f and g at P_i. Then $\sum_{i=1}^{pq} \cot \phi_i$ depends only on the (complex) points at infinity of f and g. If g is shifted in a "parallel" manner, this changes the individual ϕ_i, but the sum of the cotangents of the intersection angles is unchanged.

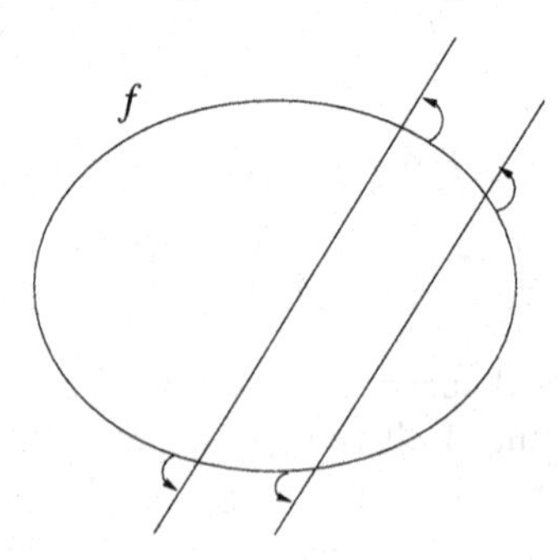

The same is true if g is subjected to a "similarity transformation," where we are always assuming that f and g intersect in $p \cdot q$ distinct real points. See the next figure.

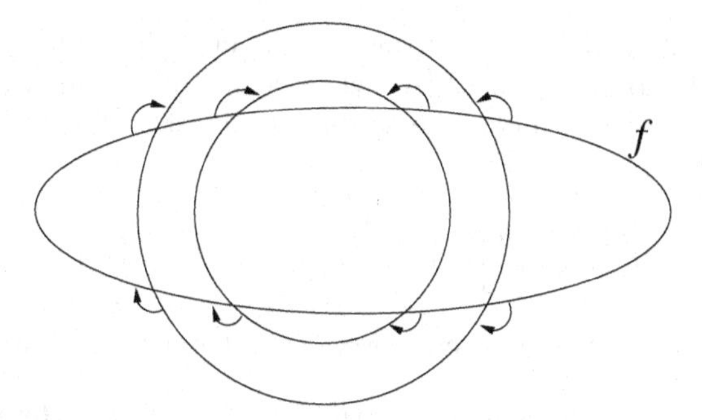

Now we come to another invariant of the intersections of two algebraic curves.

Definition 12.10. For two curves f and g, that have no points at infinity in common, we call

$$\sum(f \cap g) = \sum_P \mu_P(f,g) \cdot P$$

the *centroid* of $f \cap g$. Here the expression on the right is to be interpreted as a vector sum in K^2 (and not for example as an intersection cycle 5.6).

In order to interpret the centroid in a physical sense, when all the intersection points of f and g have real coordinates, one must divide by the number pq of intersection points. This division is not possible if pq is divisible by the characteristic of K, and then we have to give up on this interpretation. The statements that we prove about $\sum(f \cap g)$ are valid in arbitrary characteristic, and nothing essential changes if one divides by pq, insofar as this is possible. However, $\frac{1}{pq}\sum(f \cap g)$ is invariant under an arbitrary coordinate transformation, while $\sum(f \cap g)$ is invariant under coordinate transformations that fix the origin.

Next we have an "integral formula" for the centroid.

Lemma 12.11.

$$\sum(f \cap g) = \left(\int \begin{bmatrix} X df dg \\ f, g \end{bmatrix}, \int \begin{bmatrix} Y df dg \\ f, g \end{bmatrix}\right).$$

Proof. For $P = (\xi, \eta) \in \mathbb{A}^2(K)$, by the linearity of residues, we get

$$\operatorname{Res}_P \begin{bmatrix} X df dg \\ f, g \end{bmatrix} = \xi \cdot \operatorname{Res}_P \begin{bmatrix} df dg \\ f, g \end{bmatrix} + \operatorname{Res}_P \begin{bmatrix} (X-\xi) df dg \\ f, g \end{bmatrix}.$$

Here $\operatorname{Res}_P \begin{bmatrix} df dg \\ f, g \end{bmatrix} = \mu_P(f,g) \cdot 1_K$ by 12.1, and as there,

$$\operatorname{Res}_P \begin{bmatrix} (X-\xi) df dg \\ f, g \end{bmatrix} = \sigma_{\mathcal{O}/K}(x - \xi).$$

Since $x - \xi$ is a nilpotent element of $\mathcal{O}$, multiplication by $x - \xi$ yields a nilpotent endomorphism of $\mathcal{O}/K$, whose trace of course vanishes. Consequently, in the above formula, the second residue therefore vanishes. Hence this shows that

$$\left(\operatorname{Res}_P\begin{bmatrix} X\,df\,dg \\ f,g \end{bmatrix}, \operatorname{Res}_P\begin{bmatrix} Y\,df\,dg \\ f,g \end{bmatrix}\right) = \mu_P(f,g)\cdot P.$$

The statement of the lemma then follows from Chapter 11 (13).

The integral formula will be reformulated using the Residue Theorem 11.19. Let

$$f = \sum_{i=0}^{p} f_i, \quad g = \sum_{j=0}^{q} g_j$$

be decompositions of f and g into homogeneous polynomials, in particular, $Gf = f_p$, $Gg = g_q$. In the following J will denote the Jacobian determinant $\frac{\partial(f,g)}{\partial(X,Y)}$. By the Euler formula,

$$(1)\qquad \begin{aligned} Xf_X + Yf_Y &= p\cdot f - \sum_{k=0}^{p-1}(p-k)f_k = p\cdot f - f_{p-1} + \phi, \\ Xg_X + Yg_Y &= q\cdot g - \sum_{k=0}^{q-1}(q-k)g_k = q\cdot g - g_{q-1} + \psi, \end{aligned}$$

where $\deg\phi \le p-2$, $\deg\psi \le q-2$. We can calculate $X\cdot J$ by multiplying the first column of J by X and then replace this column by the column formed by the right side of equations (1). Then we get

$$X\cdot J \equiv D_1 \bmod (f,g), \quad Y\cdot J \equiv D_2 \bmod (f,g),$$

with

$$D_1 := \begin{vmatrix} \phi - f_{p-1} & f_Y \\ \psi - g_{q-1} & g_Y \end{vmatrix}, \quad D_2 := \begin{vmatrix} f_X & \phi - f_{p-1} \\ g_X & \psi - g_{q-1} \end{vmatrix},$$

and therefore by Chapter 11 (10),

$$(2)\qquad \sum(f\cap g) = \left(\int\begin{bmatrix} D_1\,dX\,dY \\ f,g \end{bmatrix}, \int\begin{bmatrix} D_2\,dX\,dY \\ f,g \end{bmatrix}\right).$$

Since $\deg\phi < p-1$, $\deg\psi < q-1$, we have either $\deg D_1 < p+q-2$ or

$$GD_1 = \begin{vmatrix} -f_{p-1} & (f_p)_Y \\ -g_{q-1} & (g_q)_Y \end{vmatrix}.$$

Similarly for D_2. From (2) and Theorem 11.19 we deduce the following formula:

Lemma 12.12. $\sum(f \cap g) =$

$$\left(\mathrm{Res}_O\begin{bmatrix}(g_{q-1}f_{pY} - f_{p-1}g_{qY})\,dXdY \\ f_p,\ g_q\end{bmatrix}, \mathrm{Res}_O\begin{bmatrix}(f_{p-1}g_{qX} - g_{q-1}f_{pX})\,dXdY \\ f_p,\ g_q\end{bmatrix}\right).$$

This lemma shows that the centroid of $f \cap g$ depends only on the degree forms and the forms of the second-highest degree of f and g. It allows answers to questions about how the centroid changes when the intersection scheme is changed.

Next we subject the curves f and g to two independent parallel displacements; i.e., we substitute for f and g the polynomials r, s, where

$$\begin{aligned} r(X,Y) &:= f(X+\alpha, Y+\beta) = f(X,Y) + \alpha f_X(X,Y) + \beta f_Y(X,Y) + \cdots, \\ s(X,Y) &:= g(X+\gamma, Y+\delta) = g(X,Y) + \gamma g_X(X,Y) + \delta g_Y(X,Y) + \cdots, \end{aligned}$$

and $(\alpha,\beta), (\gamma,\delta) \in K^2$. It is clear that

$$\begin{aligned} r_p &:= f_p, \quad r_{p-1} = f_{p-1} + \alpha(f_p)_X + \beta(f_p)_Y, \\ s_q &:= g_q, \quad s_{q-1} = g_{q-1} + \gamma(g_q)_X + \delta(g_q)_Y, \end{aligned}$$

and therefore by 12.12,

$$\sum(r \cap s) - \sum(f \cap g) = \mathrm{Res}_O\begin{bmatrix}\omega \\ f,\ g\end{bmatrix},$$

where

$$\omega := (\alpha(f_p)_X + \beta(f_p)_Y, \gamma(g_q)_X + \delta(g_q)_Y) \cdot \begin{pmatrix} -(g_q)_Y, & (g_q)_X \\ (f_p)_Y, & -(f_p)_X \end{pmatrix} \cdot dXdY$$

and where the residue is to be applied to a vector componentwise. If we set $(\gamma,\delta) = (0,0)$, and consider instead of (α,β) all multiples $\lambda \cdot (\alpha,\beta)$ ($\lambda \in K$), then we see that $\sum(r \cap s) - \sum(f \cap g)$ consists of all multiples of a vector dependent only on (f_p, g_q) and (α,β). Therefore this gives the first generalization of Newton's theorem (Chapter 5, Exercise 4):

Theorem 12.13. *Let $(\alpha,\beta) \in K^2$ be a fixed vector and let $\lambda \in K$. Furthermore, let f^λ be the curve that arises from f by a parallel displacement by the vector $\lambda \cdot (\alpha,\beta)$. Then the centroids $\sum(f^\lambda \cap g)$ ($\lambda \in K$) lie on a line.*

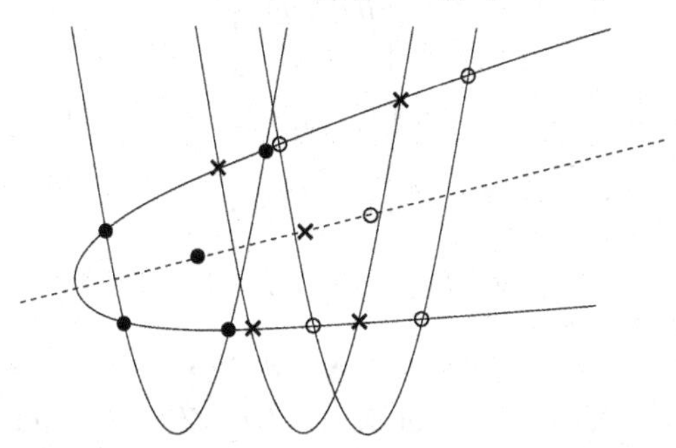

We now subject f and g to similarity transformations; i.e., we substitute for f and g the polynomials r, s, where

$$r(X,Y) := f(\lambda X, \lambda Y), \quad s(X,Y) := g(\mu X, \mu Y) \quad (\lambda, \mu \in K^*).$$

Then we have

$$r_p = \lambda^p f_p, \quad r_{p-1} = \lambda^{p-1} f_{p-1},$$
$$s_q = \mu^q g_q, \quad s_{q-1} = \mu^{q-1} g_{q-1},$$

and by Lemma 12.12 the following formula follows using Chapter 11 (17):
$\sum(r \cap s) - \sum(f \cap g) =$

$$\mathrm{Res}_O \begin{bmatrix} \\ f_p \ g_q \end{bmatrix} \left(\left[\left(\frac{1}{\lambda} - 1\right) f_{p-1}, \left(\frac{1}{\mu} - 1\right) g_{q-1} \right] \begin{bmatrix} -(g_q)_Y & (g_q)_X \\ (f_p)_Y & -(f_p)_X \end{bmatrix} dXdY \right).$$

Setting $\mu = 1$ in this gives us a second generalization of Newton's theorem:

Theorem 12.14. *Let $\lambda \in K^*$ and let f^λ be the curve given by $f^\lambda(X,Y) = f(\lambda X, \lambda Y)$. Then the centroids of $\sum(f^\lambda \cap g)$ lie on a line.*

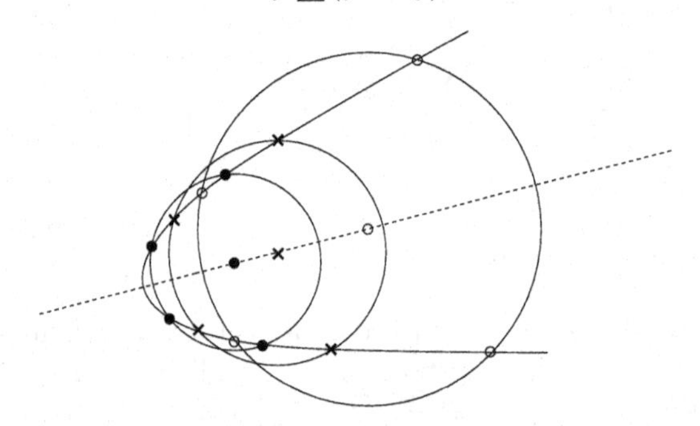

The remaining considerations of this chapter refer to the "curvature" of an algebraic curve. At the intersection points of two curves there is likewise a "residue theorem" for curvature.

For two curves f and g that intersect transversally at O, we will first compute

$$\mathrm{Res}_O \begin{bmatrix} h\, dXdY \\ f,\ g^2 \end{bmatrix}.$$

In the formula that arises, the values f_x, f_{xx}, f_{xy}, etc., of the partial derivatives at the point O will occur.

For the leading forms of f and g at O we have

$$Lf = f_x \cdot X + f_y \cdot Y,$$

$$Lg = g_x \cdot X + g_y \cdot Y,$$

where $j := f_x g_y - f_y g_x \neq 0$. We set $X' := Lf$, $Y' := Lg$ and write

$$f = X' + a_{20}X'^2 + a_{11}X'Y' + a_{02}Y'^2 + \cdots,$$

$$g = Y' + b_{20}X'^2 + b_{11}X'Y' + b_{02}Y'^2 + \cdots,$$

where $a_{ij},\ b_{ij} \in K$. Then we have

$$f = c_{11}X' + c_{12}Y',$$

$$g = c_{21}X' + c_{22}Y',$$

with

$$\begin{aligned} c_{11} &= 1 + a_{20}X' + a_{11}Y' + a_{30}X'^2 + \cdots, \\ c_{12} &= a_{02}Y' + a_{03}Y'^2 + \cdots, \\ c_{21} &= b_{20}X' + b_{11}Y' + \cdots, \\ c_{22} &= 1 + b_{02}Y' + b_{03}Y'^2 + \cdots. \end{aligned}$$

Observe that $2a_{20} = f_{x'x'},\ a_{11} = f_{x'y'},\ 2a_{02} = f_{y'y'}$, etc. If we write

$$\begin{aligned} f &= c_{11}X' + c_{12}(Y')^{-1}Y'^2, \\ g^2 &= (c_{21}^2 X' + 2c_{21}c_{22}Y')X' + c_{22}^2 Y'^2, \end{aligned}$$

then the determinant belonging to this system,

$$\Delta := c_{11}c_{22}^2 - c_{12}(Y')^{-1}(c_{21}^2 X' + 2c_{21}c_{22}Y'),$$

is a unit in $\mathcal{O}'_O$. By the transformation formula 11.17 we get

$$\operatorname{Res}_O \begin{bmatrix} h\,dX dY \\ f,\ g^2 \end{bmatrix} = \operatorname{Res}_O \begin{bmatrix} \Delta^{-1} h\,dX dY \\ X',\ Y'^2 \end{bmatrix} = \frac{1}{j} \operatorname{Res}_O \begin{bmatrix} \Delta^{-1} h\,dX' dY' \\ X',\ Y'^2 \end{bmatrix}.$$

To explicitly calculate this residue, Δ^{-1} and h can be reduced modulo $(X',\ Y'^2)\,\mathcal{O}'_O$. In the following, $\equiv$ denotes congruence mod $(X',\ Y'^2)\,\mathcal{O}'_O$. We have

$$\begin{aligned} h &\equiv h(0) + h_{Y'} \cdot Y', \\ \Delta &\equiv 1 + (f_{x'y'} + g_{y'y'}) \cdot Y', \\ \Delta^{-1} &\equiv 1 - (f_{x'y'} + g_{y'y'}) \cdot Y', \end{aligned}$$

and

$$\Delta^{-1} h \equiv h(0) + [h_{y'} - h(0)(f_{x'y'} + g_{y'y'})]Y'.$$

By 11.20 we get

$$\operatorname{Res}_O \begin{bmatrix} h\,dX dY \\ f,\ g^2 \end{bmatrix} = \frac{1}{j}(h_{y'} - h(0)(f_{x'y'} + g_{y'y'})).$$

Calculating with (X, Y)-coordinates and using the chain rules leads us, with some patience, to the following formula:

$$\operatorname{Res}_O \begin{bmatrix} h\,dXdY \\ f,\ g^2 \end{bmatrix} = \frac{1}{j^2} \cdot \frac{\partial(f,h)}{\partial(x,y)} - \frac{h(0)}{j^3} \Bigg[f_x \left(\frac{\partial(f_Y, g)}{\partial(x,y)} + \frac{\partial(f, g_Y)}{\partial(x,y)} \right) - f_y \left(\frac{\partial(f, g_X)}{\partial(x,y)} + \frac{\partial(f_X, g)}{\partial(x,y)} \right) \Bigg],$$

where the Jacobian determinants are to be taken at the point O. From the Residue Theorem 11.19 the following formula follows with $J := f_X g_Y - f_Y g_X$.

Theorem 12.15 (Formula of B. Segre [Se]). *Assume that f and g intersect transversally at every point of intersection and that they have no points at infinity in common. Let $\{P_1, \ldots, P_{pq}\} = \mathcal{V}(f) \cap \mathcal{V}(g)$ and let h be a polynomial of degree $\leq p + 2q - 3$. Then*

$$\sum_{i=1}^{pq} \left[\frac{\partial(f,g)}{\partial(X,Y)} \cdot J^{-2} \right]_{P_i} = \sum_{i=1}^{pq} \Bigg[hJ^{-3} \Bigg(\frac{\partial f}{\partial X} \left(\frac{\partial(f_Y, g)}{\partial(X,Y)} + \frac{\partial(f, g_Y)}{\partial(X,Y)} \right) - \frac{\partial f}{\partial Y} \left(\frac{\partial(f, g_X)}{\partial(X,Y)} + \frac{\partial(f_X, g)}{\partial(X,Y)} \right) \Bigg) \Bigg]_{P_i}.$$

Consider now, as did Segre, the special case

$$h := (g \cdot g_{YY} - g_Y^2) f_Y.$$

Then we have $\deg h \leq p + 2q - 3$, and yet another calculation shows that

$$\operatorname{Res}_O \begin{bmatrix} h\,dXdY \\ f,\ g^2 \end{bmatrix}$$

$$= \frac{1}{j^3} \left(f_y^3 (g_y^2 g_{xx} - 2 g_x g_y g_{xy} + g_x^2 g_{yy}) - g_y^3 (f_y^2 f_{xx} - 2 f_x f_y f_{xy} + f_x^2 f_{yy}) \right).$$

Now in the real case we can interpret the formula of B. Segre in the following way: Assume that f and g are real and that all the points of intersection P_i $(i = 1, \ldots, pq)$ are real. Furthermore, no tangents of f or g at a point P_i should be parallel to the Y-axis, which can always be arranged by a coordinate transformation. Then $f_Y(P_i) \neq 0$ and $g_Y(P_i) \neq 0$ $(i = 1, \ldots, pq)$. Write $P_i = (a_i, b_i)$ $(i = 1, \ldots, pq)$. Then, by the implicit function theorem, in a neighborhood of a_i two C^∞-functions ϕ_i and ψ_i are defined such that

$$f(X, \phi_i(X)) = g(X, \psi_i(X)) = 0,$$

and we have

$$\phi_i'(a_i) = -\frac{f_X(P_i)}{f_Y(P_i)}, \quad \psi_i'(a_i) = -\frac{g_X(P_i)}{g_Y(P_i)}$$

as well as

$$\phi_i''(a_i) = -\left[\frac{f_Y^2 f_{XX} - 2 f_X f_Y f_{XY} + f_X^2 f_{YY}}{f_Y^3} \right] (P_i)$$

and similarly for ψ_i. Also,

$$\left[\frac{J}{f_Y g_Y}\right](P_i) = \psi_i'(a_i) - \phi_i'(a_i).$$

By 12.15 we conclude

Corollary 12.16. *Under the above assumptions we have*

$$\sum_{i=1}^{pq} \frac{\psi_i''(a_i) - \phi_i''(a_i)}{(\psi_i'(a_i) - \phi_i'(a_i))^3} = 0.$$

The first derivatives are the slopes of the curves f and g at the points P_i, and the second obviously have something to do with the curvature. Let $\kappa_i(f)$ be the curvature of f at P_i, and α_i the oriented angle between f and a line through P_i parallel to the X-axis. Let $\kappa_i(g)$ and β be similarly defined. By a well-known formula we have

$$\kappa_i(f) = \frac{\phi_i''(a_i)}{\phi_i'(a_i)^3} \cdot \sin^3 \alpha_i.$$

Using 12.16 and a further calculation we get the formula

$$\sum_{i=1}^{pq} \frac{\kappa_i(g) \cdot \cos^3 \alpha_i - \kappa_i(f) \cdot \cos^3 \beta_i}{\sin^3(\beta_i - \alpha_i)} = 0.$$

In the special case that f is the X-axis we have the following:

Formula of Reiss 12.17. *If a curve g of degree q intersects the X-axis in q distinct points, then*

$$\sum_{i=1}^{q} \frac{\kappa_i(g)}{\sin^3 \beta_i} = 0,$$

where $\kappa_i(g)$ is the curvature and β_i the angle the curve makes with the X-axis at each intersection point.

If one specifies q different points on the X-axis and at each point a nonzero slope, then one can always find an algebraic curve of degree q that intersects the X-axis at the given points and has the specified slopes. By the formula of Reiss, however, one cannot also specify a curvature arbitrarily at all the points: One of the curvatures is always determined by the others and by the other slopes.

Exercises

1. Carry out the explicit calculations that were suppressed in connection with 12.15–12.17.

2. Show that the following identity holds in $\mathbb{Q}(X_1, \ldots, X_n)$:

$$\sum_{i=1}^{n} \frac{X_i^\rho}{\prod_{k \neq i} (X_i - X_k)} = \begin{cases} 0 & \text{for } \rho \leq n-2, \\ 1 & \text{for } \rho = n-1. \end{cases}$$

What do you get when $\rho = n$?

13

The Riemann–Roch Theorem

This theorem deals with the existence of rational functions on algebraic curves or on the corresponding abstract Riemann surfaces with prescribed orders at the points on the curves (or on the abstract Riemann surface). Using the methods of Appendix L we will derive two versions of the Riemann–Roch theorem, one for the curve itself and one for its Riemann surface (its function field). The theorem leads to an important birational invariant of irreducible curves, namely the genus of the associated function field. An excellent presentation of the corresponding complex-analytic theory is given by Forster [Fo].

Suppose we are given an irreducible curve F with function field $L = \mathcal{R}(F)$. Further, let $\mathfrak{X} := \mathfrak{X}(F)$ be the corresponding abstract Riemann surface, i.e., the set of all discrete valuation rings of L/K. We call the elements of $\mathfrak{X}$ "points" and denote them in general by P. The discrete valuation ring corresponding to P will be denoted by V_P, and we will let ν_P denote the discrete valuation corresponding to V_P. For the regular points P of F the local ring $\mathcal{O}_{F,P}$ is a discrete valuation ring of L/K, so we will always think of $\operatorname{Reg}(F)$ as part of $\mathfrak{X}$. For the singular points of F there are finitely many points of $\mathfrak{X}$ lying over them by the mapping $\pi : \mathfrak{X} \to \mathcal{V}_+(F)$ introduced in 6.12.

We already know that a nonconstant function $r \in L$ on $\mathfrak{X}$ has at least one zero and at least one pole, and that the number of zeros is equal to the number of poles when these are counted with their orders (7.3). The functions $r \in L$ are hence subject to strong conditions with respect to their orders at the outset. This considerably limits the possibility of constructing functions with prescribed orders at the points of $\mathfrak{X}$.

Let $\operatorname{Div}(\mathfrak{X})$ denote the divisor group on $\mathfrak{X}$. Giving a divisor $\sum \alpha_P \cdot P$ ($\alpha_P \in \mathbb{Z}$) means that one can give an order $\alpha_P \neq 0$ on finitely many points $P \in \mathfrak{X}$. Instead of α_P we will write $\nu_P(D)$ and call this number the *order of D at the point P*. We define the *support* of D by

$$\operatorname{Supp} D := \{P \in \mathfrak{X} \mid \nu_P(D) \neq 0\}.$$

For D, $D' \in \operatorname{Div}(\mathfrak{X})$ we write $D \geq D'$ in case $\nu_P(D) \geq \nu_P(D')$ for all $P \in \mathfrak{X}$. If $\nu_P(D) \geq 0$ for all $P \in \mathfrak{X}$, then we call D *effective*. For a function $r \in L^*$ we will denote, as we did earlier, by

$$(r) = \sum_{P \in \mathfrak{X}} \nu_P(r) \cdot P$$

the *principal divisor* belonging to r. Furthermore, we call

$$(r)_0 = \sum_{\nu_P(r)>0} \nu_P(r) \cdot P \quad \text{the } \textit{zero divisor},$$

$$(r)_\infty = \sum_{\nu_P(r)<0} \nu_P(r) \cdot P \quad \text{the } \textit{pole divisor},$$

of r. We will denote the subgroup of $\mathrm{Div}(\mathfrak{X})$ of all principal divisors by $\mathcal{H}(\mathfrak{X})$.

If $\phi \subset \mathbb{P}^2(K)$ is another curve, for which F is not a component, then in every local ring $\mathcal{O}_{F,Q}$ $(Q \in \mathcal{V}_+(F))$ there is a principal ideal corresponding to ϕ. The extension ideal of this ideal in V_P $(P \in \mathfrak{X},\ \pi(P) = Q)$ is a principal ideal (φ_P). We define the divisor (ϕ) of ϕ by

$$(\phi) = \sum \nu_P(\varphi_P) \cdot P.$$

Since ϕ and F intersect in only finitely many points, $\nu_P(\varphi_P) \neq 0$ for only finitely many $P \in \mathfrak{X}$. From the divisor (ϕ) we get the intersection cycle $\phi * F$ of the form

$$\phi * F = \sum_Q \left(\sum_{\pi(P)=Q} \nu_P(\varphi_P) \right) \cdot Q$$

because we have $\sum_{\pi(P)=Q} \nu_P(\varphi_P) = \mu_Q(\phi, F)$ by 7.2. The divisor (ϕ) is hence a finer invariant of the intersection $\phi \cap F$ than the intersection cycle. Of course, $\deg(\phi) = \deg \phi \cdot \deg F$ by Bézout.

Two divisors $D, D' \in \mathrm{Div}(\mathfrak{X})$ are called *linearly equivalent* if there exists an $r \in L^*$ such that $D - D' = (r)$. We write $D \equiv D'$ in this case.

Remark 13.1. The following are equivalent:

(a) $D \equiv D'$.
(b) D and D' represent the same divisor class in $\mathrm{Cl}(\mathfrak{X}) := \mathrm{Div}(\mathfrak{X})/\mathcal{H}(\mathfrak{X})$.
(c) There are curves ϕ, ψ in $\mathbb{P}^2(K)$ of the same degree, of which F is not a component, such that $D + (\phi) = D' + (\psi)$.

In particular, then (ϕ) and (ψ) are linearly equivalent.

The Riemann–Roch theorem is concerned with the dimension of the K-vector space introduced in the following definition.

Definition 13.2. For $D \in \mathrm{Div}(\mathfrak{X})$ we call

$$\mathcal{L}(D) := \{r \in L \mid \nu_P(r) \geq \nu_P(-D) \quad \text{for all } P \in \mathfrak{X}\}$$

the *vector space of multiples of* $-D$.

It consists of all function r whose orders at all points P of $\mathfrak{X}$ are no "worse" than the orders of $-D$. Evidently,

$$\mathcal{L}(D) = \bigcap_{P \in \mathfrak{X}} \mathfrak{m}_P^{-\nu_P(D)} V_P,$$

where $\mathfrak{m}_P = (\pi_P)$ denotes the maximal ideal of V_P and $\mathfrak{m}_P^{-\nu_P(D)} V_P = \pi_P^{-\nu_P(D)} \cdot V_P$.

Remark 13.3. If D and D' are linearly equivalent divisors, then $\mathcal{L}(D)$ and $\mathcal{L}(D')$ are isomorphic K-vector spaces. That is, if $D' = D + (r)$ with $r \in L^*$, then $\mathcal{L}(D') \to \mathcal{L}(D)$ $(u \mapsto ru)$ is a K-linear mapping with inverse given by r^{-1}.

We will give a description of $\mathcal{L}(D)$ that will allow us to use the results of Appendix L. However, before we do this, we should first clarify the problem of the Riemann–Roch theorem for curves.

We have already explained that $\operatorname{Reg}(F)$ embeds in $\mathfrak{X}$. A divisor D with $\operatorname{Supp} D \subset \operatorname{Reg}(F)$ will be called an *F-divisor*, and $\operatorname{Div}^F(\mathfrak{X})$ will be the group of all F-divisors. Furthermore, let

$$\Sigma := \bigcap_{P \in \operatorname{Sing}(F)} \mathcal{O}_{F,P},$$

where we think of $\Sigma := L$ if $\operatorname{Sing}(F) = \emptyset$. If $\operatorname{Sing}(F) \neq \emptyset$, then $\sum \subset V_Q$ for all $Q \in \mathfrak{X}$ that lie over a singularity of F. For $D \in \operatorname{Div}^F(\mathfrak{X})$ define

$$\mathcal{L}^F(D) := \{r \in \Sigma \mid \nu_P(r) \geq \nu_P(-D) \quad \text{for all } P \in \operatorname{Reg}(F)\}.$$

The Riemann–Roch theorem for F is concerned with the dimension of this K-vector space. If F is smooth, then of course $\mathcal{L}^F(D) = \mathcal{L}(D)$. In general we have

$$\mathcal{L}^F(D) = \Sigma \cap \bigcap_{P \in \operatorname{Reg}(F)} \mathfrak{m}_{F,P}^{-\nu_P(D)} \mathcal{O}_{F,P}\,.$$

Two F-divisors D, D' are called *linearly equivalent with respect to F* (and we write $D \equiv_F D'$) if there exists a unit $r \in \Sigma$ such that $D' - D = (r)$.

Remark 13.4. For $D, D' \in \operatorname{Div}^F(\mathfrak{X})$ we have

(a) If $D' \equiv_F D$, then $\mathcal{L}^F(D') \cong \mathcal{L}^F(D)$.
(b) If $D' \leq D$, then $\mathcal{L}^F(D') \subset \mathcal{L}^F(D)$.
(c) $\mathcal{L}^F(O) = K$.
(d) If $\deg D < 0$, then $\mathcal{L}^F(D) = \{0\}$.

Similarly, these statements also apply to the vector spaces $\mathcal{L}(D)$.

Proof. One shows (a) as in 13.3. Statement (b) is trivial, and (c) follows from the fact that every nonconstant function has at least one pole. If $\mathcal{L}^F(D)$ contains a function $r \neq 0$, then $0 = \deg r \geq -\deg D$ and hence $\deg D \geq 0$. This shows (d).

We will now establish a few properties of the ring Σ.

Lemma 13.5. *Suppose A is an integral domain. For $\mathfrak{p}_1, \dots, \mathfrak{p}_s \in \operatorname{Spec} A$, suppose $\mathfrak{p}_i \not\subset \mathfrak{p}_j$ for $i \neq j$ $(i, j = 1, \dots, s)$ and let $N := A \setminus \bigcup_{i=1}^{s} \mathfrak{p}_i$. Then*

$$A_{\mathfrak{p}_1} \cap \dots \cap A_{\mathfrak{p}_s} = A_N$$

and $\operatorname{Max} A_N = \{\mathfrak{p}_1 A_N, \dots, \mathfrak{p}_s A_N\}$. Furthermore, $A_{\mathfrak{p}_i}$ is the localization of A_N at $\mathfrak{p}_i A_N$ $(i = 1, \dots, s)$.

Proof. Let $H := A_{\mathfrak{p}_1} \cap \dots \cap A_{\mathfrak{p}_s}$. It is clear that $A_N \subset H$. To see the opposite inclusion, for $z \in H$ consider the ideal J of all $a \in A$ such that $az \in A$. Then $J \not\subset \mathfrak{p}_i$ $(i = 1, \dots, s)$ and, as one easily shows, $J \not\subset \bigcup_{i=1}^{s} \mathfrak{p}_i$. Therefore $J \cap N \neq \emptyset$ and consequently $z \in A_N$. The remaining statements of the lemma follow from C.9 and Appendix C, Exercise 3.

Since $\operatorname{Sing}(F)$ is a finite subset of $\mathcal{V}_+(F)$, one can assume that the singularities are points at finite distance. Then let $A := K[f] = K[x, y]$ be the coordinate ring of the affine curve f corresponding to F. If $\mathfrak{p}_1, \dots, \mathfrak{p}_s$ are the maximal ideals of A corresponding to the singularities of F, then by the lemma,

$$\Sigma = A_{\mathfrak{p}_1} \cap \dots \cap A_{\mathfrak{p}_s} = A_N \qquad \text{where } N := A \setminus \bigcup_{i=1}^{s} \mathfrak{p}_i, \tag{1}$$

and $\operatorname{Max}(\Sigma) = \{\mathfrak{p}_1 \Sigma, \dots, \mathfrak{p}_s \Sigma\}$. Also, for $\mathfrak{P}_i := \mathfrak{p}_i \Sigma$ $(i = 1, \dots, s)$,

$$\Sigma_{\mathfrak{P}_i} = A_{\mathfrak{p}_i}.$$

Let S be the integral closure of A in L. As was shown in Chapter 6, the points of $\mathfrak{X}$ lying over the singularities of F correspond one-to-one with the maximal ideals $\mathfrak{Q}$ of S with $\mathfrak{Q} \cap A \in \{\mathfrak{p}_1, \dots, \mathfrak{p}_s\}$, and the $S_{\mathfrak{Q}}$ are precisely the discrete valuation rings belonging to these points. In particular, $\Sigma \subset S_{\mathfrak{Q}}$ for all these $\mathfrak{Q} \in \operatorname{Max}(S)$.

Lemma 13.6. (a) $A = \Sigma \cap S$.
(b) *$r \in \Sigma$ is a unit of Σ if and only if $\nu_Q(r) = 0$ for all $Q \in \mathfrak{X}$ with $\pi(Q) \in$* $\operatorname{Sing}(F)$.

Proof. (a) Using the notation of (1) we have by F.12

$$A = \bigcap_{\mathfrak{p} \in \operatorname{Max}(A)} A_{\mathfrak{p}}, \quad \Sigma = \bigcap_{\mathfrak{p} \cap N = \emptyset} A_{\mathfrak{p}}, \quad \text{and} \quad S = \bigcap_{\mathfrak{P} \in \operatorname{Max}(S)} S_{\mathfrak{P}}.$$

For $\mathfrak{p} \cap N \neq \emptyset$ we have $A_{\mathfrak{p}} = S_{\mathfrak{P}}$ for some $\mathfrak{P} \in \operatorname{Max}(S)$, since $A_{\mathfrak{p}}$ is already a discrete valuation ring. For $\mathfrak{p} \cap N = \emptyset$ we have $A_{\mathfrak{p}} \subset S_{\mathfrak{Q}}$ for all $\mathfrak{Q} \in \operatorname{Max}(S)$ such that $\mathfrak{Q} \cap A = \mathfrak{p}$. From this (a) follows immediately.

(b) $r \in \Sigma$ is a unit of Σ if it lies in no maximal ideal $\mathfrak{p}_i \Sigma$ $(i = 1, \dots, s)$ of Σ. This is equivalent to saying that r is a unit in every $S_{\mathfrak{Q}}$ where $\mathfrak{Q} \cap A \in \{\mathfrak{p}_1, \dots, \mathfrak{p}_s\}$.

The following theorem demands that there be a unit $r \in \Sigma$ (a function $r \in L$) whose order at finitely many points of $\mathrm{Reg}(F)$ (of $\mathfrak{X}$) is prescribed, but otherwise allows for complete freedom.

Theorem 13.7. *Let $P_1, \dots, P_t \in \mathrm{Reg}(F)$ and let $\alpha_1, \dots, \alpha_t \in \mathbb{Z}$ be given. Then there exists a unit $r \in \Sigma$ such that*

$$\nu_{P_i}(r) = \alpha_i \qquad (i = 1, \dots, t).$$

The theorem holds also for arbitrary points $P_i \in \mathfrak{X}$ with an element $r \in L$.

Proof. We can assume that the affine curve f also contains the points $P_1, \dots, P_t$. Let $\mathfrak{P}_1, \dots, \mathfrak{P}_t$ be the maximal ideals of A corresponding to these points.

It is enough to prove the theorem in the case that all $\alpha_i \geq 0$. In the general case, one substitutes 0 for the α_i that are negative and then solves the problem with a unit $r_1 \in \Sigma$. After this, one substitutes 0 for the α_i that are positive and solves the problem for the $-\alpha_i$ with a unit $r_2 \in \Sigma$. Then $r := r_1 r_2^{-1}$ is a solution of the general problem. Hence suppose $\alpha_i \geq 0$ $(i = 1, \dots, t)$. Choose $z \in \mathfrak{p}_1 \cdots \mathfrak{p}_s \cdot \mathfrak{P}_1^{\alpha_1+1} \cdots \mathfrak{P}_t^{\alpha_t+1}$, $z \neq 0$, and consider the decomposition of A/zA into the direct product of its localizations by the Chinese remainder theorem. It is then clear that there is an $r \in A$ such that

$$\begin{aligned} r - 1 &\in zA_{\mathfrak{p}_i} \quad (i = 1, \dots, s),\\ r - \pi_j^{\alpha_j} &\in zA_{\mathfrak{P}_j} \quad (j = 1, \dots, t), \end{aligned}$$

where π_j is a generator of $\mathfrak{P}_j A_{\mathfrak{P}_j}$. Then r is a unit of Σ. Since $\nu_{P_j}(z) \geq \alpha_j + 1$, it follows that $\nu_{P_j}(r) = \alpha_j$ $(j = 1, \dots, t)$.

We shall now give a description of $\mathcal{L}^F(D)$, for $D \in \mathrm{Div}^F(\mathfrak{X})$, which will allow us to use the determination of $\dim \mathcal{L}^F(D)$ as given in Appendix L. Let $A = K[x, y]$ be given as in (1). By a suitable choice of coordinates, one can assume that A is a finitely generated $K[x]$-module, and that L is separable algebraic over $K(x)$. Let $[L : K(x)] = n$. As in Appendix L we set

$$R := K[x], \qquad R_\infty := K[x^{-1}]_{(x^{-1})},$$

and denote by S (by S_∞) the integral closure of R (of R_∞) in L. By L.1, S is a free R-module (S_∞ is a free R_∞-module) of rank n and we have $S \cap S_\infty = K$ by L.2. The abstract Riemann surface $\mathfrak{X}$ consists of the infinitely many points that correspond to the localizations of S at its maximal ideals, and the finitely many points one gets by localizing S_∞ at the maximal ideals of this ring. We denote the first set by $\mathfrak{X}^f$, the second by $\mathfrak{X}^\infty$ (the finite and infinite parts of $\mathfrak{X}$).

Lemma 13.8. *To each $D \in \mathrm{Div}^F(\mathfrak{X})$ there is a $D' \in \mathrm{Div}^F(\mathfrak{X})$ such that $D' \equiv_F D$ and $\mathrm{Supp}\, D' \subset \mathfrak{X}^f$. A similar statement holds for an arbitrary divisor of $\mathfrak{X}$ using linear equivalence on $\mathfrak{X}$.*

This is an immediate consequence of 13.7. For investigating $\mathcal{L}^F(D)$ (or $\mathcal{L}(D)$) one can always assume using 13.4(a) that $\operatorname{Supp} D \subset \mathfrak{X}^f$. Then we have

(2) $$\mathcal{L}^F(D) = I_D^F \cap S_\infty,$$

where

$$I_D^F := \{r \in \Sigma \mid \nu_P(r) \geq \nu_P(-D) \text{ for all } P \in \operatorname{Reg}(F) \cap \mathfrak{X}^f\}$$

and

(3) $$\mathcal{L}(D) = I_D \cap S_\infty,$$

where

$$I_D := \{r \in L \mid \nu_P(r) \geq \nu_P(-D) \text{ for all } P \in \mathfrak{X}^f\}.$$

The investigation of the dimensions of $\mathcal{L}(D)$ and $\mathcal{L}^F(D)$ can, on the basis of (2) and (3), be simultaneously carried through using Appendix L. In the following we will consider $\mathcal{L}^F(D)$, yet one needs only to substitute S for A, L for Σ, and $S\setminus\{0\}$ for N in order to get the statements for $\mathcal{L}(D)$ corresponding to the statements for $\mathcal{L}^F(D)$.

Theorem 13.9. *If an F-divisor D satisfies* $\operatorname{Supp} D \subset \mathfrak{X}^f$, *then I_D^F is a finitely generated A-module and $I_D^F \cdot \Sigma = \Sigma$.*

Proof. It is immediate from the definition of I_D^F that it is an A-module. Let $P_1, \dots, P_s$ be the points of $\operatorname{Reg}(F) \cap \mathfrak{X}^f$ with $\nu_{P_i}(D) < 0$, let $\alpha_i := -\nu_{P_i}(D)$, and let $\mathfrak{p}_i$ be the maximal ideal of A corresponding to P_i $(i = 1, \dots, s)$. Since $\mathfrak{p}_i \cap N \neq \emptyset$, it follows that $(\prod_{i=1}^s \mathfrak{p}_i^{\alpha_i}) \cap N \neq \emptyset$. For an element a in this intersection we have

$$\begin{aligned} \nu_{P_i}(a) &\geq \alpha_i \qquad (i = 1, \dots, s), \\ \nu_P(a) &\geq 0 \qquad \text{for } P \in \operatorname{Reg}(F) \cap \mathfrak{X}^f, \end{aligned}$$

and a is a unit of $\Sigma = A_N$. Therefore $a \in I_D^F$ and $I_D^F \cdot \Sigma = \Sigma$.

Consider now the points $Q_1, \dots Q_r \in \operatorname{Reg}(F) \cap \mathfrak{X}^f$ with $\beta_i := \nu_{Q_i}(D) > 0$ and the corresponding maximal ideals $\mathfrak{q}_1, \dots, \mathfrak{q}_r$ of A. Let $b \in (\prod_{i=1}^r \mathfrak{q}_i^{\beta_i}) \cap N$. For each $r \in I_D^F$ and each point $P \in \operatorname{Reg}(F) \cap \mathfrak{X}^f$ we have $\nu_P(rb) \geq 0$. Furthermore, $rb \in \Sigma$ and b is a unit of Σ. From 13.6(a) it follows that $rb \in A$. Hence $b \cdot I_D^F$ is an ideal of A. Since A is Noetherian, the ideal $b \cdot I_D^F$ is finitely generated, and so I_D^F is finitely generated as an A-module.

From L.6 it follows that we now already have the following important facts, sometimes called the *finiteness theorems*, namely that

$$\dim_K \mathcal{L}^F(D) < \infty \quad \text{and} \quad \dim_K \mathcal{L}(D) < \infty,$$

a first step toward the Riemann–Roch theorem.

Theorem 13.10. *To each finitely generated A-submodule $I \subset \Sigma$ with $I \cdot \Sigma = \Sigma$ there is exactly one divisor D with* $\operatorname{Supp}(D) \subset \operatorname{Reg}(F) \cap \mathfrak{X}^f$ *such that $I = I_D^F$. We have*

$$I \cdot \mathcal{O}_{F,P} = \mathfrak{m}_{F,P}^{-\nu_P(D)} \cdot \mathcal{O}_{F,P} \quad \textit{for all } P \in \operatorname{Reg}(F) \cap \mathfrak{X}^f.$$

Proof. (a) First let $I \subset A$ be an ideal with $I \cdot \Sigma = \Sigma$. Then $I \cdot A_\mathfrak{p} = A_\mathfrak{p}$ for $\mathfrak{p} \in \operatorname{Max}(A)$ with $\mathfrak{p} \cap N = \emptyset$. Let $I \cdot A_\mathfrak{p} = \mathfrak{p}^{\alpha_\mathfrak{p}} A_\mathfrak{p}$ for $\mathfrak{p} \in \operatorname{Max}(A)$ with $\mathfrak{p} \cap N = \emptyset$. Now $\alpha_\mathfrak{p} > 0$ for only finitely many $\mathfrak{p}$, since for $z \in I \setminus \{0\}$ there are only finitely many maximal ideals in $A/(z)$, and I is contained in the inverse images of at most these maximal ideals.

It follows that

$$I = \prod \mathfrak{p}^{\alpha_\mathfrak{p}}, \tag{4}$$

since the localizations of I and $\prod \mathfrak{p}^{\alpha_\mathfrak{p}}$ agree at each of the maximal ideals of A, and every ideal of A is the intersection of all its localizations at the maximal ideals of A (F.12).

Now set $\alpha_P := \alpha_\mathfrak{p}$ and $D = \sum(-\alpha_P) \cdot P$, where $P \in \operatorname{Reg}(F)$ is the point corresponding to $\mathfrak{p}$. Since $\nu_P(a) \geq \alpha_P$ for each $a \in I$ and $P \in \operatorname{Reg}(F) \cap \mathfrak{X}^f$, we have $I \subset I_D^F$. Conversely, for $P \in \operatorname{Reg}(F) \cap \mathfrak{X}^f$ we also have

$$\mathfrak{m}_{F,P}^{\alpha_P} \mathcal{O}_{F,P} = I \mathcal{O}_{F,P} \subset I_D^F \cdot \mathcal{O}_{F,P} = \mathfrak{m}_{F,P}^{\beta_P} \cdot \mathcal{O}_{F,P}, \tag{5}$$

where $\beta_P \geq \alpha_P$ by definition of I_D^F. It follows that $\alpha_P = \beta_P$ for all $P \in \operatorname{Reg}(F) \cap \mathfrak{X}^f$. Since I_D^F contains a unit of Σ (13.9), all the localizations of I and I_D^F coincide, and hence $I = I_D^F$.

From (5) we get the last statement of the theorem, and hence the fact that D is given uniquely by I.

(b) Now if $I \subset \Sigma$ is a finitely generated A-module with $I \cdot \Sigma = \Sigma$, then there is an element $a \in N$ with $aI \subset A$. From (a) we have $aI = I_D^F$ for some F-divisor D. Then $I = a^{-1} I_D^F = I_{D'}^F$ with $D' := \sum(\nu_P(D) + \nu_P(a)) \cdot P$. The remaining statements of the theorem are also clear in this situation.

Remark 13.11. In case $A = S$, formula (4) shows that every ideal $I \neq \{0\}$ of S is a product of powers of maximal ideals (S is a "Dedekind ring"). Furthermore, for $\mathfrak{p}_1, \dots, \mathfrak{p}_t \in \operatorname{Max}(S)$ the ring

$$H := S_{\mathfrak{p}_1} \cap \dots \cap S_{\mathfrak{p}_t}$$

is a principal ideal ring. In fact, by 13.5 the $S_{\mathfrak{p}_i}$ are the localizations of H at its maximal ideals. Let $I \subset H$ be an ideal, $I \neq \{0\}$, and let $IS_{\mathfrak{p}_j} = \mathfrak{p}_j^{\alpha_j} S_{\mathfrak{p}_j}$ ($\alpha_j \in \mathbb{N}$; $j = 1, \dots, t$). Using 13.7 there is an $r \in L$ with $IS_{\mathfrak{p}_j} = rS_{\mathfrak{p}_j}$ ($j = 1, \dots, t$). In particular, $r \in \bigcap_{j=1}^t IS_{\mathfrak{p}_j} = I$, and it follows that $I = (r)$.

Since S_∞ has only finitely many maximal ideals, it follows similarly that S_∞ is a principal ideal ring.

Theorem 13.12. *For two F-divisors D, D' with $D \geq D'$ and* $\operatorname{Supp} D \cup \operatorname{Supp} D' \subset \mathfrak{X}^f$ *we have*

$$\dim_K I_D^F / I_{D'}^F = \deg D - \deg D'.$$

Proof. By multiplying I_D^F and $I_{D'}^F$ by a suitable unit from Σ, we can assume that both are ideals in A. Then

$$I_D^F = \prod \mathfrak{p}^{\alpha_\mathfrak{p}}, \quad I_{D'}^F = \prod \mathfrak{p}^{\alpha'_\mathfrak{p}} \quad (\alpha_\mathfrak{p} = -\nu_P(D),\ \alpha'_\mathfrak{p} = -\nu_P(D')),$$

where the $\mathfrak{p}$ are the maximal ideals of A with $\mathfrak{p} \cap N \neq \emptyset$ and P is the point corresponding to $\mathfrak{p}$. We need to show that

$$\dim_K \left(\prod \mathfrak{p}^{\alpha_\mathfrak{p}} / \prod \mathfrak{p}^{\alpha'_\mathfrak{p}} \right) = \sum (\alpha'_\mathfrak{p} - \alpha_\mathfrak{p}).$$

Consider $\prod \mathfrak{p}^{\alpha_\mathfrak{p}} / \prod \mathfrak{p}^{\alpha'_\mathfrak{p}}$ as an ideal in

$$A / \prod \mathfrak{p}^{\alpha'_\mathfrak{p}} = \prod_\mathfrak{p} A_\mathfrak{p} / \mathfrak{p}^{\alpha'_\mathfrak{p}} A_\mathfrak{p}.$$

It is the direct product $\prod_\mathfrak{p} \mathfrak{p}^{\alpha_\mathfrak{p}} A_\mathfrak{p} / \mathfrak{p}^{\alpha'_\mathfrak{p}} A_\mathfrak{p}$. From E.13 it follows that

$$\dim_K (\mathfrak{p}^{\alpha_\mathfrak{p}} A_\mathfrak{p} / \mathfrak{p}^{\alpha'_\mathfrak{p}} A_\mathfrak{p}) = \alpha'_\mathfrak{p} - \alpha_\mathfrak{p},$$

and therefore the theorem is proved.

As in Appendix L we now consider the filtration $\mathcal{F} = \{\mathcal{F}_\alpha\}$ of L/R_∞ with $\mathcal{F}_\alpha = x^\alpha S_\infty$ ($\alpha \in \mathbb{Z}$). Further, let $\sigma = \sigma_{L/K(x)}$ be the canonical trace of $L/K(x)$. Then

$$\operatorname{Hom}_R(I_D^F, R) = (I_D^F)^* \cdot \sigma$$

for some finitely generated A-module $(I_D^F)^*$, and in particular,

$$\operatorname{Hom}_R(A, R) = \mathfrak{C}_{A/R} \cdot \sigma,$$

where $\mathfrak{C}_{A/R}$ is the Dedekind complementary module of A/R with respect to σ. By L.8,

$$\ell_*^F(D) := \dim_K((I_D^F)^* \cap x^{-2}\mathfrak{C}_{S_\infty/R_\infty}) < \infty,$$

and in particular,

$$g^F := \ell_*^F(0) = \dim_K(\mathfrak{C}_{A/R} \cap x^{-2}\mathfrak{C}_{S_\infty/R_\infty}) < \infty. \tag{6}$$

We now set

$$\ell^F(D) := \dim_K \mathcal{L}^F(D) \quad \text{and} \quad \chi^F(D) := \ell^F(D) - \ell_*^F(D).$$

By L.9 we have the formula

(7) $$\chi^F(D) = n - \sum_{i=1}^{n} \operatorname{ord}_{\mathcal{F}} a_i$$

if $\{a_1, \ldots, a_n\}$ is a standard basis of I_D^F. If D and D' are two F-divisors as in 13.12, then it follows from L.10 and 13.12 that

(8) $$\chi^F(D) - \chi^F(D') = \deg D - \deg D'.$$

If D and D' are two arbitrary F-divisors with support in $\mathfrak{X}^f$, then there is also such an F-divisor D'' with $D \geq D''$ and $D' \geq D''$, and two applications of formula (8) show that (8) holds for arbitrary F-divisors D, D' with $\operatorname{Supp} D \cup \operatorname{Supp} D' \subset \mathfrak{X}^f$.

For $D' = 0$ we have $\chi^F(0) = \ell^F(0) - g^F = 1 - g^F$. From (8) we therefore get the Riemann–Roch formula

(9) $$\ell^F(D) = \ell_*^F(D) + \deg D + 1 - g^F.$$

According to its definition, the number g^F, and also $\ell_*^F(D)$, could depend on the choice of x. Using (8) we will show that g^F and $\ell_*^F(D)$ actually do not depend on the choice of the element x that we used in the definitions of S, S_∞, and $\mathfrak{X}^f$.

Theorem 13.13.

(a) *There is a number $c \in \mathbb{Z}$ such that $\ell_*^F(D) = 0$ for all $D \in \operatorname{Div}^F(\mathfrak{X})$ with* $\operatorname{Supp} D \subset \mathfrak{X}^f$ *and* $\deg D > c$.
(b) *g^F depends only on F (and not on x).*
(c) *For an arbitrary $D \in \operatorname{Div}^F(\mathfrak{X})$, the number $\ell_*^F(D)$ depends only on F and D.*

Proof. (a) By L.12 we have

$$(I_D^F)^* = \mathfrak{C}_{A/R} : I_D^F,$$

and S_∞ is a principal ideal ring by 13.11. Therefore $\mathfrak{C}_{S_\infty/R_\infty} = z^{-1} S_\infty$ for some $z \in L^*$. Hence

$$\ell_*^F(D) = \dim_K(x^2 z(\mathfrak{C}_{A/R} : I_D^F) \cap S_\infty).$$

There is an element $a \in A \setminus \{0\}$ with $a x^2 z \mathfrak{C}_{A/R} \subset A$, and so $x^2 z \mathfrak{C}_{A/R} \subset a^{-1} A$ and

$$x^2 z(\mathfrak{C}_{A/R} : I_D^F) = (x^2 z(\mathfrak{C}_{A/R}) : I_D^F) \subset a^{-1} A : I_D^F.$$

By 13.10, for each $P \in \operatorname{Reg}(F) \cap \mathfrak{X}^f$ there is a $b \in I_D^F$ with $\nu_P(b) = \nu_P(-D)$. Now if $r \in L^*$ is such that $r I_D^F \subset a^{-1} A$, then $rba \in A$ and hence

$$\nu_P(r) - \nu_P(D) \geq -\nu_P(a)$$

for every $P \in X^f$. Consequently,

$$\sum_{P \in \mathfrak{X}^f} \nu_P(r) \geq \deg D - \sum_{P \in \mathfrak{X}^f} \nu_P(a).$$

Now let $c := \sum_{P \in \mathfrak{X}^f} \nu_P(a)$ and $\deg D > c$. Then $\sum_{P \in \mathfrak{X}^f} \nu_P(r) > 0$, and it cannot be the case that $r \in S_\infty$, because $\deg(r) = 0$. Therefore $\ell_*^F(D) = 0$.

(b) Let $\tilde{x} \in L$ be an element with similar properties as x and let $\tilde{\mathfrak{X}}^f$ be the finite part of $\mathfrak{X}$ with respect to $\tilde{x}$. We will write $g^F(x)$ and $g^F(\tilde{x})$ for the quantities (6) made from x respectively $\tilde{x}$. For a divisor D with $\operatorname{Supp} D \subset \mathfrak{X}^f \cap \tilde{\mathfrak{X}}^f$ we will write $\ell_*^{F,x}(D)$ and $\ell_*^{F,\tilde{x}}(D)$ for the quantity $\ell_*^F(D)$ formed from x respectively $\tilde{x}$. By (a), we can find a D such that $\ell_*^{F,\tilde{x}}(D) = \ell_*^{F,x}(D) = 0$. But then the above formula (9) shows that

$$g^F(x) = \deg D + 1 - \ell^F(D) = g^F(\tilde{x}).$$

(c) If D is now an arbitrary F-divisor and x is chosen as above, so that $\operatorname{Supp} D \subset X^f$, then the following follows from (9):

$$\ell_*^{F,x}(D) = \ell^F(D) - \deg D - 1 + g^F.$$

Since the right side does not depend on x by (b), the claim (c) has been proved.

Applying the preceding considerations to $A = S$ and $\Sigma = L$ we can also conclude that the number

$$g^L := \dim_K(\mathfrak{C}_{S/R} \cap x^{-2}\mathfrak{C}_{S_\infty/R_\infty})$$

is independent of x and depends only on L.

Definition 13.14. The number g^F is called the *genus* of F. The number g^L is called the *genus of the function field* L/K (or of the associated abstract Riemann surface $\mathfrak{X}$).

With a few basic facts about differential modules it can be shown that the vector spaces

$$\omega(F) := \mathfrak{C}_{A/R} dx \cap \mathfrak{C}_{S_\infty/R_\infty} dx^{-1} \subset \Omega^1_{L/K}$$

and

$$\omega(\mathfrak{X}) := \mathfrak{C}_{S/R} dx \cap \mathfrak{C}_{S_\infty/R_\infty} dx^{-1} \subset \Omega^1_{L/K}$$

are independent of x. One calls these the vector spaces of *global regular differentials of F respectively $\mathfrak{X}$*. According to the definition the genus is the dimension of these vector spaces.

By (9) we have proved

Theorem 13.15.

(a) *Riemann–Roch theorem for the function field L/K (or its abstract Riemann surface $\mathfrak{X}$) : For each $D \in \mathrm{Div}(\mathfrak{X})$ there exists a number $\ell_*(D) \geq 0$ depending only on D such that*

$$\ell(D) = \ell_*(D) + \deg D + 1 - g^L.$$

(b) *Riemann–Roch Theorem for the curve F : For each $D \in \mathrm{Div}^F(\mathfrak{X})$ there exists a number $\ell_*^F(D) \geq 0$ depending only on D and F such that*

$$\ell^F(D) = \ell_*^F(D) + \deg D + 1 - g^F.$$

For a divisor $D \in \mathrm{Div}(\mathfrak{X})$ we call

$$\chi(D) = \ell(D) - \ell_*(D)$$

the *Euler–Poincaré characteristic of* D and for an F-divisor D we call

$$\chi^F(D) = \ell^F(D) - \ell_*^F(D)$$

the *Euler–Poincaré characteristic of D with respect to F*. In the classical literature $\ell_*(D)$ is called the *index* (index of speciality) of D. A divisor D is called *special* if $\ell_*(D) > 0$. Clearly, for nonspecial divisors we have

$$\ell(D) = \deg D + 1 - g^L.$$

By Theorem 13.13 divisors with sufficiently large degree are nonspecial.

The Riemann–Roch theorem for F is equivalent to the formula (8), and the Riemann–Roch theorem for L/K (for $\mathfrak{X}$) to the corresponding formula

$$\chi(D) - \chi(D') = \deg D - \deg D'$$

for D, $D' \in \mathrm{Div}(\mathfrak{X})$.

To calculate the genus it is sometimes useful to use the following formula, which follows from (7) with $D = 0$:

Formula 13.16. If $\{a_1, \ldots, a_n\}$ is a standard basis for A over R, then

$$g^F = \sum_{i=1}^{n} \mathrm{ord}_{\mathcal{F}}\, a_i - n + 1.$$

Similarly for g^L using a standard basis for S over R.

The Riemann–Roch theorem for the function field L/K depends only on the discrete valuations of L/K and can therefore be recast completely independently of the theory of algebraic curves and purely in a valuation-theoretic framework, where K need not be algebraically closed. In addition to the work of F. K. Schmidt [Sch], this standpoint is taken for example by Chevalley [C] and Roquette [R].

Exercises

1. For a divisor $D \in \mathrm{Div}(\mathfrak{X})$ let $L(D)$ be the set of all effective divisors linearly equivalent to D. Show that
   (a) There is a bijective mapping
   $$L(D) \to \mathbb{P}(\mathcal{L}(D)),$$
   where $\mathbb{P}(\mathcal{L}(D))$ is the projective space associated with the vector space $\mathcal{L}(D)$ (Chapter 2).
   (b) If $D, D' \in \mathrm{Div}(\mathfrak{X})$ and if $L(D) = L(D') \neq \emptyset$, then $D \equiv D'$.
   (c) If we set $\dim L(D) := \dim_K \mathcal{L}(D) - 1$, then $\dim L(D)$ depends only on the set $L(D)$ and not on the divisor D.
2. Let F be a smooth curve of degree p in $\mathbb{P}^2(K)$ and for $q > 0$ let $L_q = K[F]_q$ be the homogeneous component of degree q of the coordinate ring $K[F]$. For $\varphi \in L_q \setminus \{0\}$ suppose the divisor (φ) of φ is defined as
   $$(\varphi) = \phi * F \quad \text{(intersection cycle)}$$
   with a preimage $\phi \in K[X_0, X_1, X_2]_q$ of φ. Let $D := (\varphi_0)$ for a fixed $\varphi_0 \in L_q \setminus \{0\}$.
   (a) Give a vector space isomorphism $\mathcal{L}(D) \to L_q$.
   (b) How large is $\dim_K \mathcal{L}(D)$?

14

The Genus of an Algebraic Curve and of Its Function Field

Here we mainly give rules for explicitly determining the genus. These originate from formula 13.16. By a function field we always understand an algebraic function field of one variable. We say that it is rational if it is generated by one transcendental element.

The genus g^F of an irreducible curve F in $\mathbb{P}^2(K)$ is not as interesting as the genus g^L of its function field $L := \mathcal{R}(F)$, because we have

Theorem 14.1. *If* $\deg F = p$*, then* $g^F = \binom{p-1}{2}$.

Proof. We assume that the coordinates have been chosen so that $\mathrm{Sing}(F)$ lies in the affine plane with respect to the line at infinity $X_0 = 0$. The corresponding affine curve is then given by

$$f(X,Y) = F(1,X,Y) = F\left(1, \frac{X_1}{X_0}, \frac{X_2}{X_0}\right) \qquad \left(X := \frac{X_1}{X_0},\ Y := \frac{X_2}{X_0}\right).$$

We can also assume that f is monic of degree p as a polynomial in Y and that $\frac{\partial f}{\partial Y} \neq 0$. Then for the coordinate ring $A = K[f] = K[X,Y]/(f) = K[x,y]$ we have

$$A = \bigoplus_{i=0}^{p-1} Ry^i,$$

with $R = K[x]$, and $L/K(x)$ is separable.

We will show that $\{1, y, \dots, y^{p-1}\}$ is a standard basis of A over R with respect to the filtration $\mathcal{F}$ used in Appendix L and Chapter 13, and that

$$\mathrm{ord}_{\mathcal{F}}\, y^i = i \qquad (i = 0, \dots, p-1).$$

To do this we consider the dehomogenization of F with respect to X_1,

$$\frac{1}{X_1^p} F(X_0, X_1, X_2) = F\left(\frac{X_0}{X_1}, 1, \frac{X_2}{X_1}\right) = F(X^{-1}, 1, X^{-1}Y) = g(U,V),$$

where $U := X^{-1}$ and $V := X^{-1}Y$. If

$$f = \sum_{i=0}^{p} \varphi_i Y^i \qquad (\varphi_i \in K[X],\ \deg \varphi_i \leq p - i),$$

then by an easy calculation we have

$$g(X^{-1}, X^{-1}Y) = \sum_{i=0}^{p} \varphi_i(X) \cdot X^{-p+i}(X^{-1}Y)^i.$$

Hence g is also monic of degree p as a polynomial in V. Note that because f is monic with respect to Y, no points at infinity (with respect to X_0) of F with X_1-coordinate 0 can exist. The coordinate ring $A' = K[g]$ of g is identified with the subalgebra $K[x^{-1}, x^{-1}y]$ of L. We have

$$A' = \bigoplus_{i=0}^{p-1} K[x^{-1}](x^{-1}y)^i,$$

and the points at infinity of f are in one-to-one correspondence with the maximal ideals of A' lying over $(x^{-1}) \cdot K[x^{-1}]$. Hence, with the notation of Appendix L,

$$S_\infty = \bigoplus_{i=0}^{p-1} R_\infty (x^{-1}y)^i,$$

and $\{1, y, \ldots, y^{p-1}\}$ is a standard basis of A (L.5). Furthermore, $\operatorname{ord}_{\mathcal{F}} y^i = i$ $(i = 0, \ldots, p-1)$.

By 13.16 it follows that

$$g^F = \sum_{i=0}^{p-1} i - p + 1 = \sum_{i=0}^{p-2} i = \binom{p-1}{2}.$$

For smooth curves F, the genus g^F is equal to the genus of its associated function field, and because of this we can derive a few statements about the genus of an algebraic function field. Recall that a model of a function field L/K is a curve F with $L = \mathcal{R}(F)$ (4.7).

Corollary 14.2. (a) *Smooth rational curves (rational function fields) have genus* 0.
(b) *Elliptic curves (elliptic function fields) have genus* 1.
(c) *If an algebraic function field L/K has a smooth plane projective curve of degree p as a model, then*

$$g^L = \binom{p-1}{2}.$$

(d) *Smooth plane curves of genus* 1 *are elliptic.*

Proof. (a) Every line has genus 0 and hence so does the field of rational functions over K. Since the smooth rational curves are birationally equivalent to lines, they also have genus 0.

(b) Elliptic curves are smooth curves of degree 3. By 14.1 they have genus 1.

(c) follows directly from 14.1, and (d) follows from (c).

This gives a new way to see that smooth quadrics are rational, because their genus is 0. Smooth curves of degree $p > 2$ cannot have a rational parametrization; in particular, this holds for elliptic curves. The example of the Fermat curves shows that there exist function fields with genus $g = \binom{p-1}{2}$ ($p \in \mathbb{N}_+$). In fact, there are function fields of genus g for every $g \in \mathbb{N}$, as we will show in 14.6. If the function fields of two curves have different genera, then of course the curves cannot be birationally equivalent.

We will show that every function field of genus 0 is rational. To do this, we will use

Lemma 14.3. *Let L/K be a function field and $x \in L$ a nonconstant function. Then for the zero divisor $(x)_0$ and the pole divisor $(x)_\infty$ of x on the abstract Riemann surface of L/K we have*

$$\deg(x)_0 = \deg(x)_\infty = [L : K(x)].$$

Proof. Consider R_∞ and S_∞ as in Appendix L and use the fact (L.1) that S_∞ is a free R_∞-module of rank $[L : K(x)]$. We then have

$$S_\infty/(x^{-1}) = \prod_{\mathfrak{P}\in \operatorname{Max}(S_\infty)} (S_\infty)_\mathfrak{P}/(x^{-1})(S_\infty)_\mathfrak{P},$$

and the $\mathfrak{P} \in \operatorname{Max}(S_\infty)$ correspond one-to-one with the poles of x. If P is the pole corresponding to $\mathfrak{P}$, then

$$-\nu_P(x) = \nu_P(x^{-1}) = \dim_K (S_\infty)_\mathfrak{P}/(x^{-1})(S_\infty)_\mathfrak{P}$$

and hence

$$\deg(x)_\infty = -\sum_{\nu_P(x)<0} \nu_P(x) = \dim_K(S_\infty/(x^{-1})S_\infty) = [L : K(x)].$$

Theorem 14.4. *For a function field L/K the following statements are equivalent.*

(a) $g^L = 0$.
(b) *There is a nonconstant function $x \in L$ with* $\deg(x)_\infty = 1$.
(c) *L is a rational function field over K.*

Proof. (a) $\Rightarrow$ (b). Let $\mathfrak{X}$ be the abstract Riemann surface of L/K and let $P \in \mathfrak{X}$. By the Riemann–Roch theorem (13.15(a)),

$$\dim_K \mathcal{L}(P) \geq \deg P + 1 = 2.$$

Therefore there is a nonconstant function x in $\mathcal{L}(P)$. This function has only P as a pole, and indeed this pole is of order 1.

(b) $\Rightarrow$ (c). For an $x \in L$ as in (b) we have $L = K(x)$ by 14.3.

(c) $\Rightarrow$ (a) was already shown in 14.2.

Corollary 14.5. *A curve has a rational parametrization if and only if its function field has genus* 0.

A function field L/K is called *hyperelliptic* if it has as a model an affine curve with equation

$$f := Y^2-(X-a_1)\cdots(X-a_p) = 0 \quad (p \geq 3,\ a_1,\dots,a_p \in K,\ a_i \neq a_j \text{ for } i \neq j),$$

that is, $L := Q(K[f])$. We call the projective closure of f a *hyperelliptic curve.*

Theorem 14.6. *Suppose* $\operatorname{Char} K \neq 2$. *Then for any such function field*

$$g^L = \begin{cases} \frac{p}{2} - 1 & \text{if } p \text{ is even,} \\ \frac{p-1}{2} & \text{if } p \text{ is odd.} \end{cases}$$

In particular, there are algebraic function fields with arbitrary genus $g \in \mathbb{N}$.

Proof. We have $A := K[f] = K[x] \oplus K[x] \cdot y$ and $L/K(x)$ is separable, because $\operatorname{Char} K \neq 2$. Using the Jacobian criterion one easily sees that f has no singularities. Therefore A is integrally closed in L. With the earlier notation, $R = K[x]$ and $S = A = R \oplus Ry$.

We now determine the integral closure S_∞ of $R_\infty := K[x^{-1}]_{(x^{-1})}$. The minimal polynomial of an element $t = a + by$ $(a, b \in K(x),\ b \neq 0)$ over $K(x)$ is

$$(T - (a + by))(T - (a - by)) = T^2 - 2aT + (a^2 - b^2(x - a_1)\cdots(x - a_p)).$$

Hence in order that t be integral over R_∞ by F.14, it is necessary and sufficient that

$$a \in R_\infty \quad \text{and} \quad a^2 - b^2(x - a_1)\cdots(x - a_p) \in R_\infty.$$

The conditions

$$a \in R_\infty \quad \text{and} \quad 2\nu_\infty(b) - p \geq 0$$

are equivalent to these, or in other words, to

$$a \in R_\infty \quad \text{and} \quad \nu_\infty(b) \geq \frac{p}{2}.$$

The last condition is synonymous with

$$b = \begin{cases} x^{-\frac{p}{2}} b' & \text{if } p \text{ is even,} \\ x^{-\frac{p+1}{2}} b' & \text{if } p \text{ is odd,} \end{cases}$$

with some $b' \in S_\infty$. We get therefore

$$S_\infty = R_\infty \oplus R_\infty x^{-\frac{p}{2}} y \qquad \text{if } p \text{ is even,}$$

$$S_\infty = R_\infty \oplus R_\infty x^{-\frac{p+1}{2}} y \quad \text{if } p \text{ is odd.}$$

We see from this that

$$\operatorname{ord}_{\mathcal{F}} y = \begin{cases} \frac{p}{2} & \text{if } p \text{ is even,} \\ \frac{p+1}{2} & \text{if } p \text{ is odd.} \end{cases}$$

In particular, $\{1, y\}$ is a standard basis of S, and by 13.16 we have

$$g^L = \begin{cases} \frac{p}{2} - 1 & \text{if } p \text{ is even,} \\ \frac{p+1}{2} - 1 & \text{if } p \text{ is odd,} \end{cases}$$

which is what we wanted to show.

Comparing 14.6 with 14.2(c), we see that not every function field has a smooth plane projective curve as a model. How is the genus of an irreducible singular curve F related to that of its function field $L := \mathcal{R}(F)$?

Theorem 14.7. *Suppose* $\deg F =: p$ *and* $\operatorname{Sing}(F) = \{P_1, \dots, P_s\}$. *Denote by* $\overline{\mathcal{O}_{F,P_i}}$ *the integral closure of* $\mathcal{O}_{F,P_i}$ *in* L. *Then we have*

$$g^L = g^F - \sum_{i=1}^{s} \dim_K \overline{\mathcal{O}_{F,P_i}}/\mathcal{O}_{F,P_i} = \binom{p-1}{2} - \sum_{i=1}^{s} \dim_K \overline{\mathcal{O}_{F,P_i}}/\mathcal{O}_{F,P_i}.$$

Here $\overline{\mathcal{O}_{F,P_i}}/\mathcal{O}_{F,P_i}$ *is to be understood as the residue class vector space of* $\overline{\mathcal{O}_{F,P_i}}$ *modulo* $\mathcal{O}_{F,P_i}$.

Proof. We consider the situation that underlies Formula 13.16. Let $\{a_1, \dots, a_n\}$ be a standard basis for A over R, and let $\{b_1, \dots, b_n\}$ be one for S over R, where S is the integral closure of A in L. By 13.16 it follows that

$$g^F - g^L = \sum_{i=1}^{n} \operatorname{ord}_{\mathcal{F}} a_i - \sum_{i=1}^{n} \operatorname{ord}_{\mathcal{F}} b_i.$$

On the other hand, the last difference is equal to $\dim_K S/A$ by L.7.

Let $\mathfrak{p}_1, \dots, \mathfrak{p}_s \in \operatorname{Max}(A)$ be the prime ideals corresponding to $P_1, \dots, P_s$. Then

$$\mathcal{O}_{F,P_i} = A_{\mathfrak{p}_i}, \quad \overline{\mathcal{O}_{F,P_i}} = S_{\mathfrak{p}_i} \quad (i = 1, \dots, s),$$

where $S_{\mathfrak{p}_i}$ denotes the localization of S at $A \setminus \mathfrak{p}_i$. For $\mathfrak{p} \in \operatorname{Max}(A) \setminus \{\mathfrak{p}_1, \dots, \mathfrak{p}_s\}$ we have

$$A_{\mathfrak{p}} = S_{\mathfrak{p}}.$$

Since S is a finitely generated A-module, there is an element $a \in A$, $a \neq 0$, with $aS \subset A$. We then have

$$\dim_K(S/A) = \dim_K(aS/aA),$$

where aS and aA are ideals in A and hence aS/aA is an ideal in A/aA. Write

$$A/aA = \prod_{\mathfrak{p}} A_{\mathfrak{p}}/aA_{\mathfrak{p}}$$

by the Chinese remainder theorem and identify aS/aA with $\prod_{\mathfrak{p}} aS_{\mathfrak{p}}/aA_{\mathfrak{p}}$. Then we get

$$\dim_K S/A = \sum_{\mathfrak{p}} \dim_K(aS_{\mathfrak{p}}/aA_{\mathfrak{p}}) = \sum_{\mathfrak{p}} \dim_K(S_{\mathfrak{p}}/A_{\mathfrak{p}}) = \sum_{i=1}^{s} \dim_K \overline{\mathcal{O}_{F,P_i}}/\mathcal{O}_{F,P_i}\,.$$

Definition 14.8. For $P \in \mathcal{V}_+(F)$ we call

$$\delta(P) := \dim_K \overline{\mathcal{O}_{F,P}}/\mathcal{O}_{F,P}$$

the *singularity degree of F at the point P*. If $\operatorname{Sing}(F) = \{P_1, \ldots, P_s\}$, then we call

$$\delta(F) := \sum_{P \in \mathcal{V}_+(F)} \delta(P) = \sum_{i=1}^{s} \delta(P_i)$$

the *singularity degree of F*.

One can calculate the genus of the function field of a curve F using the formula in 14.7 if one can succeed in determining the singularity degree of F. We will go into this further in Chapter 17.

Corollary 14.9. *A rational curve of degree p has singularity degree*

$$\binom{p-1}{2}.$$

Exercises

1. Deduce the following from 14.7: An irreducible curve F of degree p has at most $\binom{p-1}{2}$ singularities. If F has this number of singularities, then F is rational. In particular, every singular irreducible cubic is rational.
2. Show that for an irreducible curve F and $P \in \mathcal{V}_+(F)$ we have
$$\delta(P) \geq m_P(F) - 1.$$
3. What is the singularity degree of the projective closure of
   (a) the curve $Y^n + XY + X = 0$ $(n \geq 2)$?
   (b) the curve with the parametrization
$$x = \frac{1}{1+t^n}, \quad y = \frac{t}{1+t^n} \ ?$$
4. A hyperelliptic curve of degree p has only one singularity P. Determine $\delta(P)$.

15

The Canonical Divisor Class

This chapter complements the Riemann–Roch theorem. It will be shown that for every irreducible projective algebraic curve F, an F-divisor C exists (a "canonical divisor") such that $\ell_*^F(D) = \ell^F(C - D)$ for every $D \in \operatorname{Div}^F(\mathfrak{X})$. The corresponding fact is also valid for the Riemann–Roch theorem on the abstract Riemann surface $\mathfrak{X}$ associated with F, and it has important applications.

The Dedekind complementary module occurs in the definition of the genus of a curve (and of its function field). We want to give a more precise description of this module in a special situation:

Let R be an integral domain with quotient field Z and let A be an integral domain of the form

$$A = R[Y]/(f) = R[y],$$

where $f \in R[Y]$ is a monic polynomial of degree $p > 0$. Suppose the quotient field L of A is separable over Z and let σ be the canonical trace of L/Z.

Theorem 15.1. *For the complementary module $\mathfrak{C}_{A/R}$ of A/R with respect to σ we have*

$$\mathfrak{C}_{A/R} = (f'(y))^{-1} \cdot A.$$

Proof. Because of the monic assumption on f we have

$$(1) \qquad A = \bigoplus_{i=0}^{p-1} Ry^i$$

and

$$(2) \qquad L = Z[Y]/(f) = \bigoplus_{i=0}^{p-1} Zy^i.$$

Hence f is the minimal polynomial of y over Z. Let $L^e := L \otimes_Z L$ and let I be the kernel of the canonical ring homomorphism $\mu : L \otimes_Z L \to L$ ($a \otimes b \mapsto ab$). By H.20 the traces of L/Z are in one-to-one correspondence with the generating elements of the L-module $\operatorname{Ann}_{L^e} I$. We will first describe this L-module more precisely. By (2),

$$L^e = L \otimes_Z L \cong L[Y]/(f),$$

and I can be identified with the principal ideal $(Y-y)/(f)$ in $L[Y]/(f)$. Write $f=(Y-y)\cdot\varphi$ with $\varphi\in L[Y]$. Then $\operatorname{Ann}_{L^e} I$ is identified with the principal ideal $(\varphi)/(f)$ in $L[Y]/(f)$. If

$$f=Y^p+r_1Y^{p-1}+\cdots+r_p \quad (r_i\in R),$$

then in $L[Y]$ we have

$$\begin{aligned} f=f-f(y)&=(Y^p-y^p)+r_1(Y^{p-1}-y^{p-1})+\cdots+r_{p-1}(Y-y)\\ &=(Y-y)\cdot\left[\sum_{\alpha=0}^{p-1}y^\alpha Y^{p-1-\alpha}+r_1\sum_{\alpha=0}^{p-2}y^\alpha Y^{p-2-\alpha}+\cdots+r_{p-1}\right], \end{aligned}$$

and hence φ is equal to the expression in the square brackets. Its image in L^e is

$$\Delta_y^f:=\sum_{\alpha=0}^{p-1}y^\alpha\otimes y^{p-1-\alpha}+r_1\sum_{\alpha=0}^{p-2}y^\alpha\otimes y^{p-2-\alpha}+\cdots+r_{p-1}. \tag{3}$$

Since this element generates the ideal $\operatorname{Ann}_{L^e} I$, by H.20 this element corresponds to a trace $\tau_f^y\in\omega_{L/Z}$ with the property

$$1=\sum_{\alpha=0}^{p-1}\tau_f^y(y^\alpha)y^{p-1-\alpha}+r_1\cdot\sum_{\alpha=0}^{p-2}\tau_f^y(y^\alpha)y^{p-2-\alpha}+\cdots+r_{p-1}\tau_f^y(1).$$

From this formula it follows by comparing coefficients with respect to the basis $\{1,y,\dots,y^{p-1}\}$ of L/Z that

$$\tau_f^y(y^i)=\begin{cases}0 & \text{for } i=0,\dots,p-2,\\ 1 & \text{for } i=p-1.\end{cases} \tag{4}$$

In particular, $\tau_f^y(A)\subset R$.

On the other hand, the formula (3) shows that $\mu(\Delta_y^f)=f'(y)$, and from this it follows by H.20(c) that

$$\sigma=f'(y)\cdot\tau_f^y.$$

From (4) we therefore get the following general formula for the canonical trace of L/Z:

$$\sigma\left(\frac{y^i}{f'(y)}\right)=\begin{cases}0 & \text{for } i=0,\dots,p-2,\\ 1 & \text{for } i=p-1.\end{cases} \tag{5}$$

Let $u\in L$ be given. One can write u in the form

$$u=\frac{1}{f'(y)}\sum_{i=0}^{p-1}a_iy^i \qquad (a_i\in Z).$$

Then $u \in \mathfrak{C}_{A/R}$ precisely when

$$\sigma(uy^j) = \tau_f^y\left(\left(\sum_{i=0}^{p-1} a_i y^i\right)\cdot y^j\right) \in R \quad \text{for } j = 0,\dots,p-1. \tag{6}$$

If $a_i \in R$ for $i = 0,\dots,p-1$, then this condition is certainly satisfied. Conversely, assume that (6) holds for an arbitrary $u \in L$. Then from (4) we see that $a_{p-1} \in R$. Applying the formula (6) for $j = 1$, it follows that $a_{p-2} \in R$. By induction we get $a_i \in R$ for $i = 0,\dots,p-1$. This shows that

$$\mathfrak{C}_{A/R} = (f'(y))^{-1} \cdot A,$$

and this concludes the proof.

Now let F be an irreducible curve in $\mathbb{P}^2(K)$ of degree p. We assume that the singularities of F are at finite distance, and that $A = K[X,Y]/(f) = K[x,y]$ is the corresponding affine coordinate ring. We can also assume, as we did earlier, that f is monic in Y of degree p and $L := Q(A)$ is separable over $K(x)$. With the notation introduced in Chapter 13, we have for the genus of F,

$$g^F = \dim_K(\mathfrak{C}_{A/R} \cap x^{-2}\mathfrak{C}_{S_\infty/R_\infty}). \tag{7}$$

By 15.1, $\mathfrak{C}_{A/R} = (\frac{\partial f}{\partial y})^{-1} \cdot A$ with $\frac{\partial f}{\partial y} := \frac{\partial f}{\partial Y}(x,y)$. Hence we also have the formula

$$g^F = \dim_K\left(A \cap x^{-2}\frac{\partial f}{\partial y}\mathfrak{C}_{S_\infty/R_\infty}\right).$$

Since S_∞ is a principal ideal ring (13.11), by 13.6 it is clear that a divisor C with $\operatorname{Supp} C \subset \mathfrak{X}^\infty$ exists such that

$$\mathcal{L}^F(C) = A \cap x^{-2}\frac{\partial f}{\partial y} \cdot \mathfrak{C}_{S_\infty/R_\infty},$$

and for this divisor we have

$$g^F = \ell^F(C).$$

If D is an F-divisor with $\operatorname{Supp} F \subset \mathfrak{X}^f$, then by L.12,

$$(I_D^F)^* = \mathfrak{C}_{A/R} : I_D^F = \left(\frac{\partial f}{\partial y}\right)^{-1}(A : I_D^F).$$

If $\mathfrak{p} \in \operatorname{Max}(A)$ corresponds to a singular point of F, then $I_D^F \cdot A_\mathfrak{p} = A_\mathfrak{p}$ (13.9), and we have

$$(A : I_D^F) \cdot A_\mathfrak{p} = A_\mathfrak{p} : A_\mathfrak{p} = A_\mathfrak{p}.$$

On the other hand, if $\mathfrak{p} \in \operatorname{Max}(A)$ corresponds to a regular point P of F, then $I_D^F \cdot A_\mathfrak{p} = \mathfrak{p}^{-\nu_P(D)}A_\mathfrak{p}$, and so

$$(A : I_D^F) \cdot A_\mathfrak{p} = A_\mathfrak{p} : \mathfrak{p}^{-\nu_P(D)} A_\mathfrak{p} = \mathfrak{p}^{\nu_P(D)} A_\mathfrak{p}.$$

Therefore by 13.10,

$$A : I_D^F = I_{-D}^F$$

and

$$(I_D^F)^* \cap x^{-2} \mathfrak{C}_{S_\infty/R_\infty} \cong I_{-D}^F \cap x^{-2} \frac{\partial f}{\partial y} \mathfrak{C}_{S_\infty/R_\infty} = \mathcal{L}^F(C-D).$$

Hence

$$\ell_*^F(D) = \ell^F(C-D).$$

The Riemann–Roch theorem for F can now be completed as follows.

Theorem 15.2. *There is an F-divisor C, unique up to linear equivalence with respect to F, such that for an arbitrary F-divisor D,*

$$\ell^F(D) = \ell^F(C-D) + \deg D + 1 - g^F. \tag{8}$$

Moreover,

$$\ell^F(C) = g^F, \qquad \deg C = 2g^F - 2, \tag{9}$$

and C is also uniquely determined by these properties up to linear equivalence.

Proof. We have already shown that an F-divisor C exists such that (8) is true for all $D \in \mathrm{Div}^F(\mathfrak{X})$ with $\mathrm{Supp}\, D \subset \mathfrak{X}^f$. By 13.8 and 13.4(a) equation (8) holds for arbitrary F-divisors D. Let C be chosen such that (8) holds.

Setting $D=0$ in (8), we see that $\ell^F(C) = g^F$. Setting $D = C$ we get that $\deg C = 2g^F - 2$.

Now suppose C' is an F-divisor for which (9) is satisfied. Then we have $\deg(C - C') = 0$, and from (8), it follows by substituting $D = C - C'$ that $\ell^F(C - C') = 1$. For an F-divisor D of degree 0, we can have $\ell^F(D) = 1$ only when D is a principal divisor: For $r \in \mathcal{L}^F(D)$, $r \neq 0$, we have $(r) \geq -D$ and $0 = \deg(r) \geq -\deg D = 0$, hence $(r) = -D$ and $D = (r^{-1})$. Since (r) is an F-divisor, r must be a unit of Σ (13.6(b)). This shows that $C' \equiv_F C$ and the theorem is proved.

In order to complete the Riemann–Roch heorem for the algebraic function fields, a simpler way is possible. As in Chapter 13, we choose an element x in the function field L/K that is transcendental over K such that $L/K(x)$ is separable. The symbols R, R_∞, S, S_∞ as well as X^f and X^∞ are to have their earlier meaning. In contrast to 15.1, the S-module $\mathfrak{C}_{S/R}$ need not be generated by one element. However, $\mathfrak{C}_{S_\infty/R_\infty} = z^{-1} S_\infty$ with an element $z \in L^*$, since S_∞ is a principal ideal ring. Hence we have the formula

$$g^L = \dim_K(\mathfrak{C}_{S/R} \cap x^{-2} \mathfrak{C}_{S_\infty/R_\infty}) = \dim_K(x^2 z \mathfrak{C}_{S/R} \cap S_\infty). \tag{10}$$

Here $x^2 z \mathfrak{C}_{S/R}$ is a nonzero finitely generated S-module, and by 13.10 there is a divisor C with $\mathrm{Supp}\, C \subset \mathfrak{X}^f$ and $x^2 z \mathfrak{C}_{S/R} = I_C$. From (10) we have

$$g^L = \ell(C).$$

If D is an arbitrary divisor with $\operatorname{Supp} D \subset X^f$, then

$$\begin{aligned}\ell_*(D) &= \dim_K(I_D^* \cap x^{-2}\mathfrak{C}_{S_\infty/R_\infty}) = \dim_K(x^2 z I_D^* \cap S_\infty) \\ &= \dim_K((x^2 z \mathfrak{C}_{S/R} : I_D) \cap S_\infty) = \dim_K((I_C : I_D) \cap S_\infty).\end{aligned}$$

As above, we have $I_C : I_D = I_{C-D}$ and hence

$$\ell_*(D) = \dim_K(I_{C-D} \cap S_\infty) = \ell(C - D).$$

Analagous to 15.2, we now obtain the following.

Theorem 15.3. *There is a divisor $C \in \operatorname{Div}(\mathfrak{X})$, unique up to linear equivalence, such that for all divisors $D \in \operatorname{Div}(\mathfrak{X})$,*

$$\ell(D) = \ell(C - D) + \deg D + 1 - g^L.$$

Moreover,

$$\ell(C) = g^L, \qquad \deg C = 2g^L - 2,$$

and C is also uniquely determined by these properties up to linear equivalence.

Definition 15.4. (a) An F-divisor C with $\ell^F(C) = g^F$, $\deg C = 2g^F - 2$ is called a *canonical divisor of* F. The corresponding divisor class in $\operatorname{Div}^F(\mathfrak{X})$ is called the *canonical class* of F.
(b) A divisor $C \in \operatorname{Div}(\mathfrak{X})$ with $\ell(C) = g^L$, $\deg C = 2g^L - 2$ is called a *canonical divisor of* $\mathfrak{X}$ (or of L/K), and the corresponding divisor class in $\operatorname{Div}(\mathfrak{X})$ is called the *canonical class* of $\mathfrak{X}$ (of L/K).

We give now some applications of 15.2 and 15.3.

Theorem 15.5 (Riemann's Theorem). *For every F-divisor D such that $\deg D > 2g^F - 2$ we have*

$$\ell^F(D) = \deg D + 1 - g^F,$$

and for every divisor $D \in \operatorname{Div}(\mathfrak{X})$ with $\deg D > 2g^L - 2$ we have

$$\ell(D) = \deg D + 1 - g^L.$$

Proof. If C is a canonical divisor, then $\deg(C - D) < 0$ by assumption. Therefore $\ell^F(C - D) = 0$ respectively $\ell(C - D) = 0$. Now apply 15.2 and 15.3.

Corollary 15.6. *For every F-divisor D with $\deg D \geq 2g^F$ and every $P \in \operatorname{Reg}(F)$ we have*

$$\ell^F(D - P) = \ell^F(D) - 1.$$

For every divisor $D \in \operatorname{Div}(\mathfrak{X})$ with $\deg D \geq 2g^L$ and every $P \in \mathfrak{X}$ we have

$$\ell(D - P) = \ell(D) - 1.$$

Another application of Riemann's theorem gives the following.

Theorem 15.7. *Every function field L/K of genus 1 has an elliptic curve as a model.*

Proof. Let P be a point on the abstract Riemann surface $\mathfrak{X}$ associated with L/K. Since $g^L = 1$, Riemann's theorem (15.5) shows that

$$\ell(\nu P) = \nu \quad \text{for all } \nu \in \mathbb{N}_+.$$

Let $\{1, x\}$ be a K-basis of $\mathcal{L}(2P)$. Then $\nu_P(x) = -2$, because if we had $\nu_P(x) = -1$, then we would have $L = K(x)$ by 14.3, hence $g^L = 0$. Again by 14.3 we have $[L : K(x)] = 2$.

Now let $\{1, x, y\}$ be a K-basis of $\mathcal{L}(3P)$. Then $\nu_P(y) = -3$, for otherwise we would have $\mathcal{L}(3P) = \mathcal{L}(2P)$. From $[L : K(y)] = 3$ and $[L : K(x)] = 2$ it follows that $y \notin K(x)$, and so $L = K(x, y)$.

We have $1, x, y, xy, x^2, y^2, x^3 \in \mathcal{L}(6P)$ and $\ell(6P) = 6$. Therefore there must be a nontrivial relation between these functions. Consequently, the function field L/K has an irreducible curve F of degree ≤ 3 as a model. If F were of degree ≤ 2, or if F were of degree 3 and singular, then we would have $g^L = 0$ by 14.7. Thus F must be an elliptic curve.

Now let F again be an arbitrary irreducible curve. For $P \in \text{Reg}(F)$ one can consider the functions $r \in \Sigma := \bigcap_{Q \in \text{Sing}(F)} \mathcal{O}_{F,Q}$ that have a pole only at the one point P. Similarly, for $P \in \mathfrak{X}$, one can ask which functions $r \in L$ have a pole only at P. What orders of the pole are then possible?

We discuss the problem for F. The solution for $\mathfrak{X}$ is similar. By 15.5 we have

$$\ell^F((2g^F - 1) \cdot P) = g^F, \tag{11}$$

and by 15.6 for $\nu \geq 2g^F$,

$$\ell^F(\nu \cdot P) = \ell^F((\nu - 1) \cdot P) + 1.$$

For $\nu \geq 2g^F$ there always exists a function $r_\nu \in \Sigma$ with pole divisor $(r_\nu)_\infty = \nu \cdot P$. By (11), besides the constant functions there are still $g^F - 1$ linearly independent functions with P as their only pole and with pole order $< 2g^F$. If two functions r, r' have the same order at P, then one can always find $\kappa \in K$ with $\nu_P(r - \kappa r') > \nu_P(r)$. From this it follows that there are integers $0 < \nu_1 < \cdots < \nu_{g^F - 1} < 2g^F$ such that for every $\nu \in \{0, \nu_1, \ldots, \nu_{g^F-1}\}$ there exists a function r_ν with

$$(r_\nu)_\infty = \nu \cdot P,$$

while if $\nu < 2g^F$ with $\nu \notin \{0, \nu_1, \ldots, \nu_{g^F-1}\}$, no such function exists. The functions r_ν are linearly independent over K, since they have different orders. If there were other functions, then we would have $\ell^F((2g^F - 1)P) > g^F$, contradicting (11). We have therefore shown:

Theorem 15.8 (Weierstraß Gap Theorem). *For every $P \in \operatorname{Reg}(F)$ there are natural numbers ℓ_i $(i = 1, \dots, g^F)$ with*

$$0 < \ell_1 < \cdots < \ell_{g^F} < 2g^F$$

such that the following is true: For each $\nu \in \mathbb{N} \setminus \{\ell_1, \dots, \ell_{g^F}\}$ there exists a function $r_\nu \in \Sigma$ with pole divisor

$$(r_\nu)_\infty = \nu \cdot P.$$

For $\nu \in \{\ell_1, \dots, \ell_{g^F}\}$ there is no such function. Also, the corresponding theorem is true for $\mathfrak{X}$ and the functions from L.

The integers $\ell_1, \dots, \ell_{g^F}$ are called the *Weierstraß gaps* of the point P and

$$H_P^F := \mathbb{N} \setminus \{\ell_1, \dots, \ell_{g^F}\}$$

is called the *Weierstraß semigroup* of the point P. The Weierstraß semigroup H_P^L is similarly defined. It is clear that we are dealing with a subsemigroup of $(\mathbb{N}, +)$, since the order of a product of two functions equals the sum of their orders.

It is an unsolved problem to determine which subsemigroups $H \subset \mathbb{N}$ with only finitely many gaps ("numerical" semigroups) are Weierstraß semigroups H_P^L. By a result of Buchweitz (cf. Exercise 5) there are numerical semigroups that are not of this form. On the other hand, it is known that large classes of numerical semigroups do occur as Weierstraß semigroups. The works of Eisenbud–Harris [EH] and Waldi [Wa$_1$] contain collections of results of this kind, and in them are detailed references to the literature of this research area.

The rest of this chapter is concerned with the canonical class and the genus of a function field L/K. We will use the notation introduced in connection with 15.3. For $\mathfrak{P} \in \operatorname{Max}(S)$ we have

$$S_{\mathfrak{P}} \cdot \mathfrak{C}_{S/R} = \mathfrak{P}^{-d_{\mathfrak{P}}} S_{\mathfrak{P}} \quad \text{for some } d_{\mathfrak{P}} \in \mathbb{Z},$$

and similarly, for $\mathfrak{P} \in \operatorname{Max}(S_\infty)$,

$$(S_\infty)_{\mathfrak{P}} \cdot \mathfrak{C}_{S_\infty/R_\infty} = \mathfrak{P}^{-d_{\mathfrak{P}}} (S_\infty)_{\mathfrak{P}} \quad (d_{\mathfrak{P}} \in \mathbb{Z}).$$

Here $d_{\mathfrak{P}} \neq 0$ for only finitely many $\mathfrak{P}$ (see 13.10). We show that $d_{\mathfrak{P}} \geq 0$ for all $\mathfrak{P}$. This follows immediately from

Lemma 15.9. $S \subset \mathfrak{C}_{S/R}$ *and* $S_\infty \subset \mathfrak{C}_{S_\infty/R_\infty}$.

Proof. By definition, $\mathfrak{C}_{S/R} = \{u \in L \mid \sigma(Su) \subset R\}$. In order to show that $S \subset \mathfrak{C}_{S/R}$, one has only to show that for every $u \in S$ the trace $\sigma(u)$ belongs to R. Since R is integrally closed in its quotient field Z, all the coefficients of a minimal polynomial of u over Z are contained in R (F.14). Since $-\sigma(u)$ is the second-highest coefficient of the minimal polynomial, $\sigma(u) \in R$. The same proof shows that $S_\infty \subset \mathfrak{C}_{S_\infty/R_\infty}$.

If $P \in \mathfrak{X}$ is the point corresponding to $\mathfrak{P}$, then set $d_P := d_{\mathfrak{P}}$. The effective divisor

$$D_x := \sum_{P \in \mathfrak{X}} d_P \cdot P$$

is called the *different divisor* of L with respect to x, and d_P is called the *different exponent* at the point P.

For the canonical divisor C with $I_C = x^2 z \mathfrak{C}_{S/R}$ that we considered in connection with (10) we have

$$\begin{aligned} \nu_P(C) &= d_P - (2\nu_P(x) + \nu_P(z)) && (P \in \mathfrak{X}^f), \\ \nu_P(C) &= 0 && (P \in \mathfrak{X}^\infty), \end{aligned}$$

and hence

$$\begin{aligned} \nu_P(C + (x^2 z)) &= d_P && (P \in \mathfrak{X}^f), \\ \nu_P(C + (x^2 z)) &= d_P + 2\nu_P(x) && (P \in \mathfrak{X}^\infty). \end{aligned}$$

Consequently, $C \equiv D_x - 2(x)_\infty$, and $D_x - 2(x)_\infty$ is also a canonical divisor.

Theorem 15.10 (Hurwitz Formula). *Let $D_x = \sum_{P \in \mathfrak{X}} d_P \cdot P$ be the different divisor of L with respect to x. Then*

$$g^L = \frac{1}{2} \deg D_x - [L : K(x)] + 1 = \frac{1}{2} \sum_{P \in \mathfrak{X}} d_P - [L : K(x)] + 1.$$

Proof. Since $\deg(x)_\infty = [L : K(x)]$ by 14.3 and $D_x - 2(x)_\infty$ is a canonical divisor, it follows from 15.3 that

$$2g^L - 2 = \deg C = \deg D_x - 2[L : K(x)],$$

and from this the assertion follows.

In particular, the different divisor always has even degree. In order to apply the formula, more exact knowledge of the different exponents d_P is needed. This will be provided by the Dedekind different theorem.

For $\mathfrak{P} \in \operatorname{Max}(S)$ and $\mathfrak{p} := \mathfrak{P} \cap R$ let

$$\mathfrak{p} S_{\mathfrak{P}} = \mathfrak{P}^{e_{\mathfrak{P}}} \cdot S_{\mathfrak{P}} \qquad (e_{\mathfrak{P}} \in \mathbb{N}_+). \tag{12}$$

If $P \in \mathfrak{X}^f$ is the point corresponding to $\mathfrak{P}$, then we set $e_P = e_{\mathfrak{P}}$. Similarly, e_P is defined for $P \in \mathfrak{X}^\infty$. The number e_P is called the *ramification index* of $L/K(x)$ at the point P.

Theorem 15.11 (Dedekind Different Theorem). *The different exponent d_P and the ramification index e_P satisfy the following relations:*

$$\begin{aligned} d_P &= e_P - 1, && \textit{in case } e_P \cdot 1_K \neq 0, \\ d_P &\geq e_P, && \textit{in case } e_P \cdot 1_K = 0. \end{aligned}$$

In particular, there are only finitely many ramification points of $L/K(x)$, i.e., points $P \in \mathfrak{X}$ with $e_P > 1$.

In the special case that K is a field of characteristic 0, the Hurwitz formula can be written in the following form, according to the different theorem:

$$g^L = \frac{1}{2} \sum_{P \in \mathfrak{X}} (e_P - 1) - [L : K(x)] + 1. \tag{13}$$

The number $\sum_{P \in \mathfrak{X}} (e_P - 1)$ is called the *total ramification number* of $L/K(x)$. It is necessarily an even number. Formula (13) also holds if Char K is larger than $[L : K(x)]$, because $e_P \leq [L : K(x)]$ for all $P \in \mathfrak{X}$ by F.13.

To prove the different theorem it is enough to consider a $\mathfrak{P} \in \operatorname{Max}(S)$ and $\mathfrak{p} := \mathfrak{P} \cap R$. We pass to the completions $\hat{R}_\mathfrak{p}$ and $\hat{S}_\mathfrak{p}$ of $R_\mathfrak{p}$ with respect to $\mathfrak{p}R_\mathfrak{p}$, and $S_\mathfrak{p}$ with respect to $\mathfrak{p}S_\mathfrak{p}$, where $S_\mathfrak{p}$ is the quotient ring of S with denominator set $R \setminus \mathfrak{p}$. To simplify the notation we write R for $R_\mathfrak{p}$ and S for $S_\mathfrak{p}$ and denote $\mathfrak{p}R_\mathfrak{p}$ by $\mathfrak{m}$. The proof makes use of the following five lemmas.

Lemma 15.12. *The canonical mapping*

$$\alpha : \hat{R} \otimes_R S \longrightarrow \hat{S}$$

is a ring isomorphism.

Proof. Let $\mathfrak{m} = (a_1, \ldots, a_n)$. Then

$$\hat{R} = R[[X_1, \ldots, X_n]]/(X_1 - a_1, \ldots, X_n - a_n),$$

$\hat{S} = S[[X_1, \ldots, X_n]]/(X_1 - a_1, \ldots, X_n - a_n)$ (cf. K.15), and α is induced by

$$R[[X_1, \ldots, X_n]] \longrightarrow S[[X_1, \ldots, X_n]].$$

Since S is a finitely generated R-module, one sees easily that $S[[X_1, \ldots, X_n]] \cong R[[X_1, \ldots, X_n]] \otimes_R S$ in a canonical way, and the assertion follows by passing to the residue class ring modulo $(X_1 - a_1, \ldots, X_n - a_n)$.

Lemma 15.13. *Let* $\operatorname{Max}(S) = \{\mathfrak{P}_1, \ldots, \mathfrak{P}_h\}$. *Then there is a canonical ring isomorphism*

$$\hat{S} \cong \widehat{S_{\mathfrak{P}_1}} \times \cdots \times \widehat{S_{\mathfrak{P}_h}}.$$

Proof. Since $\hat{S}$ is integral over $\hat{R}$ (15.12), the maximal ideals of $\hat{S}$ are precisely the prime ideals of $\hat{S}$ lying over $\mathfrak{m}\hat{R}$ (F.9). They are in one-to-one correspondence with the prime ideals of $\hat{S}/\mathfrak{m}\hat{S} \cong \hat{R}/\mathfrak{m}\hat{R} \otimes_{R/\mathfrak{m}} S/\mathfrak{m}S \cong S/\mathfrak{m}S$. In other words, the maximal ideals of $\hat{S}$ are given by

$$\mathfrak{M}_i := \mathfrak{m}\hat{R} \otimes S + \hat{R} \otimes \mathfrak{P}_i = \hat{R} \otimes \mathfrak{m}S + \hat{R} \otimes \mathfrak{P}_i = \hat{R} \otimes \mathfrak{P}_i \quad (i = 1, \ldots, h).$$

Here $\mathfrak{M}_i$ has $\mathfrak{P}_i$ as its preimage in S $(i = 1, \ldots, h)$. By K.11 there exists a canonical isomorphism

$$\hat{S} \cong \hat{S}_{\mathfrak{M}_1} \times \cdots \times \hat{S}_{\mathfrak{M}_h},$$

and the $\hat{S}_{\mathfrak{M}_i}$ are complete local rings ($i = 1,\dots,h$). There is a canonical homomorphism $S_{\mathfrak{P}_i} \to \hat{S}_{\mathfrak{M}_i}$, and the ideal $\mathfrak{M}_i\hat{S}_{\mathfrak{M}_i} = \mathfrak{P}_i\hat{S}_{\mathfrak{M}_i}$ is generated by one element $t_i \in \mathfrak{P}_i$. Thus the $\hat{S}_{\mathfrak{M}_i}$ are complete discrete valuation rings:

$$\hat{S}_{\mathfrak{M}_i} = K[[T_i]],$$

where T_i is an indeterminate corresponding to t_i. It is also clear that $\hat{S}_{\mathfrak{M}_i}$ is the completion of the discrete valuation ring $S_{\mathfrak{P}_i}$:

$$\hat{S}_{\mathfrak{M}_i} = \widehat{S_{\mathfrak{P}_i}} \quad (i = 1,\dots,h).$$

By 15.13 the ring $\hat{S}$ has no nonzero nilpotent elements, but does have zerodivisors for $h > 1$. However, it is clear that $\hat{L} := Q(\hat{S})$ is the direct product of the fields $\hat{L}_i := Q(\widehat{S_{\mathfrak{P}_i}})$. We set $Z := K(x)$ and $\hat{Z} := Q(\hat{R})$.

Lemma 15.14. *The canonical homomorphism*

$$\beta : \hat{Z} \otimes_Z L \longrightarrow \hat{L}$$

is an isomorphism.

Proof. By 15.12 there is a canonical isomorphism $\hat{R} \otimes_R S \cong \hat{S}$. The elements $x \in \hat{R} \setminus \{0\}$ are therefore nonzerodivisors in $\hat{S}$ (G.4(b)). Because $\hat{S}$ is finitely generated as an R-module, every element of $\hat{L}$ can be written with a numerator from $\hat{S}$ and a denominator from $\hat{R} \setminus \{0\}$. This shows that β is surjective. Because of dimension considerations, β must be bijective.

In the following we identify $\hat{L}$ with $\hat{Z} \otimes_Z L$ and $\hat{S}$ with $\hat{R} \otimes_R S$. In $\hat{L}$ we then have $\hat{L} = \hat{Z} \cdot L$, and $\hat{S} = \hat{R} \cdot S$. If σ is the canonical trace of L/Z, then $1 \otimes \sigma$ is the canonical trace of $\hat{L}/\hat{Z}$ and the complementary module $\mathfrak{C}_{\hat{S}/\hat{R}}$ with respect to $1 \otimes \sigma$ is defined. According to the identifications, $\mathfrak{C}_{\hat{S}/\hat{R}}$ and $\mathfrak{C}_{S/R}$ are contained in $\hat{L}$.

Lemma 15.15. $\mathfrak{C}_{\hat{S}/\hat{R}} \cong \hat{S} \cdot \mathfrak{C}_{S/R}$.

Proof. Let $B = \{a_1,\dots,a_n\}$ be a basis of S as an R-module. Then $1 \otimes B := \{1 \otimes a_1,\dots,1 \otimes a_n\}$ is a basis of $\hat{S} = \hat{R} \otimes_R S$ as an $\hat{R}$-module. If $\{a_1^*,\dots,a_n^*\}$ is the complementary basis to B with respect to σ, then obviously $\{1 \otimes a_1^*,\dots,1 \otimes a_n^*\}$ is the complementary basis to $1 \otimes B$ with respect to $1 \otimes \sigma$. Hence

$$\mathfrak{C}_{\hat{S}/\hat{R}} = \bigoplus_{i=1}^{n} \hat{R}(1 \otimes a_i^*) = \hat{R} \otimes_R \left(\bigoplus_{i=1}^{n} Ra_i^* \right) = \hat{R}\mathfrak{C}_{S/R} = \hat{S}\mathfrak{C}_{S/R}.$$

As shown above,

$$\hat{L} = \hat{L}_1 \times \cdots \times \hat{L}_h,$$

and by 15.14 the $\hat{L}_i$ are finite and separable field extensions of $\hat{Z}$, because L/Z is such. If σ_i is the canonical trace of $\hat{L}_i/\hat{Z}$, then $1 \otimes \sigma = (\sigma_1,\dots,\sigma_h)$ by H.3. Using 15.13, the next result follows easily (cf. H.2):

Lemma 15.16. $\mathfrak{C}_{\hat{S}/\hat{R}} = \mathfrak{C}_{\widehat{S_{\mathfrak{P}_1}}/\hat{R}} \times \cdots \times \mathfrak{C}_{\widehat{S_{\mathfrak{P}_h}}/\hat{R}}$.

If d is the different exponent corresponding to $\mathfrak{P} \in \{\mathfrak{P}_1, \ldots, \mathfrak{P}_h\}$, so that $S_{\mathfrak{P}}\mathfrak{C}_{S/R} = \mathfrak{P}^{-d}S_{\mathfrak{P}}$, then 15.15 and 15.16 show that

$$\mathfrak{C}_{\widehat{S_{\mathfrak{P}}}/\hat{R}} = \widehat{S_{\mathfrak{P}}}\mathfrak{C}_{S/R} = \mathfrak{P}^{-d}\widehat{S_{\mathfrak{P}}}. \tag{14}$$

Thus the different exponent can also be calculated in the completion. As we will see, this has the advantage that the formula in 15.1 can be used.

If e is the ramification index at the point $\mathfrak{P}$, so $\mathfrak{p}S_{\mathfrak{P}} = \mathfrak{P}^e S_{\mathfrak{P}}$, then also

$$\mathfrak{p}\widehat{S_{\mathfrak{P}}} = \mathfrak{P}^e \widehat{S_{\mathfrak{P}}}.$$

To prove the Dedekind different theorem, we have to show that $d = e-1$ if $e \cdot 1_K \neq 0$, and that $d \geq e$ if $e \cdot 1_K = 0$. Identify $\hat{R}$ and $\widehat{S_{\mathfrak{P}}}$ with power series rings $K[[t]] \subset K[[T]]$. Then $t = \varepsilon \cdot T^e$ for some unit $\varepsilon \in K[[T]]$. Hence $K[[T]]/tK[[T]]$ is a K-algebra of dimension e, and therefore

$$\widehat{S_{\mathfrak{P}}} = \hat{R} \oplus \hat{R}T \oplus \cdots \oplus \hat{R}T^{e-1}.$$

Let f be the minimal polynomial of T over $\hat{Z}$. It is of the form

$$f = Y^e + r_1 Y^{e-1} + \cdots + r_e \quad (r_i \in \hat{R},\ i = 1, \ldots, e),$$

and no r_i can be a unit in $\hat{R}$. Suppose that this is not the case. Let j be the largest index for which r_j is a unit. Then $r_j T^{e-j}$ in $K[[T]]$ would have smaller order in $K[[T]]$ than T^e and $r_i T^{e-i}$ for $i \neq j$, which cannot be, since $T^e + \sum_{i=1}^{e} r_i T^{e-i} = 0$.

From $\widehat{S_{\mathfrak{P}}} = \hat{R}[Y]/(f)$, it follows by 15.1 that $\mathfrak{C}_{\widehat{S_{\mathfrak{P}}}/\hat{R}} = (f'(T))^{-1} \cdot \widehat{S_{\mathfrak{P}}}$ where

$$f'(T) = eT^{e-1} + (e-1)r_1 T^{e-2} + \cdots + r_{e-1}.$$

If $e \cdot 1_K \neq 0$, then $f'(T)$ has order $e-1$ and it follows that $d = e-1$. If $e \cdot 1_K = 0$, then $f'(T)$ has order $\geq e$, since all r_i have such an order. The different theorem has therefore been proved.

Differents (ideals defined with the help of differentiation) are well known in various situations (cf. [Ku$_2$], Chapter 10, and Appendix G). Theorem 15.1 is a special case of an important connection between several concepts of differents ([Ku$_2$], 10.17).

Exercises

1. (Reciprocity theorem of Brill–Noether). Let F be a curve and let D, D' be F-divisors such that $D + D'$ is a canonical divisor for F. Show that

$$\deg D - 2\ell^F(D) = \deg D' - 2\ell^F(D').$$

2. Let F be a smooth curve of degree $p > 3$ and let ϕ be a curve of degree $p-3$. Then the divisor (ϕ) of ϕ is a canonical divisor on $\mathfrak{X} = \mathcal{V}_+(F)$.

In the following exercises let $\mathfrak{X}$ be the abstract Riemann surface of a function field L/K. Let C be a canonical divisor of $\mathfrak{X}$. Assume that $\mathfrak{X}$ has genus $g \geq 2$.

3. Show that $\ell(\nu \cdot C) = (2\nu - 1) \cdot (g-1)$ for $\nu \geq 2$.
4. (Weierstraß gaps and canonical class). Show that:
   (a) A natural number $\nu \geq 1$ is a Weierstraß gap of $\mathfrak{X}$ at the point $P \in \mathfrak{X}$ if and only if there exists a $g_\nu \in \mathcal{L}(C)$ with $\nu_P(g_\nu) = \nu - 1 - \nu_P(C)$.
   (b) Let $\Lambda := \{\nu + \mu \mid \nu \text{ and } \mu \text{ are Weierstraß gaps of } \mathfrak{X} \text{ at } P\}$. Then for every $\lambda \in \Lambda$ there is an $f_\lambda \in \mathcal{L}(2C)$ with
   $$\nu_P(f_\lambda) = \lambda - 2 - 2\nu_P(C).$$
   (c) Λ has at most $3g - 3 = \ell(2C)$ elements.
5. (Buchweitz). Let H be a numerical semigroup with $g \geq 2$ gaps and let h be the number of integers $\ell_1 + \ell_2$, where ℓ_1 and ℓ_2 are gaps of H. Show that:
   (a) If H is a Weierstraß semigroup, then $h \leq 3g - 3$.
   (b) The semigroup with the gaps
   $$1, 2, 3, 4, 5, 6, 7, 8, 9, 10, 11, 12, 19, 21, 24, 25$$
   is not a Weierstraß semigroup.

16

The Branches of a Curve Singularity

Irreducible polynomials in $K[X,Y]$ can decompose in the power series ring $K[[X,Y]]$. In the geometry over $\mathbb{C}$, this fact corresponds to the possibility of decomposing curves into "analytic" branches "in a neighborhood" of a singularity, and thereby allowing us to analyze them more precisely. Also, a similar theory will be discussed for curves over an arbitrary algebraically closed field.

The local ring $\mathcal{O}_{F,P}$ of a point P on a curve F in $\mathbb{P}^2(K)$ has the affine description (without loss of generality $P=(0,0)$)

$$\mathcal{O}_{F,P} \cong K[X,Y]_{(X,Y)}/(f), \tag{1}$$

where $f \in K[X,Y]$ is a polynomial with homogenization F. The completion $\widehat{\mathcal{O}_{F,P}}$ of $\mathcal{O}_{F,P}$ with respect to its maximal ideal $\mathfrak{m}_{F,P}$ has by K.17 the presentation

$$\widehat{\mathcal{O}_{F,P}} \cong K[[X,Y]]/(f). \tag{2}$$

Thus $\widehat{\mathcal{O}_{F,P}}$ is a complete Noetherian local ring (K.7) with maximal ideal $\widehat{\mathfrak{m}_{F,P}} = \mathfrak{m}_{F,P} \cdot \widehat{\mathcal{O}_{F,P}}$. The canonical map $\mathcal{O}_{F,P} \to \widehat{\mathcal{O}_{F,P}}$ is injective, because $\mathcal{O}_{F,P}$ is $\mathfrak{m}_{F,P}$-adically separated (Krull intersection theorem).

Let Lf be the leading form of f, considered as a power series in $K[[X,Y]]$. Unlike the presentation in Appendix B, here we give the leading form the usual degree ≥ 0.

Lemma 16.1. (a) *We have* $\dim \widehat{\mathcal{O}_{F,P}} = 1$ *and* $\operatorname{edim} \widehat{\mathcal{O}_{F,P}} \leq 2$.

(b) *The following statements are equivalent:*

(α) $\operatorname{edim} \widehat{\mathcal{O}_{F,P}} = 1$ *(i.e., $\widehat{\mathcal{O}_{F,P}}$ is a complete discrete valuation ring).*

(β) $\deg Lf = 1$.

(γ) *P is a regular point of F.*

(c) *If X is not a factor of Lf and $\deg Lf =: m$, then $\widehat{\mathcal{O}_{F,P}}$ is a free $K[[X]]$-module of rank m.*

Proof. We show (c) first. By assumption, f is Y-general of order m (K.18) and so by the Weierstraß preparation theorem (K.19),

$$\widehat{\mathcal{O}_{F,P}} = K[[X]] \oplus K[[X]] \cdot y \oplus \cdots \oplus K[[X]] \cdot y^{m-1},$$

where y denotes the residue class of Y in $\widehat{\mathcal{O}_{F,P}}$.

The assumption that X does not divide the leading form Lf is of course satisfied by a suitable choice of coordinate system; thus $\widehat{\mathcal{O}_{F,P}}$ is always a finitely generated module over a power series ring in one variable. From this, and from F.10, it follows easily that $\dim \widehat{\mathcal{O}_{F,P}} = 1$. The presentation (2) shows that $\operatorname{edim} \widehat{\mathcal{O}_{F,P}} \leq 2$, and $\operatorname{edim} \widehat{\mathcal{O}_{F,P}} = 1$ is equivalent to $\deg Lf = 1$. Hence, statement (b) is clear.

While the completion $\widehat{\mathcal{O}_{F,P}}$ for $P \in \operatorname{Reg}(F)$ already played a role in Chapter 15, these complete rings are now used to study the singular points of F.

Since $K[[X,Y]]$ is a unique factorization domain (K.22), a power series f in $K[[X,Y]]$ decomposes into a product of powers

$$f = c \cdot f_1^{\alpha_1} \cdots f_h^{\alpha_h} \quad (c \in K[[X,Y]] \text{ a unit, } \alpha_i \in \mathbb{N}_+) \tag{3}$$

of (pairwise nonassociate) irreducible power series f_i $(i = 1, \ldots, h)$. As we will see in 16.6, it is possible for an irreducible f in $K[X,Y]$ to properly decompose in $K[[X,Y]]$.

Lemma 16.2. *The ring* $R := \widehat{\mathcal{O}_{F,P}}$ *has exactly* h *minimal prime ideals, namely* $\mathfrak{p}_i := (f_i)/(f)$ $(i = 1, \ldots, h)$. *Here* α_i *is the smallest natural number with* $\mathfrak{p}_i^{\alpha_i} R_{\mathfrak{p}_i} = (0)$; *in particular, the* α_i *are invariants of* $\widehat{\mathcal{O}_{F,P}}$ *and so are also invariants of* P.

Proof. $R/\mathfrak{p}_i \cong K[[X,Y]]/(f_i)$ is of course an integral domain, and hence $\mathfrak{p}_i$ is a prime ideal of R. Since $\dim R = 1$, it must be a minimal prime ideal. Every prime ideal of $K[[X,Y]]$ containing f contains one of the factors f_i, and its image in R then contains $\mathfrak{p}_i$. Therefore the $\mathfrak{p}_i$ $(i = 1, \ldots, h)$ are all the minimal prime ideals of R.

Because of the permutability of localization and residue class ring construction,

$$R_{\mathfrak{p}_i} = K[[X,Y]]_{(f_i)}/fK[[X,Y]]_{(f_i)} = K[[X,Y]]_{(f_i)}/f_i^{\alpha_i}K[[X,Y]]_{(f_i)},$$

and the statement about α_i follows.

Definition 16.3. The rings

$$Z_i := \widehat{\mathcal{O}_{F,P}}/\mathfrak{p}_i^{\alpha_i} \cong K[[X,Y]]/(f_i^{\alpha_i}) \quad (i = 1, \ldots, h)$$

are called the (analytic) *branches of* F *at the point* P.

These are complete Noetherian local rings of Krull dimension 1 and of embedding dimension ≤ 2. In addition to the maximal ideal, these rings have only one other prime ideal, namely $\mathfrak{p}_i/\mathfrak{p}_i^{\alpha_i} = (f_i)/(f_i^{\alpha_i})$. As in 16.1 (c) one sees that each Z_i is a finite free module over a power series algebra $K[[X]]$.

The number $m_i := \deg Lf_i^{\alpha_i}$ is an invariant of the branch Z_i, since $\mathrm{gr}_{\mathfrak{m}_i} Z_i \cong K[X,Y]/(Lf_i^{\alpha_i})$, where $\mathfrak{m}_i$ is the maximal ideal of Z_i, and m_i gives the place where the Hilbert function of $\mathrm{gr}_{\mathfrak{m}_i} Z_i$ differs from the Hilbert function of the polynomial ring for the first time.

We call m_i the *multiplicity of the branch* Z_i. It is clear that

$$m_P(F) = \sum_{i=1}^{h} m_i. \tag{4}$$

The branch Z_i is called *integral* if Z_i is an integral domain (so $\alpha_i = 1$), and *regular* if Z_i is a discrete valuation ring. This is the case if and only if $m_i = 1$, and then there is a canonical K-isomorphism

$$Z_i \cong K[[T_i]]$$

onto a power series ring in one variable T_i.

Definition 16.4. The curve F is called *reduced at* P (*irreducible at* P) if no multiple components of F pass through P (if F is reduced at P and only one component of F passes through P).

The curve F is reduced (irreducible) at P if and only if $\mathcal{O}_{F,P}$ has no nonzero nilpotent elements (is an integral domain). If F is irreducible at P, then for every branch Z_i of F at P, the canonical mapping

$$\mathcal{O}_{F,P} \longrightarrow \widehat{\mathcal{O}_{F,P}}/\mathfrak{p}_i^{\alpha_i} = Z_i$$

is injective. If it had a kernel $\mathfrak{a}_i \neq (0)$, then $\mathcal{O}_{F,P}/\mathfrak{a}_i$ would be a finite-dimensional K-algebra; then this would also be the case for its completion. So $(\mathcal{O}_{F,P}/\mathfrak{a}_i)^\wedge = \mathfrak{a}_i\widehat{\mathcal{O}_{F,P}}/\widehat{\mathcal{O}_{F,P}}$ would be a finite-dimensional K-algebra, and then Z_i would also be a finite-dimensional K-algebra, a contradiction.

Theorem 16.5. *If F is reduced at P, then all the branches of F at P are integral. For the minimal prime ideals $\mathfrak{p}_i$ $(i = 1, \dots, h)$ of $\widehat{\mathcal{O}_{F,P}}$ we have*

$$\bigcap_{i=1}^{h} \mathfrak{p}_i = (0).$$

Proof. Without loss of generality we can assume that in (1) the polynomial f in $K[X,Y]$ has a factorization of the form

$$f = f_1 \cdots f_m$$

with pairwise nonassociate irreducible polynomials f_i $(i = 1, \dots, m)$ that are all contained in (X,Y). Then the f_i are also pairwise relatively prime as power series: If $i \neq j$, then $A = K[X,Y]_{(X,Y)}/(f_i, f_j)$ is a finite-dimensional local

K-algebra. Then $A = \hat{A} = K[[X,Y]]/(f_i, f_j)$. If f_i and f_j had a nonunit g from $K[[X,Y]]$ as a divisor, then $B := K[[X,Y]]/(g)$ would be a homomorphic image of A. However, this cannot be the case, because B (by the Weierstraß preparation theorem) is certainly not finite-dimensional over K.

It is enough to show that if f in $K[X,Y]$ is irreducible, then f has no multiple factors in $K[[X,Y]]$. If f is irreducible, then $\frac{\partial f}{\partial X}$ and $\frac{\partial f}{\partial Y}$ do not vanish simultaneously. Suppose $\frac{\partial f}{\partial Y} \neq 0$. The power series f and $\frac{\partial f}{\partial Y}$ are then relatively prime, and the K-algebra $K[[X,Y]]/(f, \frac{\partial f}{\partial Y})$ is finite-dimensional.

Suppose it were the case that $f = g^2 \cdot \varphi$ ($g, \varphi \in K[[X,Y]]$, g irreducible). Then because of

$$\frac{\partial f}{\partial Y} = 2g\frac{\partial g}{\partial Y}\varphi + g^2 \cdot \frac{\partial \varphi}{\partial Y},$$

g would also be a divisor of $\frac{\partial f}{\partial Y}$. As above, this gives a contradiction, since $K[[X,Y]]/(g)$ is not finite-dimensional over K.

Under the assumptions of the theorem, in equation (3) we must have

$$\alpha_1 = \cdots = \alpha_h = 1.$$

Since $\mathfrak{p}_i = (f_i)/(f)$, it follows that $\bigcap_{i=1}^h \mathfrak{p}_i = (0)$.

The following theorem gives a sufficient condition for the decomposability of a power series in two variables. It uses a variation of Hensel's lemma. We consider the grading on $K[X,Y]$ in which $\deg X =: p > 0$ and $\deg Y =: q > 0$. For $f \in K[[X,Y]] \setminus \{0\}$, let the leading form Lf be the homogeneous polynomial of smallest degree that occurs in f using the above grading, and $\operatorname{ord} f := \deg Lf$. There is a trivial, but useful, *irreducibility criterion*: If by a suitable choice of p and q the leading form Lf is an irreducible polynomial, then f is an irreducible power series. In particular, homogeneous irreducible polynomials are also irreducible as power series.

The following theorem is a partial converse to this criterion.

Theorem 16.6. *For $f \in K[[X,Y]] \setminus \{0\}$, suppose Lf has a decomposition $Lf = \varphi_1 \cdot \varphi_2$ with nonconstant relatively prime (homogeneous) polynomials $\varphi_j \in K[X,Y]$ $(j = 1,2)$. Then there are power series $f_j \in K[[X,Y]]$ with $Lf_j = \varphi_j$ $(j = 1,2)$ and*

$$f = f_1 \cdot f_2.$$

Proof. Set $G := K[X,Y]$, $\alpha_j := \deg \varphi_j$ $(j = 1,2)$, and $\alpha := \operatorname{ord} f = \deg Lf = \alpha_1 + \alpha_2$. Using the relative primeness of φ_1 and φ_2, we will first show that

(5) $G_k = G_{k-\alpha_1} \cdot \varphi_1 + G_{k-\alpha_2} \cdot \varphi_2$ for $k > (\alpha_1 - p) + (\alpha_2 - q) = \alpha - p - q$.

If $p = q = 1$, then this statement follows from the considerations of the Hilbert function in A.12(b). We can deduce the general case from this as follows: Let $H := K[U,V]$ be the polynomial ring in two variables U and V of degree 1. Using $X \mapsto U^p$, $Y \mapsto V^q$ there is an embedding $K[X,Y] \hookrightarrow K[U,V]$, and this is homogeneous of degree 0. The monomials U^iV^j $(0 \leq i \leq p-1, 0 \leq j \leq q-1)$ form a basis of H over G. Let $\overline{G} = G/(\varphi_1, \varphi_2)$, $\overline{H} = H/(\varphi_1, \varphi_2)H$ and let u, v be the residue classes of U, V in $\overline{H}$. Then $\{u^iv^j \mid 0 \leq i \leq p-1, 0 \leq j \leq q-1\}$ is a (homogeneous) basis of $\overline{H}$ as a $\overline{G}$-module. Let $\overline{G}_\rho$ be the homogeneous component of $\overline{G}$ of largest degree. Then $\overline{G}_\rho \cdot u^{p-1}v^{q-1}$ is the homogeneous component of $\overline{H}$ of largest degree. By A.12(b) this has degree $\alpha_1 + \alpha_2 - 2$. It follows that

$$\rho = \alpha_1 + \alpha_2 - 2 - (p-1) - (q-1) = (\alpha_1 - p) + (\alpha_2 - q) = \alpha - p - q,$$

and so we have shown formula (5).

Using this formula a decomposition $f = f_1 \cdot f_2$ will be constructed step by step. Set

$$f^{(1)} := f - Lf = f - \varphi_1\varphi_2 \quad \text{and} \quad \alpha^{(1)} := \operatorname{ord} f^{(1)}.$$

Then $\alpha^{(1)} \geq \alpha + 1$. Suppose, for an $i \geq 1$, that we have already found polynomials $p_j \in G$ with $Lp_j = \varphi_j$ $(j = 1, 2)$ such that $f^{(i)} := f - p_1p_2$ has order $\alpha^{(i)} \geq \alpha + i$. The formula (5) can be applied with $k = \alpha^{(i)}$. We can therefore write

$$Lf^{(i)} = \psi_2\varphi_1 + \psi_1\varphi_2$$

with $\psi_2 \in G_{\alpha^{(i)}-\alpha_1}$, $\psi_1 \in G_{\alpha^{(i)}-\alpha_2}$, and then we have

$$\deg(\psi_1\psi_2) = 2\alpha^{(i)} - \alpha_1 - \alpha_2 = 2\alpha^{(i)} - \alpha \geq \alpha + i + 1.$$

For

$$f^{(i+1)} := f - (p_1 + \psi_1)(p_2 + \psi_2),$$

we then have

$$f^{(i+1)} = f^{(i)} - (\psi_2 p_1 + \psi_1 p_2) - \psi_1\psi_2$$

and so $\operatorname{ord} f^{(i+1)} \geq \alpha + i + 1$. Furthermore, $p_j + \psi_j$ has leading form φ_j, since $\deg \psi_j > \alpha_j$ $(j = 1, 2)$.

By continuing with this method, we approximate f more and more closely by a product of two polynomials with leading forms φ_1, φ_2. Passing to the limit, we obtain the desired decomposition $f = f_1 f_2$ of f as a product of two power series f_j with $Lf_j = \varphi_j$ $(j = 1, 2)$.

Corollary 16.7. *If f is an irreducible power series, then its leading form Lf is a power of an irreducible homogeneous polynomial. In particular, for $p = q = 1$,*

$$Lf = (aX - bY)^\mu \qquad (\mu \in \mathbb{N},\ a, b \in K,\ (a, b) \neq (0, 0)).$$

In the initial situation let $Z_i = K[[X, Y]]/(f_i^{\alpha_i})$ be a branch of F at P. Then

$$Lf^{(\alpha_i)} = (a_iX - b_iY)^{\mu_i} \qquad (\mu_i \in \mathbb{N}_+,\ (a, b) \neq (0, 0)),$$

and $t_i : a_iX - b_iY = 0$ is one of the tangents of F at P. We call t_i the *tangent of the branch* Z_i. Every branch therefore has a unique tangent, and every tangent to F at P is also a tangent of a certain branch. But different branches can indeed have the same tangent.

Example 16.8 (Classification of double points). In case $m_P(F) = 2$ we have $\deg Lf = 2$. Then Lf is either the product of two linearly independent linear homogeneous polynomials or is the square of a linear homogeneous polynomial. The following cases are possible:

(a) *Normal crossings* (*nodes, ordinary double points*): F has two distinct tangents at P. Then by 16.6, F also has two branches at P, which moreover are regular. A specific example of this is the folium of Descartes $X^2-Y^2+X^3=0$ ($\operatorname{Char} K \neq 2$). Although the polynomial $X^2-Y^2+X^3$ is irreducible in $K[X,Y]$, it decomposes in $K[[X,Y]]$, since $Lf = X^2-Y^2 = (X+Y)(X-Y)$.

(b) *Ordinary cusps*: F has a (double) tangent at P and only one branch. A specific example of this kind is Neil's parabola $Y^2 - X^3 = 0$. The polynomial Y^2-X^3 is irreducible in $K[X,Y]$. It is homogeneous if we set $\deg X = 2$, $\deg Y = 3$. Then it can also not decompose as a power series.

(c) *Tacnodes*: F has a (double) tangent at P but two different branches. An example of this situation is given by

$$f = Y^2 - X^2Y^2 - X^4 = 0, \quad P = (0,0) \quad (\operatorname{Char} K \neq 2).$$

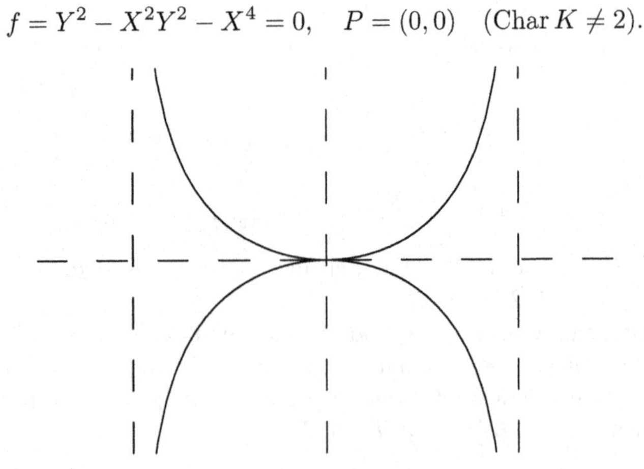

Here $Y = 0$ is a double tangent. The curve f is irreducible, since $\frac{X^4}{1-X^2}$ is not a square in $K(X)$. If we set $\deg X = 1$, $\deg Y = 2$, then $Lf = Y^2-X^4 = (Y+X^2)(Y-X^2)$ and by 16.6, there are two different regular branches to f at P.

As for algebraic curves, one obtains further invariants for the branches Z by passing to the integral closure of Z.

Theorem 16.9. *Let $Z = K[[X,Y]]/(f)$ be an integral branch and let $\overline{Z}$ be the integral closure of Z in $Q(Z)$. Then $\overline{Z}$ is finitely generated as a Z-module. There is also a K-isomorphism*

$$\overline{Z} \cong K[[T]]$$

onto a power series ring in one variable T; i.e., $\overline{Z}$ is a complete discrete valuation ring.

Proof. By 16.1(c) we may assume without loss of generality that

$$Z = \bigoplus_{i=0}^{m-1} K[[X]]y^i,$$

where y denotes the residue class of Y in Z. It is enough to show that $\overline{Z}$ is finitely generated as a $K[[X]]$-module. If $L := Q(Z)$ is separable over $K((X)) := Q(K[[X]])$, then this follows from F.7. In the inseparable case, we proceed as in L.1. Let $\operatorname{Char} K =: p > 0$ and let L_{sep} be the separable closure of $K((X))$ in L. The integral closure $\tilde{Z}$ of $K[[X]]$ in L_{sep} is in any case a finite $K[[X]]$-module. Furthermore, there is an $e \in \mathbb{N}$ with $L^{p^e} \subset L_{\text{sep}}$. Then $\overline{Z}^{p^e} \subset \tilde{Z}$, and since $K[[X]]$ is finite over $K[[X]]^{p^e} = K[[X^{p^e}]]$, it is also the case that $\tilde{Z}$ and $\overline{Z}^{p^e}$ are finite over $K[[X^{p^e}]]$. However, $\overline{Z}$ is isomorphic to $\overline{Z}^{p^e}$ and $K[[X]]$ to $K[[X^{p^e}]]$. Therefore $\overline{Z}$ is finite over $K[[X]]$.

Since Z is a complete local ring and $\overline{Z}$ is a finitely generated Z-module, $\overline{Z}$ decomposes by K.11 into the direct product of the localizations at its maximal ideals. But $\overline{Z}$ is an integral domain, so only one maximal ideal can occur, i.e., $\overline{Z}$ is a local ring. By F.8, then, $\overline{Z}$ is in fact a (complete) discrete valuation ring. Since it has K as its residue class field, it is isomorphic to $K[[T]]$.

In the situation of the theorem, let x, y be the residue classes of X, Y in Z. Under the injection $Z \hookrightarrow K[[T]]$, the elements x and y will be mapped to power series α, $\beta \in K[[T]]$ and we will have $f(\alpha,\beta) = 0$ in $K[[T]]$. We write

$$Z = K[[\alpha,\beta]], \tag{6}$$

where $K[[\alpha,\beta]]$ denotes the image under the substitution homomorphism $K[[X,Y]] \to K[[T]]$ ($X \mapsto \alpha$, $Y \mapsto \beta$). We also say that the branch Z is given by the (*analytic*) *parametric representation* (α,β).

By 16.9 the residue class vector space $\overline{Z}/Z$ over K is finite-dimensional (cf. 7.1). We call

$$\delta(Z) := \dim_K \overline{Z}/Z$$

the *singularity degree* of the (integral) branch Z.

Example 16.10. Let $p, q \in \mathbb{N}_+$ be relatively prime natural numbers with $p < q$. We consider the affine curve

$$f : X^p - Y^q = 0.$$

In case $p = 2$, $q = 3$ this is Neil's semicubical parabola (Figure 1.6).

First we will show that $X^p - Y^q$ is irreducible in $K[X, Y]$. We endow the polynomial ring $K[X, Y]$ with the grading where $\deg X := q$ and $\deg Y := p$. Then f is homogeneous of degree $p \cdot q$. If f had a proper divisor, then this divisor would have to be homogeneous (A.3) and of degree $\leq p \cdot q - p$.

But all the monomials $X^i Y^j$ with $i \leq p-1$, $j \leq q-1$ have distinct degrees $iq + jp$, for if

$$iq + jp = i'q + j'p \qquad (i' \leq p - 1,\ j' \leq q - 1)$$

and $i \geq i'$, then $(i - i')q = (j' - j)p$, and because p and q are relatively prime, we must have $i = i'$ and $j = j'$. Hence the homogeneous polynomials of degree $\leq p \cdot q - p$ are of the form cX^iY^j $(c \in K)$ and they do not divide f.

From the irreducibility of f in $K[X, Y]$, it follows that f is irreducible also in $K[[X, Y]]$, because f is a homogeneous polynomial.

The kernels of the substitution homomorphisms

$$K[X, Y] \longrightarrow K[T],$$

$$K[[X, Y]] \longrightarrow K[[T]],$$

where $X \mapsto T^q$, $Y \mapsto T^p$, contain f. Since f is irreducible in both rings, f generates the kernel. Hence we have

$$K[f] = K[X, Y]/(f) \cong K[T^q, T^p]$$

and

$$Z := K[[X, Y]]/(f) \cong K[[T^q, T^p]].$$

So f is a rational curve. It has only one branch at $P = (0, 0)$, namely Z. This branch is integral and has $X = 0$ as a p-fold tangent. Further, $\overline{Z} := K[[T]]$ is the integral closure of Z in $Q(Z)$, and (T^q, T^p) is a parametric representation of Z.

Let $H = \langle p, q \rangle = \{ip + jq \mid i, j \in \mathbb{N}\}$ be the numerical semigroup generated by p and q. Then

$$Z = \left\{ \sum_{h \in H} \kappa_h T^h \in K[[T]] \mid \kappa_h \in K \right\}.$$

The singularity degree $\delta(Z) = \dim_K \overline{Z}/Z$ is thus the number of gaps of H. It is an exercise to show that

$$\delta(Z) = \frac{1}{2}(p - 1)(q - 1).$$

The Jacobian criterion shows that f has at most one singularity at finite distance at $(0, 0)$. Besides that one, f has only one point at infinity with branch

$$Z_\infty = K[[X,Y]]/(X^{q-p} - Y^q),$$

and this has singularity degree

$$\delta(Z_\infty) = \frac{1}{2}(q-p-1)(q-1).$$

If L is the function field of the curve f, then by 14.7,

$$\begin{aligned} g^L &= \binom{q-1}{2} - \delta(Z) - \delta(Z_\infty) \\ &= \frac{1}{2}\left[(q-1)(q-2) - (p-1)(q-1) - (q-p-1)(q-1)\right] = 0 \end{aligned}$$

in accordance with the fact that f is rational.

In what follows, we call the rings $\Gamma := K[[X,Y]]/(f)$, where $f \in K[[X,Y]]$ is not a unit, (plane) *algebroid curves*. For example, the rings $\widehat{\mathcal{O}_{F,P}}$ of (2) are algebroid curves, and of course also the branches of a curve singularity are algebroid curves. For an algebroid curve, its branches are defined in exactly the same way as for a curve singularity. Lemma 16.1 also holds for algebroid curves: These are complete Noetherian local rings of Krull dimension 1. Using 16.1 one also sees that $K[[X,Y]]$ has, except for the maximal ideal and the zero ideal, only prime ideals of the form (f), where f is an irreducible power series.

If $\Gamma_i = K[[X,Y]]/(f_i)$ $(i = 1,2)$ are two algebroid curves with relatively prime power series f_1, f_2, then $A := K[[X,Y]]/(f_1,f_2)$ is a finite-dimensional K-algebra, for by the relative primeness of f_1 and f_2, the only prime ideal of $K[[X,Y]]$ that contains (f_1,f_2) is (X,Y). The elements of the maximal ideal of A are nilpotent (C.12), and it follows that $\dim_K A < \infty$.

Definition 16.11. $\mu(\Gamma_1,\Gamma_2) := \dim_K K[[X,Y]]/(f_1,f_2)$ is called the *intersection multiplicity* of the algebroid curves Γ_1 and Γ_2.

This definition in particular defines the intersection multiplicity of two branches. If $f_1, f_2 \in K[X,Y]$ and $P = (0,0)$, then $\mu(\Gamma_1,\Gamma_2) = \mu(f_1,f_2)$ is the intersection multiplicity of the curves f_1, f_2 at P, because

$$K[X,Y]_{(X,Y)}/(f_1,f_2) \cong K[[X,Y]]/(f_1,f_2),$$

since the first ring is already complete as a finite-dimensional local K-algebra. There are properties for intersection multiplicity of algebroid curves that are similar to those for algebraic curves, except for Bézout's theorem, which is of a global nature. The proofs are analogous to those for algebraic curves, and we do not include them here.

Rules 16.12.

(a) $\mu(\Gamma_1,\Gamma_2) = 1$ if and only if Γ_1 and Γ_2 are regular branches with distinct tangents.

(b) Additivity: If $\Gamma_1 = K[[X,Y]]/(\varphi_1 \cdot \varphi_2)$ with nonunits $\varphi_i \in K[[X,Y]]$ and $\Gamma_1^{(i)} = K[[X,Y]]/(\varphi_i)$ $(i = 1,2)$, then for an arbitrary branch Γ_2,

$$\mu(\Gamma_1, \Gamma_2) = \mu(\Gamma_1^{(1)}, \Gamma_2) + \mu(\Gamma_1^{(2)}, \Gamma_2).$$

(c) Let $Z_1, \dots, Z_r$ be the branches of Γ_1, and $Z'_1, \dots, Z'_s$ the branches of Γ_2. Then

$$\mu(\Gamma_1, \Gamma_2) = \sum_{\substack{i=1,\dots,r \\ j=1,\dots,s}} \mu(Z_i, Z'_j).$$

For an algebraic curve F, with branches $Z_1, \dots, Z_r$ at a point P, the intersection multiplicities $\mu(Z_i, Z_j)$ $(i \neq j)$ are interesting invariants of the singularity P.

Example 16.13. Ordinary Singularities.

A point P of a curve F is called an *ordinary singularity* if $m := m_P(F) > 1$ and if F also has m distinct tangents at P.

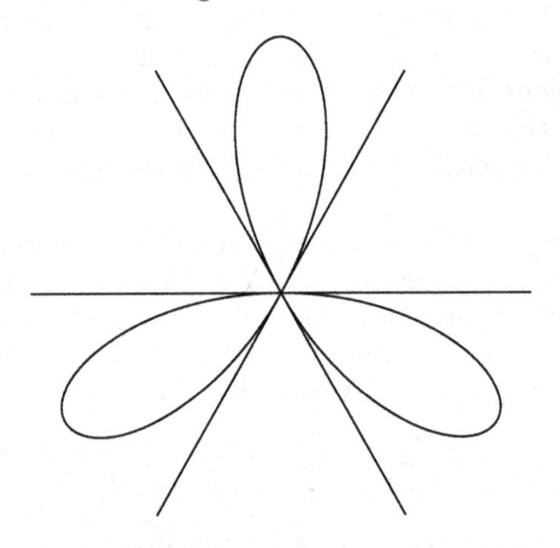

The last condition is equivalent (under the assumptions of (1)) to saying that the leading form Lf decomposes into m nonassociate linear factors. By 16.6, then, $f = f_1 \cdots f_m$ with pairwise nonassociate power series f_i of order 1 $(i = 1, \dots, m)$. At an ordinary singularity of multiplicity m the curve thus has m different branches $Z_i = K[[X,Y]]/(f_i)$, which are all regular and for which

$$\mu(Z_i, Z_j) = 1 \qquad (\text{if } i \neq j).$$

We mention that it is not especially difficult to transfer over the formulas of residue calculus from Chapters 11 and 12, as long as they are of a local nature, to algebroid curves.

For an irreducible curve F with function field $L := \mathcal{R}(F)$ and with abstract Riemann surface $\mathfrak{X}$, the canonical mapping $\pi : \mathfrak{X} \to \mathcal{V}_+(F)$ was discussed in 6.12. We will now prove the remarkable fact that for every $P \in \mathcal{V}_+(F)$ the points of $\pi^{-1}(P)$ are in one-to-one correspondence with the branches of F at P.

Let $R := \mathcal{O}_{F,P}$ and let $S := \overline{\mathcal{O}_{F,P}}$ be the integral closure of R in L. The points of $\pi^{-1}(P)$ are then the discrete valuation rings $S_{\mathfrak{P}_i}$ $(i = 1, \ldots, s)$, where $\operatorname{Max}(S) = \{\mathfrak{P}_1, \ldots, \mathfrak{P}_s\}$. On the other hand, if $\mathfrak{p}_1, \ldots, \mathfrak{p}_t$ are the minimal prime ideals of $\hat{R}$, then $Z_i := \hat{R}/\mathfrak{p}_i$ are the branches of F at P. Therefore by 16.9 the integral closure $\overline{Z}_i$ of Z in $L_i := Q(Z_i)$ is a complete discrete valuation ring $(i = 1, \ldots, t)$.

Theorem 16.14. *With the above notation, $s = t$, and by a suitable renumbering, $\mathfrak{p}_i$ is the kernel of the canonical homomorphism $\hat{R} \to \widehat{S_{\mathfrak{P}_i}}$, and so $\widehat{S_{\mathfrak{P}_i}}$ can be identified in a canonical way with $\overline{Z_i}$ $(i = 1, \ldots, s)$.*

Proof. We will first study the ring $\hat{R} \otimes_R S$. Denote by $\mathfrak{m} := \mathfrak{m}_{F,P}$ the maximal ideal of R and let $\hat{S}$ be the completion of S with respect to $\mathfrak{m}S$. We then have, as in 15.12, a canonical isomorphism

$$\hat{S} \cong \hat{R} \otimes_R S.$$

Here $\hat{S}$ is integral over the image of $\hat{R}$ in $\hat{S}$, and as in 15.13, there is a canonical isomorphism

$$\hat{S} \cong \widehat{S_{\mathfrak{P}_1}} \times \cdots \times \widehat{S_{\mathfrak{P}_s}}. \tag{7}$$

Hence $Q(\hat{S})$ is the direct product of the s fields $Q(\widehat{S_{\mathfrak{P}_i}})$:

$$Q(\hat{S}) \cong Q(\widehat{S_{\mathfrak{P}_1}}) \times \cdots \times Q(\widehat{S_{\mathfrak{P}_s}}). \tag{8}$$

We also write $\hat{R} = K[[X, Y]]/(f)$, and if $f = f_1 \cdots f_t$ is the decomposition of f into irreducible factors f_i, so $\mathfrak{p}_i = (f_i)/(f)$ $(i = 1, \ldots, t)$, then we immediately get from this representation that $\hat{R} \setminus \bigcup_{i=1}^{t} \mathfrak{p}_i$ is the set of all nonzerodivisors of $\hat{R}$. The Chinese remainder theorem then gives a canonical decomposition

$$Q(\hat{R}) = \hat{R}_{\mathfrak{p}_1}/\mathfrak{p}_1\hat{R}_{\mathfrak{p}_1} \times \cdots \times \hat{R}_{\mathfrak{p}_t}/\mathfrak{p}_t\hat{R}_{\mathfrak{p}_t} \cong Q(Z_1) \times \cdots \times Q(Z_t) \tag{9}$$

corresponding to the canonical injection

$$\hat{R} \hookrightarrow \hat{R}/\mathfrak{p}_1 \times \cdots \times \hat{R}/\mathfrak{p}_t.$$

Since the canonical mappings $R \to \hat{R}/\mathfrak{p}_i$ $(i = 1, \ldots, t)$ are injective, as we have already remarked in connection with 16.4, the elements of $N := R \setminus \{0\}$ are nonzerodivisors on $Z_1 \times \cdots \times Z_t$, and so all the more are not zerodivisors on $\hat{R}$. Similarly, from (7) we conclude that the elements of N are not zerodivisors on $\hat{S}$. By G.6 (d) there is a canonical isomorphism

$$\hat{S}_N \cong \hat{R}_N \otimes_{R_N} S_N.$$

Further, $R_N = Q(R) = Q(S) = S_N$, because S is finite over R. Therefore there is a canonical isomorphism $\hat{R}_N \cong \hat{S}_N$ induced by $\hat{R} \to \hat{S}$. Even more we have

$$Q(\hat{R}) \cong Q(\hat{S}).$$

Comparing (8) and (9), we see that $t = s$ and that there are induced isomorphisms $Q(Z_i) \cong Q(\widehat{S_{\mathfrak{P}_i}})$ (with a suitable numbering) coming from the canonical homomorphisms $\hat{R} \to \widehat{S_{\mathfrak{P}_i}}$ $(i = 1, \dots, s)$. Since $\widehat{S_{\mathfrak{P}_i}}$ is integrally closed in $Q(\widehat{S_{\mathfrak{P}_i}})$ and is finite over Z_i, we can identify $\widehat{S_{\mathfrak{P}_i}}$ with $\overline{Z_i}$.

As the theorem shows, this allows us to identify the abstract Riemann surface $\mathfrak{X}$ with the set of all branches at points of the curve F, and to identify the divisor group $\operatorname{Div}(\mathfrak{X})$ with the free abelian group on the set of all these branches.

One understands Theorem 16.14 better if one knows the fact that every plane affine algebraic curve can be obtained from a smooth curve C in a higher-dimensional affine space by projection onto the plane. At a given point of the plane curve there are as many branches at that point as there are points in C that are preimages of that given point. For each of these, according to how the projection of C behaves (transversal, tangential), various singularities arise in the plane.

Theorem 16.14 is a ring-theoretic analog of these facts from higher-dimensional geometry.

Exercises

1. Let $\operatorname{Char} K = 0$, and suppose Z is an integral branch of multiplicity m. Show that:
   (a) Every unit in $K[[T]]$ has an nth root for each $n \in \mathbb{N}$.
   (b) Z has a parametric representation of the form (T^m, β), where $\beta \in K[[T]]$ is a power series of order $> m$.
   (Set $X = T^m$, $Y = \beta(T) = \sum_{i=0}^{\infty} b_i T^i$. Then the *Newton–Puiseux series* of Z is defined by $Y = \sum_{i=0}^{\infty} b_i X^{\frac{i}{m}}$. This is the basis for the definition of further numerical and geometric invariants of the branch Z, see [BK], 8.3).
2. Determine the number of branches at the origin for the following curves from Chapter 1: the cissoid of Diocles, the conchoid of Nichomedes, the cardioid, and the four-leaf rose.

3. What can you say about the nature of the real and complex singularities of the algebraic curve represented by the olympic emblem

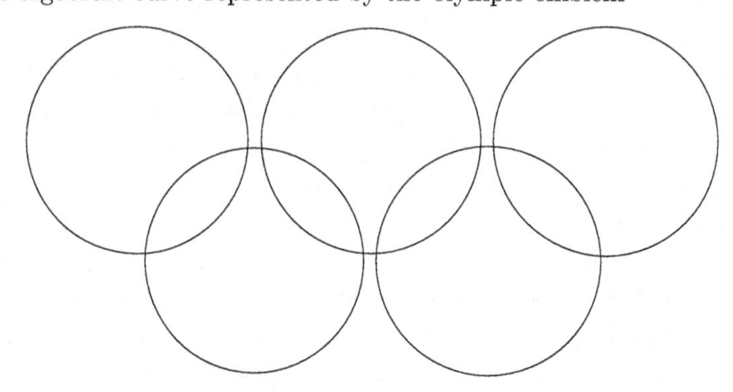

and of its projective closure?

Same question for the astroid $(X^2 + Y^2 - 1)^3 + 27X^2Y^2 = 0$.

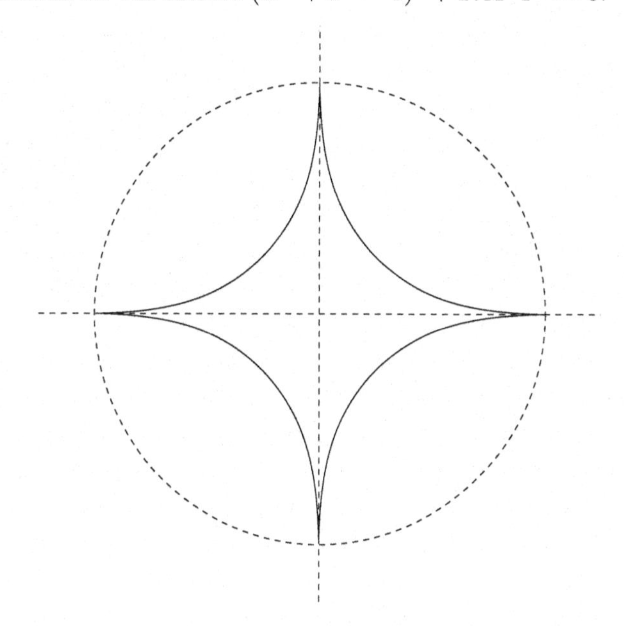

17

Conductor and Value Semigroup of a Curve Singularity

In this chapter we will relate the invariants of singularities of curves already introduced, such as "multiplicity," "tangents," "singularity degree," "branches," and "intersection multiplicity between the branches," to the conductor and value semigroup. This will allow a more precise classification of curve singularities than was possible up to now. Also, we will be led to other formulas for calculating the genus of the function field of a curve.

For two rings R, S with $R \subset S \subset Q(R)$ we call

$$(1) \qquad \mathcal{F}_{S/R} := \{z \in Q(R) \mid z \cdot S \subset R\}$$

the *conductor* of S/R. It is clear that $\mathcal{F}_{S/R}$ is an S-ideal that lies in R. Also, $\mathcal{F}_{S/R}$ is even the (uniquely determined) largest ideal of S that lies in R. Since $Q(R) = Q(S)$, we can consider $\mathcal{F}_{S/R}$ also as the complementary module $\mathfrak{C}_{S/R}$ of S/R with respect to the identity as the trace of $Q(S)/Q(R)$.

Remarks 17.1.

(a) The conductor $\mathcal{F}_{S/R}$ equals R precisely when $S = R$.
(b) If S is finitely generated as an R-module, then $\mathcal{F}_{S/R}$ contains a nonzerodivisor on R.

Proof. (a) is trivial. For (b), consider a system of generators $\{s_1, \dots, s_n\}$ of S as an R-module and write $s_i = \frac{r_i}{r}$ $(r_i, r \in R)$, where r is a nonzerodivisor. Obviously, $r \in \mathcal{F}_{S/R}$.

Lemma 17.2. *If S is finitely generated as an R-module and $N \subset R$ is multiplicatively closed, then $\mathcal{F}_{S_N/R_N}$ is defined and*

$$\mathcal{F}_{S_N/R_N} = (\mathcal{F}_{S/R})_N.$$

Proof. It is easy to see that $R_N \subset S_N \subset Q(R_N)$. Hence $\mathcal{F}_{S_N/R_N}$ is defined, and it is clear that $(\mathcal{F}_{S/R})_N \subset \mathcal{F}_{S_N/R_N}$.

If $\{s_1, \dots, s_n\}$ generates S as an R-module, then $S_N = \sum_{i=1}^n R_N \cdot \frac{s_i}{1}$. For $\frac{x}{\nu} \in \mathcal{F}_{S_N/R_N}$ $(x \in R, \nu \in N)$ we have

$$\frac{x}{\nu} \cdot \frac{s_i}{1} = \frac{x s_i}{\nu} \in R_N \qquad (i = 1, \dots, n).$$

Write $\frac{xs_i}{\nu} = \frac{r_i}{\mu}$ with $r_i \in R$, $\mu \in N$ and choose an element $\nu' \in N$ with $\nu'\mu x s_i = \nu'\nu r_i$ $(i = 1, \dots, n)$. Then $\frac{x}{\nu} = \frac{\nu'\mu x}{\nu'\mu\nu}$, and we have written $\frac{x}{\nu}$ with a numerator from $\mathcal{F}_{S/R}$. Hence we also have $\mathcal{F}_{S_N/R_N} \subset (\mathcal{F}_{S/R})_N$.

In the following, again let F be an irreducible curve in $\mathbb{P}^2(K)$ and let f be its affine part. For $P \in \mathcal{V}_+(F)$, let $\overline{\mathcal{O}_{F,P}}$ denote the integral closure of $\mathcal{O}_{F,P}$ in $L := \mathcal{R}(F)$. Furthermore, let S be the integral closure of $A := K[f] = K[x,y]$ in L, and let $\mathfrak{X}$ be the abstract Riemann surface of L/K. Also let the map $\pi : \mathfrak{X} \to \mathcal{V}_+(F)$ be defined as in 6.12.

Definition 17.3.

(a) $\mathcal{F}_P := \mathcal{F}_{\overline{\mathcal{O}_{F,P}}/\mathcal{O}_{F,P}}$ is called the *conductor of the singularity* P.
(b) $\mathcal{F}_{S/A}$ is called the *conductor of the affine curve* f.

By 17.1(a) we have $\mathcal{F}_P = \mathcal{O}_{F,P}$ precisely when $\overline{\mathcal{O}_{F,P}} = \mathcal{O}_{F,P}$, i.e., when $P \in \operatorname{Reg}(F)$. The conductor $\mathcal{F}_P$ is therefore interesting only when P is a singular point. By 17.1(b) the conductors $\mathcal{F}_P$ and $\mathcal{F}_{S/A}$ are nonzero ideals in $\mathcal{O}_{F,P}$ respectively A. If $P \in \mathcal{V}(f)$ and $\mathfrak{m}_P$ denotes the maximal ideal of A corresponding to P, then by 17.2,

$$\mathcal{F}_P = (\mathcal{F}_{S/A})_{\mathfrak{m}_P}. \tag{2}$$

By 13.11, the ring $\overline{\mathcal{O}_{F,P}}$ is a principal ideal ring; hence $\mathcal{F}_P$ is a principal $\overline{\mathcal{O}_{F,P}}$-ideal. If $\mathfrak{P}_1, \dots, \mathfrak{P}_s$ are the maximal ideals of $\overline{\mathcal{O}_{F,P}}$, then we can write

$$\mathcal{F}_P = \mathfrak{P}_1^{c_1} \cdots \mathfrak{P}_s^{c_s}, \tag{3}$$

where $c_i \in \mathbb{N}$ are uniquely determined. If $P_i \in \mathfrak{X}$ is the point corresponding to $\mathfrak{P}_i$, then we set $c_{P_i} = c_i$ $(i = 1, \dots, s)$. Furthermore, set $c_P = 0$ for $P \in \operatorname{Reg}(F)$.

Definition 17.4. The divisor $\mathcal{F}_{\mathfrak{X}/F} := \sum_{P \in \mathfrak{X}} c_P \cdot P$ is called the *conductor divisor of* F, its degree $c(F)$ is called the *conductor degree of* F, and for $P \in \operatorname{Sing}(F)$ we call $c(P) := \sum_{\pi(R)=P} c_R$ the *conductor degree of the point* P.

Using the notation of (3) we have

$$c(P) = \sum_{i=1}^{s} c_i = \dim_K \overline{\mathcal{O}_{F,P}}/\mathcal{F}_P \tag{4}$$

and therefore

$$\deg \mathcal{F}_{\mathfrak{X}/F} = \sum_{P \in \operatorname{Sing}(F)} \dim_K \overline{\mathcal{O}_{F,P}}/\mathcal{F}_P. \tag{5}$$

We define the *conductor degree of the affine curve f* by

$$c(A) := \sum_{P \in \mathrm{Sing}(f)} c(P) = \dim_K S/\mathcal{F}_{S/A}. \tag{6}$$

The connection between conductor degree, singularity degree, and genus of $\mathfrak{X}$ will be derived from the following theorem, called Dedekind's formula for conductor and complementary module.

Dedekind's Formula 17.5. Suppose we have rings $R \subset S \subset T \subset Q(S)$ and every nonzerodivisor of R is also a nonzerodivisor on S; i.e., the inclusion $R \subset S$ defines a ring homomorphism $Q(R) \to Q(S) = Q(T)$. Suppose also that $Q(S)$ is a finitely generated free $Q(R)$-module and that there exists a trace $\sigma : Q(S) \to Q(R)$. Denote the complementary modules with respect to this trace by $\mathfrak{C}_{T/R}$ and $\mathfrak{C}_{S/R}$. Suppose also that $\mathfrak{C}_{S/R}$ is generated as an S-module by a unit of $Q(S)$. Then

$$\mathfrak{C}_{T/R} = \mathcal{F}_{T/S} \cdot \mathfrak{C}_{S/R}.$$

Proof. For $z \in \mathcal{F}_{T/S}$ and $u \in \mathfrak{C}_{S/R}$ we have $\sigma(zuT) \subset R$ because $zT \subset S$. Therefore $\mathcal{F}_{T/S} \cdot \mathfrak{C}_{S/R} \subset \mathfrak{C}_{T/R}$.

Write $\mathfrak{C}_{S/R} = c \cdot S$ for some unit $c \in Q(S)$. Every $v \in \mathfrak{C}_{T/R}$ can then be written in the form $v = c \cdot w$ for some $w \in Q(S)$, and for $t \in T$ we then have

$$\sigma(twcS) = \sigma(vtS) \subset R.$$

Hence $twc \in \mathfrak{C}_{S/R} = c \cdot S$. It follows that $tw \in S$ and so $w \in \mathcal{F}_{T/S}$. Therefore $v \in \mathcal{F}_{T/S} \cdot \mathfrak{C}_{S/R}$. Thus $\mathfrak{C}_{T/R} \subset \mathcal{F}_{T/S} \cdot \mathfrak{C}_{S/R} \subset \mathfrak{C}_{T/R}$, and hence we have equality.

Corollary 17.6 (Product formula for conductors). *Suppose we have rings $R \subset S \subset T \subset Q(R)$, and $\mathcal{F}_{S/R}$ is generated as an S-module by a unit of $Q(R)$. Then*

$$\mathcal{F}_{T/R} = \mathcal{F}_{T/S} \cdot \mathcal{F}_{S/R}.$$

In order to apply the formula to a curve F, we can assume that the singularities of F lie at finite distance, that the coordinate ring A of the curve f is finitely generated as a module over $R := K[x]$, and that L is separable over $K(x)$. Because by 15.1 the A-module $\mathfrak{C}_{A/R}$ is generated by one element $\neq 0$, it follows from 17.5 that

$$\mathfrak{C}_{S/R} = \mathcal{F}_{S/A} \cdot \mathfrak{C}_{A/R}. \tag{7}$$

Let $\mathfrak{X}^f$, $\mathfrak{X}^\infty$ as well as $R_\infty := K[x^{-1}]_{(x^{-1})}$ and S_∞ have their earlier meanings (see Chapter 13). For $P \in \mathfrak{X}^f$ let $\mathfrak{P}$ be the corresponding maximal ideal of S and let $\kappa_P \in \mathbb{Z}$ be the number defined by $S_\mathfrak{P} \cdot \mathfrak{C}_{S/R} = \mathfrak{P}^{-\kappa_P} S_\mathfrak{P}$. Similarly, let $(S_\infty)_\mathfrak{P}(x^{-2}\mathfrak{C}_{S_\infty/R_\infty}) = \mathfrak{P}^{-\kappa_P}(S_\infty)_\mathfrak{P}$ if $P \in X^\infty$. Then $C :=$

$\sum_{P\in\mathfrak{X}} \kappa_P \cdot P$ is a canonical divisor of $\mathfrak{X}$, as was shown in the proof of 15.3. If we set $S_{\mathfrak{P}}\mathfrak{C}_{A/R} = \mathfrak{P}^{-\lambda_P} S_{\mathfrak{P}}$ for $P \in \mathfrak{X}^f$ and $\lambda_P = \kappa_P$ for $P \in \mathfrak{X}^\infty$, then the divisor $C' := \sum_{P\in\mathfrak{X}} \lambda_P \cdot P$ is linearly equivalent on $\mathfrak{X}$ to a canonical divisor of F. This follows from the preparatory remarks to 15.2. Now from (7) we have

Theorem 17.7. *Let F be a curve of degree d.*

(a) (*Connection between the canonical classes of $\mathfrak{X}$ and F*)

$$C = C' - \mathcal{F}_{X/F}.$$

(b) $g^L = g^F - \frac{1}{2} \cdot c(F) = \binom{d-1}{2} - \frac{1}{2} \cdot c(F).$

Statement (a) is a direct consequence of (7). Passing to the degree of the divisors, we get $2g^L - 2 = 2g^F - 2 - c(F)$ by 15.2 and 15.3, and (b) follows.

Comparing with 14.7 and 14.8 we have the following.

Corollary 17.8. $c(F) = 2\delta(F)$, *where $\delta(F)$ denotes the singularity degree of F.*

The formula in 17.8 also holds locally, and the global formula follows of course also from the local proof:

Theorem 17.9 (Gorenstein[Go]). *For every $P \in \mathcal{V}_+(F)$ we have*

$$c(P) = 2\delta(P).$$

Proof. Consider a maximal chain of ideals (composition series)

$$\mathcal{O}_{F,P} = I_0 \supsetneq I_1 \supsetneq \cdots \supsetneq I_\delta = \mathcal{F}_P. \tag{8}$$

That is, we suppose that this chain cannot be properly refined by the insertion of any further $\mathcal{O}_{F,P}$-ideal. We therefore have $I_j/I_{j+1} \cong K$ $(j = 0, \dots, \delta - 1)$ and hence $\delta = \dim_K \mathcal{O}_{F,P}/\mathcal{F}_P$. Dualizing, as in Appendix L, formula (9), we get a chain of $\mathcal{O}_{F,P}$-modules

$$\mathcal{O}_{F,P} = \mathcal{O}'_{F,P} = I'_0 \subset I'_1 \subset \cdots \subset I'_\delta = \mathcal{F}'_P. \tag{8'}$$

In L.13 it was shown, under the assumptions there, that dualizing an ideal twice returns us to the original ideal. This fact also holds locally if one localizes such a ring A at a maximal ideal. The assumption that the complementary module $\mathcal{C}_{A/R}$ is generated by one element is in any case satisfied in our situation by 15.1 We can therefore apply L.13 here, and get $I''_j = I_j$ for $j = 0, \dots, \delta$. Hence (8′) cannot be refined.

Furthermore, $\mathcal{F}_P = \mathcal{O}_{F,P} :_L \overline{\mathcal{O}}_{F,P} = \overline{\mathcal{O}}'_{F,P}$ and therefore $\mathcal{F}'_P = \overline{\mathcal{O}}''_{F,P} = \overline{\mathcal{O}}_{F,P}$. From (8) it follows that $\delta = \dim_K \overline{\mathcal{O}}_{F,P}/\mathcal{O}_{F,P} = \delta(P)$. Because $\mathcal{F}_P \subset \mathcal{O}_{F,P} \subset \overline{\mathcal{O}}_{F,P}$, we have

$$c(P) = \dim_K \overline{\mathcal{O}}_{F,P}/\mathcal{F}_P = \dim_K \overline{\mathcal{O}}_{F,P}/\mathcal{O}_{F,P} + \dim_K \mathcal{O}_{F,P}/\mathcal{F}_P = 2\delta = 2\delta(P).$$

Next we want to derive a formula for the conductor degree that will take into consideration the branches of a curve singularity. To do that we pass to the completion $\widehat{\mathcal{O}_{F,P}}$ of $\mathcal{O}_{F,P}$. As in 16.4 we set $R := \mathcal{O}_{F,P}$ and $S := \overline{\mathcal{O}_{F,P}}$. Let $Z_1, \ldots, Z_s$ be the branches of F at P corresponding to the maximal ideals $\mathfrak{P}_1, \ldots, \mathfrak{P}_s$ of S. If $\overline{Z_i}$ is the integral closure of Z_i, then $\overline{Z_i} \cong K[[T_i]]$ is a power series ring and

$$(9) \quad \hat{R} \otimes_R S \cong \widehat{S_{\mathfrak{P}_1}} \times \cdots \times \widehat{S_{\mathfrak{P}_s}} \cong \overline{Z_1} \times \cdots \times \overline{Z_s} \cong K[[T_1]] \times \cdots \times K[[T_s]]$$

is the integral closure of $\hat{R}$ in $Q(\hat{R})$, as Theorem 16.14 and its proof show. We will denote by $\hat{\mathcal{F}}_P$ the conductor of $\hat{R} \otimes_R S$ over $\hat{R}$.

Lemma 17.10 (Compatibility of Conductors with Completions).

$$\hat{\mathcal{F}}_P = \mathcal{F}_P \cdot \widehat{\mathcal{O}_{F,P}}.$$

Proof. Let $S = \sum_{i=1}^n R s_i$, with $s_i = \frac{r_i}{r}$ ($r_i, r \in R$; $i = 1, \ldots, n$). Then $\hat{R} \otimes_R S = \sum_{i=1}^n \hat{R} \cdot (1 \otimes s_i)$ and $r \in \mathcal{F}_P$. It is clear that $\mathcal{F}_P \subset \hat{\mathcal{F}}_P$ and hence $\mathcal{F}_P \cdot \widehat{\mathcal{O}_{F,P}} \subset \hat{\mathcal{F}}_P$.

To prove the opposite inclusion observe that

$$\mathcal{F}_P = \{u \in R \mid u r_i \in rR \ (i = 1, \ldots, n)\}$$

and similarly,

$$\hat{\mathcal{F}}_P = \{z \in \hat{R} \mid z r_i \in r\hat{R} \ (i = 1, \ldots, n)\}.$$

Since R is a 1-dimensional local ring, we have $\hat{R}/r\hat{R} \cong \widehat{R/rR} \cong R/rR$, and it follows that

$$\hat{\mathcal{F}}_P = \mathcal{F}_P + r\hat{R}.$$

Hence $\hat{\mathcal{F}}_P = \mathcal{F}_P \cdot \hat{R}$, since $r \in \mathcal{F}_P$.

If $\mathcal{F}_P = \mathfrak{P}_1^{c_1} \cdots \mathfrak{P}_s^{c_s}$ as in (3), then 17.10 in connection with (9) yields the formula

$$(10) \qquad \hat{\mathcal{F}}_P = (T_1^{c_1}, \ldots, T_s^{c_s}) \cdot (K[[T_1]] \times \cdots \times K[[T_s]]).$$

We want to apply the product formula 17.6 to $\mathcal{F}_P$ using the ring extensions

$$\hat{R} \subset Z_1 \times \cdots \times Z_s \subset \overline{Z}_1 \times \cdots \times \overline{Z}_s.$$

As abbreviations we set

$$\tilde{R} := Z_1 \times \cdots \times Z_s \ \text{ and } \ T := \overline{Z}_1 \times \cdots \times \overline{Z}_s = K[[T_1]] \times \cdots \times K[[T_s]].$$

If the branches Z_i are given as in Chapter 16 by $Z_i = K[[X,Y]]/(f_i)$ with irreducible power series f_i ($i = 1, \ldots, s$), then let $g_i := \prod_{j \neq i} f_j$ and denote by $\tilde{g}_i$ the image of g_i in Z_i ($i = 1, \ldots, s$). We first determine $\mathcal{F}_{\tilde{R}/\hat{R}}$:

Lemma 17.11. *The conductor $\mathcal{F}_{\tilde{R}/\hat{R}}$ equals $(\tilde{g}_1, \dots, \tilde{g}_s)\tilde{R}$. In particular, $\mathcal{F}_{\tilde{R}/\hat{R}}$ is a principal ideal of $\tilde{R}$ generated by a nonzerodivisor.*

Proof. Since $\tilde{g}_i \neq 0$ for $i = 1, \dots, s$, the element $(\tilde{g}_1, \dots, \tilde{g}_s)$ is a nonzerodivisor of $\tilde{R}$. Also,

$$(0, \dots, \tilde{g}_i, \dots, 0) \cdot \tilde{R} = (0, \dots, \tilde{g}_i Z_i, \dots, 0) \subset \hat{R},$$

for if $\tilde{h} \in Z_i$ has preimage h in $K[[X, Y]]$, then $(0, \dots, \tilde{g}_i \tilde{h}, \dots, 0)$ is the image of $g_i h$ under the canonical homomorphism $K[[X, Y]] \to Z_1 \times \cdots \times Z_s$, since g_i is divisible by all f_j with $j \neq i$. Hence $(0, \dots, \tilde{g}_i, \dots, 0) \in \mathcal{F}_{\tilde{R}/\hat{R}}$ and therefore also $(\tilde{g}_1, \dots, \tilde{g}_s) \in \mathcal{F}_{\tilde{R}/\hat{R}}$.

Conversely, suppose $(z_1, \dots, z_s) \in \mathcal{F}_{\tilde{R}/\hat{R}}$. Then in particular,

$$(0, \dots, z_i, \dots, 0) \in \hat{R}.$$

That is, there is an $h \in K[[X, Y]]$ that is divisible by all f_j with $j \neq i$ and that has image z_i in Z_i. It follows that $z_i = \tilde{g}_i \cdot z_i'$ ($z_i' \in Z_i$) and therefore $(z_1, \dots, z_s) \in (\tilde{g}_1, \dots, \tilde{g}_s) \cdot \tilde{R}$.

Now set $d_i := \dim_K Z_i/(\tilde{g}_i)$. Because of the additivity of intersection multiplicities of branches (16.12b) we have

$$d_i = \sum_{j \neq i} \mu(Z_j, Z_i) \quad (i = 1, \dots, s).$$

Further, let c_i' be the conductor degree of the branch Z_i, i.e.,

$$c_i' := \dim_K \overline{Z}_i / \mathcal{F}_{\overline{Z}_i/Z_i} \quad (i = 1, \dots, s).$$

With this data we can now obtain the desired formula for the conductor degree.

Theorem 17.12.

$$\hat{\mathcal{F}}_P = (T_1^{c_1' + d_1}, \dots, T_s^{c_s' + d_s}) \cdot (K[[T_1]] \times \cdots \times K[[T_s]]).$$

In particular,

$$(11) \qquad c(P) = 2 \cdot \sum_{1 \leq i < j \leq s} \mu(Z_i, Z_j) + \sum_{i=1}^{s} c_i',$$

Proof. It is clear that

$$\mathcal{F}_{T/\tilde{R}} = \mathcal{F}_{\overline{Z}_1 \times \cdots \times \overline{Z}_s / Z_1 \times \cdots \times Z_s} = \mathcal{F}_{\overline{Z}_1/Z_1 \times \cdots \times \overline{Z}_s/Z_s}.$$

Because of 17.11 the product formula for the conductor can be applied using the ring extension $\hat{R} \subset \tilde{R} \subset T$:

$$\hat{\mathcal{F}}_P = \mathcal{F}_{T/\hat{R}} = \mathcal{F}_{T/\tilde{R}} \cdot \mathcal{F}_{\tilde{R}/\hat{R}} = (T_1^{c_1'}, \dots, T_s^{c_s'})(\tilde{g}_1, \dots, \tilde{g}_s) \cdot T.$$

Since $\tilde{g}_i$ has order d_i in $K[[T_i]]$, we get the first statement of the theorem. The second follows from 17.10.

By this theorem, the calculation of the conductor degree of a singularity is reduced to the calculation of the conductor degrees of its branches and of the intersection multiplicities between its branches.

Example 17.13. Conductor degree of ordinary singularities.

If P is an ordinary singularity of F with multiplicity m, then by (11) and by what was said in 16.13, we have

$$c(P) = m(m-1), \text{ so } \delta(P) = \tbinom{m}{2}.$$

For a curve of degree d with only ordinary singularities P_i $(i = 1, \dots, s)$, we have the formula

$$g^L = \binom{d-1}{2} - \sum_{i=1}^{s} \binom{m_{P_i}(F)}{2}. \tag{12}$$

One reason this formula is especially significant is because, according to a theorem of Max Noether, every plane algebraic curve can be transformed into a birationally equivalent curve with only ordinary singularities using a sequence of quadratic transformations. This is one of the main theorems in the theory of plane algebraic curves, for which the reader is referred to Fulton [Fu] (Chapter 7 and the Appendix). A more precise theorem of Clebsch says that every algebraic function field has a plane algebraic curve with only normal crossings as a model. However, in the proof of this theorem, one leaves plane geometry.

If a curve F of degree d has only s *normal crossings* as singularities (i.e., $m_{P_i}(F) = 2$ for $P_i \in \mathrm{Sing}(F)$ and F has two distinct tangents at each P_i), then (12) becomes the simple formula

$$g^L = \binom{d-1}{2} - s. \tag{13}$$

The conductor divisor $\mathcal{F}_{\mathfrak{X}/F}$ allows the following generalization of the fundamental theorem of Max Noether (5.14 and 7.19):

Theorem 17.14. *Let G and H be two curves in $\mathbb{P}^2(K)$ of which F is not a component and that have divisors (G) and (H) on $\mathfrak{X}$. If*

$$(H) \geq (G) + \mathcal{F}_{\mathfrak{X}/F},$$

then $F \cap G$ is a subscheme of H. In particular, $H \in (F, G)$.

Proof. For $P \in \mathcal{V}_+(F) \cap \mathcal{V}_+(G)$, let (g_P) and (h_P) be the principal ideals in $\mathcal{O}_{F,P}$ corresponding to (G) and (H). We have to show that $(h_P) \subset (g_P)$ for all these P.

If $P \in \mathrm{Reg}(F)$, then by assumption $\nu_P(h_P) \geq \nu_P(g_P)$, and the result follows. So let $P \in \mathrm{Sing}(P)$ and let $P_1, \dots, P_r$ be the points of $\mathfrak{X}$ lying over P. Then for $i = 1, \dots, r$,

$$\nu_{P_i}(h_P) \geq \nu_{P_i}(g_P) + \nu_{P_i}(\mathcal{F}_{\mathfrak{X}/F}).$$

In the principal ideal ring $\overline{\mathcal{O}_{F,P}}$ we therefore have $h_P \in g_P \cdot \overline{\mathcal{O}_{F,P}} \cdot \mathcal{F}_P \subset g_P \cdot \mathcal{O}_{F,P}$, which was to be shown.

Example 17.15. Suppose F has only ordinary singularities $P_1, \dots, P_s$. Then

$$\mathcal{F}_{\mathfrak{X}/F} = \sum_{i=1}^{s} \sum_{\pi(Q)=P_i} (m_{P_i}(F) - 1) \cdot Q.$$

In this case the condition of Theorem 17.14 is

$$\begin{array}{ll} \nu_P(h_P) \geq \nu_P(g_P) & \text{for } P \in \operatorname{Reg}(F), \\ \nu_Q(h_P) \geq \nu_Q(g_P) + m_P(F) - 1 & \text{for } P \in \operatorname{Sing}(F),\ \ Q \in \pi^{-1}(P). \end{array}$$

We will say now a little more about the calculation of the conductor degree of branches.

Under the assumptions of 17.12 consider the embedding

$$\widehat{\mathcal{O}_{F,P}} \hookrightarrow K[[T_1]] \times \cdots \times K[[T_s]] =: T.$$

An element $z \in \widehat{\mathcal{O}_{F,P}}$ is a nonzerodivisor of $\widehat{\mathcal{O}_{F,P}}$ if and only if its image $(z_1, \dots, z_s)$ in T has all its components $z_i \neq 0$. We call the s-tuple

$$\nu(z) := (\nu_1(z_1), \dots, \nu_s(z_s)) \in \mathbb{N}^s$$

the *value* of z. Here ν_i is the order function on $K[[T_i]]$.

Definition 17.16. (a) $H_P := \{\nu(z) \mid z \text{ is a nonzerodivisor of } \widehat{\mathcal{O}_{F,P}}\}$ is called the *value semigroup* of F at P.

(b) For an integral branch $Z = K[[X,Y]]/(f)$, we call

$$H_Z := \{\nu(z) \mid z \in Z \setminus \{0\}\}$$

the *value semigroup* of Z. Here ν denotes the order function on the integral closure $\overline{Z}$ of Z.

It is clear that H_P is a subsemigroup of $(\mathbb{N}^s, +)$, and H_Z is a subsemigroup of $(\mathbb{N}, +)$. The units of $\widehat{\mathcal{O}_{F,P}}$ are precisely the elements with value $(0, \dots, 0)$. The zerodivisors of $\widehat{\mathcal{O}_{F,P}}$ are not assigned any value. Since $\hat{\mathcal{F}}_P$ is a T-ideal of the form

$$\hat{\mathcal{F}}_P = (T_1^{c_1}, \dots, T_s^{c_s}) \cdot T,$$

obviously

$$(c_1, \dots, c_s) + \mathbb{N}^s \subset H_P.$$

In particular, the semigroup H_Z has only finitely many gaps; i.e., H_Z is a numerical semigroup.

If $z \in \widehat{\mathcal{O}_{F,P}}$ is a zerodivisor, say $z = (0, z_2, \ldots, z_s)$ with $z_i \neq 0$ for $i = 2, \ldots, s$ and $\nu(z_i) = \nu_i$ $(i = 2, \ldots, s)$, then $(\nu, \nu_2, \ldots, \nu_s) \in H_P$ for all $\nu \geq c_1$. Furthermore, $(\mu_1, \ldots, \mu_s) \in H_P$ if $\mu_i = \sum_{j \neq i} \mu(Z_i, Z_j)$ $(i = 1, \ldots, s)$. Value semigroups of (plane) curve singularities and numerical semigroups have been thoroughly studied by many authors. Here are some names: Barucci, V, Dobbs, D.E., Fontana, M. [BDF]; Barucci, V., D'Anna, M., Fröberg, R. [BDFr$_1$], [BDFr$_2$]; Bertin, J. and Carbonne, P. [BC]; Campillo, A., Delgado, F., Kiyek, K. [CDK]; Delgado, F. [De]; Garcia, A. [Ga]; Waldi, R. [Wa$_2$]. The lists of references to these papers and the MathSciNet will help the reader to gain more information about this area of research.

We close this chapter with observations about plane branches Z. If $\mathcal{F}_{\overline{Z}/Z} = T^c \cdot K[[T]]$, then c is the conductor degree of Z. On the other hand, $c-1$ is the largest gap of H_Z, for if there were an element $z \in Z$ with $\nu(z) = c-1$, then all $y \in \overline{Z}$ with $\nu(y) \geq c-1$ would be contained in Z: If $\nu(y) \geq c$, then $y \in \mathcal{F}_{\overline{Z}/Z} \subset Z$, and if $\nu(y) = c-1$, then $y - \kappa z \in \mathcal{F}_{\overline{Z}/Z}$ for some $\kappa \in K$, and it follows that $y \in Z$. We would then have $T^{c-1} \in \mathcal{F}_{\overline{Z}/Z}$, a contradiction.

For a numerical semigroup H we call the smallest number c with $c + \mathbb{N} \subset H$ the *conductor* of H. This is in agreement with the above observation. The calculation of the conductor degree of a branch reduces to the calculation of the conductor of its semigroup. The largest gap $c-1$ of a numerical semigroup H is called the *Frobenius number of* H. Its computation (the Frobenius problem) has also given rise to many papers.

Lemma 17.17. *If $\ell_1, \ldots, \ell_\delta$ are the gaps of a numerical semigroup H with conductor c, then $c \leq 2\delta$.*

Proof. If $h \in H$ with $h < c$, then $c-1-h \notin H$, for $(c-1-h)+h = c-1 \notin H$. Hence there are at least as many gaps as elements $h \leq c$, $h \in H$.

Definition 17.18. A numerical semigroup H with conductor c is called *symmetric* if for $z \in \mathbb{Z}$, $c-1-z \in H$ if and only if $z \notin H$.

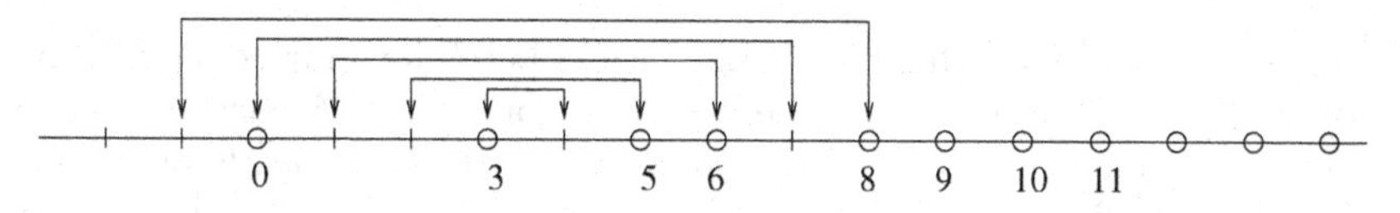

Theorem 17.19 (Apéry). *The value semigroup H_Z of an integral branch Z is symmetric.*

Proof. If c is the conductor of H_Z, then $\mathcal{F}_{\overline{Z}/Z} = T^c \cdot K[[T]]$. For each $h \in H_Z$ with $h < c$, there exists an element $z_h \in Z$ with $\nu(z_h) = h$. It is then clear that

$$Z = \bigoplus_{h \in H_Z,\ h < z} K z_h \oplus \mathcal{F}_{\overline{Z}/Z},$$

and therefore $\delta(Z) = \dim_K \overline{Z}/Z$ is equal to the number δ of gaps of H_Z. As in 17.9 one shows that $c(Z) = 2\delta(Z)$. Then $c = 2\delta$ and H_Z is symmetric.

Characterizations of numerical semigroups that occur as value semigroups of branches of irreducible plane algebroid curves are given in Angermüller [An] and Garcia–Stöhr [GSt]. These results are tied to earlier publications of Apéry [Ap], Azevedo [Az], Abhyankar–Moh [AM], and Moh [Mo]. See also [BDFr$_1$].

We conclude with an example that shows how to determine the value semigroup and conductor degree of a branch given by a parametric representation.

Example 17.20. Let $Z = \mathbb{C}[[\alpha, \beta]] \subset \mathbb{C}[[T]]$ with

$$\alpha = T^4, \quad \beta = T^6 + T^7.$$

Since $4 \cdot \mathbb{N} + 6 \cdot \mathbb{N} \subset H_Z$, all even integers ≥ 4 belong to H_Z. Furthermore,

$$\beta^2 - \alpha^3 = 2T^{13} + T^{14},$$

and so $13 \in H_Z$. It is easy to see that 16 is the conductor of H_Z and that H_Z has the following appearance:

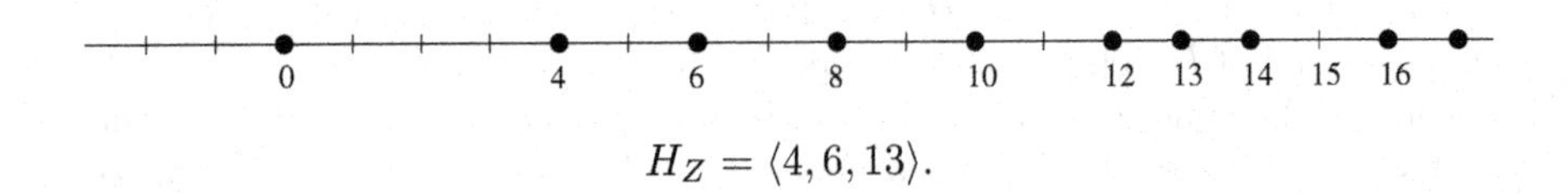

$$H_Z = \langle 4, 6, 13 \rangle.$$

Exercises

1. Sketch the value semigroup of a normal crossing.
2. Determine the number of branches at the origin of the following curves in $\mathbb{A}^2(\mathbb{C})$ and sketch their corresponding value semigroups:

$$Y^2 - X^4 + X^5,$$

$$Y^4 - X^6 + X^8,$$

3. (A generalization of Chapter 15, Exercise 1) Let F be an irreducible curve of degree $p > 3$ and let $\mathcal{F}_{\mathfrak{X}/F}$ be its conductor divisor. A curve G that is not a component of F is called *adjoint* to F if $(G) \geq \mathcal{F}_{\mathfrak{X}/F}$. Show that
   (a) If G is adjoint to F and $\deg G = p - 3$, then $C := (G) - \mathcal{F}_{\mathfrak{X}/F}$ is an effective canonical divisor of $\mathfrak{X}$.
   (b) If F has only ordinary singularities and $\mathfrak{X}$ is not rational, then F has an adjoint curve of degree $p - 3$.
   (c) If F has only ordinary singularities, then every effective canonical divisor is of the form $(G) - \mathcal{F}_{\mathfrak{X}/F}$ for some adjoint curve G of degree $p - 3$. (Observe that every nonrational abstract Riemann surface has an effective canonical divisor.)
4. Let p and q be two relatively prime natural numbers and let $H := \langle p, q \rangle$ be the numerical semigroup generated by p and q. Determine the Frobenius number of H. Show that

(a) The branch given by $X^p - Y^q$ has the semigroup H.
(b) H is symmetric (this follows from (a), but one can also easily give a direct proof).

Part II

Algebraic Foundations

Algebraic Foundations

The following list of keywords should give an indication of the parts of algebra that are assumed to be well known, so when these words appear in statements in the text, it should be clear what is meant.

From linear algebra:

- The theory of vector spaces, matrices, and linear transformations.
- The theory of determinants.
- The concepts of module, free module, and torsion-free module.
- Submodules and residue class modules.
- Linear maps between modules, and the dual module.
- The fundamental theorem for modules over a principal ideal domain (PID).
- The Hilbert basis theorem for modules.

From ring theory:

- Basic concepts of units, zerodivisors, nilpotent elements, integral domains.
- Ring homomorphisms and the homomorphism theorems.
- Ideals and residue class rings.
- Prime ideals and maximal ideals.
- Polynomial rings and power series rings in several variables.
- Basic facts about unique factorization domains (UFDs).
- The concept of a Noetherian ring and the Hilbert basis theorem for polynomial rings.

By an algebra S/R we mean a triple (R, S, ρ), where R and S are commutative rings and $\rho : R \to S$ is a ring homomorphism. The map ρ is called the structure homomorphism for the algebra.

From field theory:

- The theory of finite (field) extensions.
- Algebraic and transcendental elements of a field extension.

- Conjugate elements.
- Algebraically closed fields.
- The field of quotients of an integral domain.

What we still need besides this is collected together in the following Appendices A–L. To save time and space, many results are not stated in their most general form, but only in the form that we need. Textbooks on commutative algebra such as [B], [E], [Ku_1], and [M] will of course give a more complete presentation.

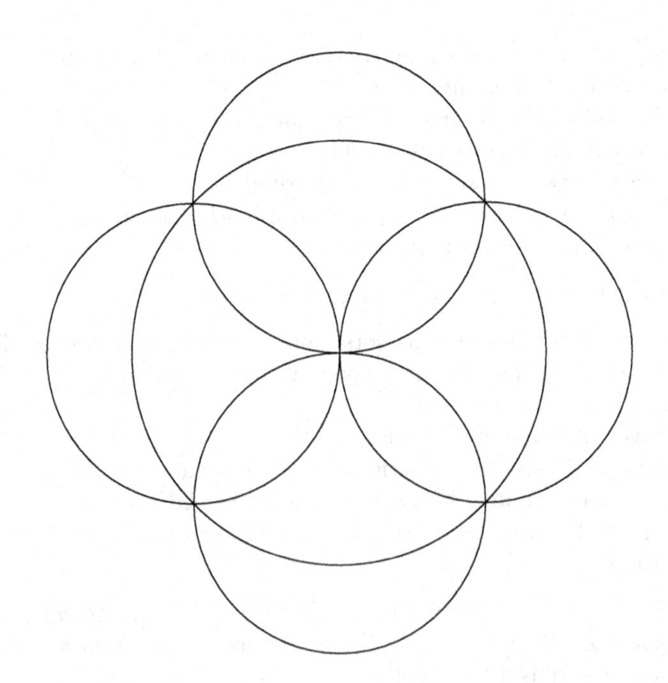

A

Graded Algebras and Modules

This is a short chapter on linear algebra. Let G/K be a given algebra, and let M be a G-module. Then G and M can be thought of as K-modules, in a natural way.

Definition A.1. A *grading* of G/K is a family $\{G_k\}_{k\in\mathbb{Z}}$ of K-submodules $G_k \subset G$ such that:

(a) $G = \bigoplus_{k\in\mathbb{Z}} G_k$.

(b) $G_k G_l \subset G_{k+l}$ for all $k, l \in \mathbb{Z}$.

G is called a *graded K-algebra* if G is furnished with a grading $\{G_k\}_{k\in\mathbb{Z}}$. The elements of G_k are called *homogeneous of degree k*. If $g \in G$ is written in the form $g = \sum_{k\in\mathbb{Z}} g_k$, with $g_k \in G_k$, then we call g_k the *homogeneous component of g of degree k*.

Example A.2. Let $G = K[X_1, \dots, X_m]$ be the polynomial algebra in the variables $X_1, \dots, X_m$ over a ring K, and let G_k, for $k \in \mathbb{Z}$, be the set of all homogeneous polynomials of degree k:

$$\sum_{\nu_1+\cdots+\nu_m=k} a_{\nu_1\dots\nu_m} X_1^{\nu_1} \cdots X_m^{\nu_m} \qquad (a_{\nu_1\dots\nu_m} \in K)$$

Here $G_k = \{0\}$ for $k < 0$. It is clear that $\{G_k\}_{k\in\mathbb{Z}}$ is a grading of G/K.

A homogeneous polynomial $F \in G$ of degree k has the following property:

$$(1) \qquad F(\lambda X_1, \dots, \lambda X_m) = \lambda^k F(X_1, \dots, X_m) \quad \text{for all } \lambda \in K.$$

Conversely, if K is an infinite field and $F \in G$ is a polynomial with the property (1), then F is homogeneous of degree k. If $F = \sum_{i\in\mathbb{N}} F_i$ is the decomposition of F into homogeneous components, then it follows from (1) applied to F and to F_i that

$$F(\lambda X_1, \dots, \lambda X_m) = \lambda^k F(X_1, \dots, X_m) = \lambda^k \sum_{i\in\mathbb{N}} F_i(X_1, \dots, X_m)$$

and

$$F(\lambda X_1, \ldots, \lambda X_m) = \sum_{i \in \mathbb{N}} F_i(\lambda X_1, \ldots, \lambda X_m) = \sum_{i \in \mathbb{N}} \lambda^i F_i(X_1, \ldots, X_m).$$

Since λ takes on infinitely many values, it follows by comparing coefficients that $F = F_k$.

Another property of a homogeneous polynomial of degree k is given by *Euler's formula*:

$$kF = \sum_{i=1}^{m} X_i \frac{\partial F}{\partial X_i}. \tag{2}$$

If $\mathbb{Q} \subset K$, then it is not difficult to show that the homogeneous polynomials of degree k are characterized by this formula.

Now let $G = \bigoplus_{k \in \mathbb{Z}} G_k$ be an arbitrary graded K-algebra. By A.1(b), G_0 is a subring of G and the G_k are G_0-modules. Thus $1 \in G_0$, for if $1 = \sum_{k \in \mathbb{Z}} l_k$ is a decomposition of the identity into homogeneous components ($l_k \in G_k$), then for every homogeneous $g \in G$, we have $g = g \cdot 1 = \sum_{k \in \mathbb{Z}} g \cdot l_k$, and then by comparison we have $g = g \cdot l_0$. Since this holds for arbitrary $g \in G$, we have $1 = l_0$.

Lemma A.3. *If G is an integral domain, then any divisor of a homogeneous element of G is again homogeneous.*

Proof. Let $g \in G$ be homogeneous. Suppose $g = ab$, where $a, b \in G$. Write

$$a = a_p + a_{p+1} + \cdots + a_q \quad (a_i \text{ homogeneous of degree } i,\ p \leq q,\ a_p \neq 0,\ a_q \neq 0)$$

and

$$b = b_m + b_{m+1} + \cdots + b_n \quad (b_j \text{ homogeneous of degree } j,\ m \leq n,\ b_m \neq 0,\ b_n \neq 0).$$

Then we have

$$g = a_p b_m + \cdots + a_q b_n.$$

Therefore, $a_p b_m \neq 0$ and $a_q b_n \neq 0$, since G is an integral domain. Also, $a_p b_m$ is the homogeneous component of g of degree $p + m$, and $a_q b_n$ is the homogeneous component of g of degree $q + n$. Since g is homogeneous, we must have $p = q$ and $m = n$. Therefore $a = a_p$ and $b = b_m$.

The lemma can be applied in the special case of a polynomial algebra $K[X_1, \ldots, X_m]$ over an arbitrary integral domain K. If K is a unique factorization domain, then the irreducible factors of a homogeneous polynomial are themselves homogeneous polynomials in $K[X_1, \ldots, X_m]$. In this regard, we shall mention the *graded version of the fundamental theorem of algebra*:

Theorem A.4. *Let K be an algebraically closed field and $F \in K[X, Y]$ a homogeneous polynomial of degree d. Then F decomposes into linear factors:*

$$F = \prod_{i=1}^{d} (a_i X - b_i Y) \qquad (a_i, b_i) \in K^2 \quad (i = 1, \ldots, d).$$

Proof. Let $F = \sum_{j=0}^{d} c_j X^j Y^{d-j}$ $(c_j \in K)$. Because K is algebraically closed, the polynomial $f := \sum_{j=0}^{d} c_j X^j$ decomposes into linear factors

$$f = \prod_{i=1}^{d} (a_i X - b_i).$$

Then $F(X,Y) = Y^d f(\frac{X}{Y}) = \prod_{i=1}^{d} (a_i X - b_i Y)$.

Now let $G = \bigoplus_{k \in \mathbb{Z}} G_k$ be a graded K-algebra. Next we define the notion of a graded G-module.

Definition A.5. A *grading on* M is a family $\{M_k\}_{k \in \mathbb{Z}}$ of K-submodules $M_k \subset M$ such that

(a) $M = \bigoplus_{k \in \mathbb{Z}} M_k$.
(b) $G_k M_l \subset M_{k+l}$ for all $k, l \in \mathbb{Z}$.

If M is furnished with a grading, we call M a *graded module* over the graded ring G.

The concepts "homogeneous element" and "homogeneous component," introduced above for graded algebras, carry over to graded modules. By A.5(b), the M_k are G_0-modules.

If $M = \oplus M_k$ is a graded module over a graded ring G and if M is finitely generated, then M has a finite set of generators consisting of homogeneous elements: one simply takes all homogeneous components of elements of a finite set of generators of M.

Now let $U \subset M$ be a submodule.

Definition A.6. A submodule U of M is called a *graded* (or *homogeneous*) *submodule* of M if whenever $u \in U$ and $u = \sum_{k \in \mathbb{Z}} u_k$ is a decomposition of u into homogeneous components $u_k \in M_k$ $(k \in \mathbb{Z})$, then $u_k \in U$ for all $k \in \mathbb{Z}$.

In particular, this definition defines a *homogeneous ideal* in a graded ring, e.g., in a polynomial ring. A homogeneous submodule $U \subset M$ is itself a graded module over G:

$$U = \bigoplus_{k \in \mathbb{Z}} U_k \qquad \text{with} \qquad U_k := U \cap M_k \quad (k \in \mathbb{Z}).$$

Lemma A.7. *For a submodule $U \subset M$, the following are equivalent:*

(a) *U is a homogeneous submodule of M.*
(b) *U is generated by homogeneous elements of M.*
(c) *The family $\{(M_k + U)/U\}_{k \in \mathbb{Z}}$ is a grading of M/U.*

Proof. (*a*) $\Rightarrow$ (*b*) The homogeneous elements of U trivially form a generating system of U.

(*b*) $\Rightarrow$ (*a*) Let $\{x_\lambda\}$ be a generating system of U consisting of homogeneous elements of M, and let $\deg x_\lambda =: d_\lambda$. For $u \in U$, write $u = \sum_\lambda g_\lambda x_\lambda$ with $g_\lambda \in G$ and decompose each g_λ into homogeneous components: $g_\lambda = \sum_i g_{\lambda_i}$ with $g_{\lambda_i} \in G_i$. Then

$$u = \sum_k \left(\sum_\lambda \sum_{i+d_\lambda = k} g_{\lambda_i} x_\lambda \right)$$

and $u_k := \sum_\lambda \sum_{i+d_\lambda=k} g_{\lambda_i} x_\lambda$ is homogeneous of degree k. Since $u_k \in U$ for all $k \in \mathbb{Z}$, this proves (a).

(*a*) $\Rightarrow$ (*c*) It is clear that $M/U = \sum_{k\in\mathbb{Z}} (M_k + U)/U$, so we need to show only that the sum is direct. For $m_k \in M_k$, denote by $\overline{m_k}$ the residue class in M/U. If $\sum_{k\in\mathbb{Z}} \overline{m_k} = 0$ for elements $m_k \in M_k$ ($k \in \mathbb{Z}$), then $\sum_{k\in\mathbb{Z}} m_k \in U$. Since U is a homogeneous submodule of M, it follows that $m_k \in U$ and $\overline{m_k} = 0$ for all $k \in \mathbb{Z}$.

(*c*) $\Rightarrow$ (*a*) Each element $u \in U$ can be written in the form $u = \sum_{k\in\mathbb{Z}} u_k$ with $u_k \in M_k$ ($k \in \mathbb{Z}$). Then in M/U,

$$0 = \overline{u} = \sum_{k\in\mathbb{Z}} \overline{u_k},$$

and therefore $\overline{u_k} = 0$ for all $k \in \mathbb{Z}$; hence $u_k \in U$.

If $U \subset M$ is a homogeneous submodule, we usually tacitly assume that M/U has the grading given by A.7(c). The canonical epimorphism $M \to M/U$ is "homogeneous of degree 0"; i.e., homogeneous elements of M are mapped to homogeneous elements of the same degree. If $I \subset G$ is a homogeneous ideal, then G/I is also a graded K-algebra with the grading $\{G_k + I/I\}_{k\in\mathbb{Z}}$.

The submodule

$$IM := \left\{ \sum_\alpha x_\alpha m_\alpha \mid x_\alpha \in I, m_\alpha \in M \right\}$$

of M is generated by homogeneous elements and is therefore a graded submodule; thus M/IM is a graded G-module. As a special case we have the residue class module M/gM, whenever $g \in G$ is a homogeneous element.

A graded K-algebra $G = \bigoplus_{k\in\mathbb{Z}} G_k$ is called *positively graded* if $G_k = \{0\}$ for $k < 0$. If $M = \bigoplus_{k\in\mathbb{Z}} M_k$ is a graded G-module, we call the grading $\{M_k\}$ *bounded below* if there is a $k_0 \in Z$ such that $M_k = \{0\}$ for $k < k_0$.

The next lemma will be used quite often.

Nakayama's Lemma for Graded Modules A.8. *Let G be a graded K-algebra, $I \subset G$ an ideal generated by homogeneous elements of positive degree, M a graded G-module, and $U \subset M$ a graded submodule. If the grading of M/U is bounded below and if*

$$M = U + IM,$$

then $M = U$.

Proof. $N := M/U$ is a graded G-module with a bounded-below grading, and we have $N = IN$. Suppose $N \neq \{0\}$. Let $n \in N \setminus \{0\}$ be a homogeneous element of smallest degree. We can write this element in the form

$$n = \sum x_\alpha n_\alpha,$$

where $x_\alpha \in I$ are homogeneous elements of positive degree and $n_\alpha \in N \setminus \{0\}$. However, we must then have $\deg(n_\alpha) < \deg(n)$, which is a contradiction. Therefore $N = \{0\}$ and $M = U$.

Now let G be a positively graded algebra and suppose

$$G = G_0[x_1, \ldots, x_m]$$

for some homogeneous elements x_i of degree $\alpha_i \in \mathbb{N}_+$ $(i = 1, \ldots, m)$. This situation occurs with polynomial algebras (A.2) and their residue class algebras.

Each G_k $(k \in \mathbb{N})$ is generated as a G_0-module by elements $x_1^{\nu_1} \cdots x_m^{\nu_m}$ with $\sum_{i=1}^m \nu_i \alpha_i = k$. If, furthermore, M is a finitely generated graded G-module,

$$M = Gm_1 + \cdots + Gm_t,$$

where m_i are homogeneous elements of degree d_i $(i = 1, \ldots, t)$, then for all $k \in \mathbb{Z}$, M_k is generated as a G_0-module by the (finitely many) elements

$$x_1^{\nu_1} \cdots x_m^{\nu_m} m_i \quad \text{with} \quad d_i + \sum_{j=1}^m \nu_j \alpha_j = k.$$

Finally, if $G_0 = K$ is a field, then the M_k are finite-dimensional K-vector spaces. The dimensions of these vector spaces play a role in many questions in algebra, algebraic geometry, and combinatorics.

Definition A.9. Under the assumptions above, the mapping $\chi_M : \mathbb{Z} \to \mathbb{N}$ defined by

$$\chi_M(k) := \dim_K M_k \quad (k \in \mathbb{Z})$$

is called the *Hilbert function* of the graded G-module M.

Example A.10 (The Hilbert function of a polynomial algebra). Let K be a field and $G = K[X_1, \dots, X_m]$ a polynomial algebra over K in m variables $X_1, \dots, X_m$ of degree 1. Then $\chi_G(k) = 0$ for $k < 0$ and

$$\chi_G(k) = \binom{m+k-1}{m-1} = \binom{m+k-1}{k} \quad \text{for } k \geq 0.$$

This is the formula for the number of monomials $X_1^{\nu_1} \cdots X_m^{\nu_m}$ of degree k.

Lemma A.11. *Under the assumptions of A.9, let $g \in G$ be a homogeneous element of degree d. Suppose that the map $\mu_g : M \to M \quad (m \mapsto gm)$ is injective. Then*

$$\chi_{M/gM}(k) = \chi_M(k) - \chi_M(k-d) \qquad (k \in \mathbb{Z}).$$

Proof. We have $\mu_g(M_{k-d}) \subset M_k$ (by A.5(b)), and therefore we have for each $k \in \mathbb{Z}$ an exact sequence of K-vector spaces

$$0 \to M_{k-d} \xrightarrow{\mu_g} M_k \to (M/gM)_k \to 0.$$

The formula above follows immediately.

Examples A.12. Let K be a field.

(a) For a homogeneous polynomial $F \in K[X_1, \dots, X_m]$ of degree $d > 0$, let $G := K[X_1, \dots, X_m]/(F)$. Then

$$\chi_G(k) = \begin{cases} \binom{m+k-1}{m-1} & \text{for } 0 \leq k < d, \\ \binom{m+k-1}{m-1} - \binom{m+k-d-1}{m-1} & \text{for } d \leq k. \end{cases}$$

(b) Let $F, G \in K[X, Y]$ be homogeneous polynomials with $\deg F = p > 0$, $\deg G = q > 0$, and let $A := K[X, Y]/(F, G)$. If F and G are relatively prime and if $p \leq q$, then the Hilbert function χ_A of A is

$$\chi_A(k) = \begin{cases} k+1 & \text{for } 0 \leq k < p, \\ p & \text{for } p \leq k < q, \\ p+q-k-1 & \text{for } q \leq k < p+q, \\ 0 & \text{for } k \geq p+q. \end{cases}$$

Figure A.1 is a sketch of the "graph" of this function. The proof follows by two applications of Lemma A.11. For $A^0 := K[X, Y]$ we easily see that

$$\chi_{A^0}(k) = k+1 \qquad (k \in \mathbb{N}),$$

and for $A^1 := K[X, Y]/(F)$, by (a) we have

$$\chi_{A^1}(k) = \begin{cases} k+1 & \text{for } 0 \leq k \leq p-1, \\ p & \text{for } p \leq k. \end{cases}$$

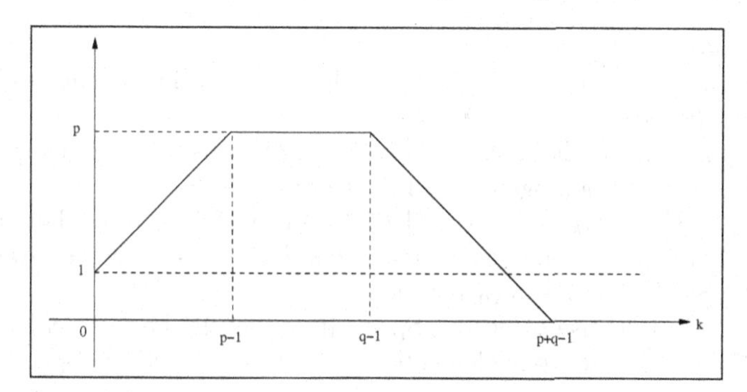

Fig. A.1. The graph of the Hilbert function χ_A of $A = K[X,Y]/(F,G)$.

Since F and G are relatively prime, the map consisting of multiplication by G on A^1 is a nonzerodivisor on A^1. Therefore we can use A.11 and get the formula above. From this we find in particular that

$$\dim_K A = \sum_{k=0}^{p+q-2} \chi_A(k) = pq, \tag{3}$$

which one can see easily for example from figure A.1.

Exercises

1. Let K be a field and $G = \bigoplus_{k \geq 0} G_k$ a positively graded K-algebra. Let $G_0 = K$ and $G = K[x_1, \dots, x_n]$, where the elements x_i are homogeneous of degree $d_i \in \mathbb{N}_+$ $(i = 1, \dots, n)$. The formal power series

$$H_G(t) = \sum_{k=0}^{\infty} \chi_G(k) t^k \in \mathbb{Z}[[t]]$$

is called the *Hilbert series* of G. Show that if $g \in G$ is homogeneous of degree d and is a nonzerodivisor on G, then

$$H_{G/gG}(t) = (1 - t^d) H_G(t).$$

2. The polynomial algebra $P = K[X_1, \dots, X_n]$ over a field K may be given the grading for which $\deg X_i = d_i \in N_+$ $(i = 1, \dots, n)$.
   (a) Prove that
   $$(1 - t^{d_1}) \cdots (1 - t^{d_n}) H_P(t) = 1.$$
   (b) Which power series do you get when $d_1 = \cdots = d_n = 1$?
3. Let I be the kernel of the K-epimorphism $\alpha : P \to G$, where $\alpha(X_i) = x_i$ $(i = 1, \dots, n)$ (G and P as in Exercises 1 and 2). Show that I is a homogeneous ideal of P.

4. Show that:
   (a) With the hypotheses of Exercise 2 let $F \in P$ be homogeneous of degree k. Then $kF = \sum_{i=1}^{n} d_i X_i \frac{\partial F}{\partial X_i}$.
   (b) If this formula holds for an $F \in P$ and if K is a field of characteristic 0, then F is homogeneous of degree k.
5. Let $G = \bigoplus_{k \in \mathbb{Z}} G_k$ be a graded ring and let $I \subset G$ be a homogeneous ideal. For homogeneous elements $a, b \notin I$, assume that we always have $ab \notin I$. Show that I is a prime ideal.
6. For G as in Exercise 5, let $\mathfrak{P} \in \operatorname{Spec}(G)$ and let $\mathfrak{P}^*$ be the ideal generated by all the homogeneous elements of $\mathfrak{P}$. Show that $\mathfrak{P}^* \in \operatorname{Spec}(G)$. Conclude that all the minimal prime ideals of G are homogeneous and that for each homogeneous ideal $I \subset G$ the minimal prime divisors of I are homogeneous.

B

Filtered Algebras

We will not attempt here to explain what filtered algebras are good for: that will be shown with the applications. We will just say that this appendix is fundamental for the entire text, and that our friends in computer algebra work with similar methods in order to do effective calculations in polynomial rings or to give explicit solutions to systems of algebraic equations (see [KR] for an excellent introduction). For this, however, it is necessary to replace the $\mathbb{Z}$-grading and $\mathbb{Z}$-filtration by a G-grading and G-filtration, where G is an ordered abelian group. There are no fundamental difficulties in transferring over the results of this section into the more general case.

Let S/R be an algebra.

Definition B.1. An (ascending) *filtration* of S/R is a family $\mathcal{F} = \{\mathcal{F}_i\}_{i\in\mathbb{Z}}$ of R-submodules $\mathcal{F}_i \subset S$ $(i \in \mathbb{Z})$ such that

(a) $\mathcal{F}_i \subset \mathcal{F}_{i+1}$ for all $i \in \mathbb{Z}$,
(b) $\mathcal{F}_i \cdot \mathcal{F}_j \subset \mathcal{F}_{i+j}$ for all $i, j \in \mathbb{Z}$,
(c) $1 \in \mathcal{F}_0$,
(d) $\bigcup_{i\in\mathbb{Z}} \mathcal{F}_i = S$.

An algebra S/R with a filtration is called a *filtered algebra.* We write $(S/R, \mathcal{F})$ for such an algebra. If $\bigcap_{i\in\mathbb{Z}} \mathcal{F}_i = \{0\}$, we call $\mathcal{F}$ *separated.*

If $(S/R, \mathcal{F})$ is a filtered algebra, it follows from B.1(b) and B.1(c) that $\mathcal{F}_0$ is a subring of S and each $\mathcal{F}_i$ is an $\mathcal{F}_0$-module. If $\mathcal{F}$ is separated, one can define the *order* of a nonzero element $f \in S$ with respect to $\mathcal{F}$ as follows:

$$\operatorname{ord}_{\mathcal{F}} f := \operatorname{Min}\{i \in \mathbb{Z} \mid f \in \mathcal{F}_i\}.$$

Also, we set $\operatorname{ord}_{\mathcal{F}} 0 = -\infty$. The following formulas follow easily from B.1.

Rules B.2. Let $f, g \in S$.

(a) $\operatorname{ord}_{\mathcal{F}}(f + g) \leq \operatorname{Max}\{\operatorname{ord}_{\mathcal{F}} f, \operatorname{ord}_{\mathcal{F}} g\}$. If $\operatorname{ord}_{\mathcal{F}} f \neq \operatorname{ord}_{\mathcal{F}} g$, then we have equality.
(b) $\operatorname{ord}_{\mathcal{F}}(f \cdot g) \leq \operatorname{ord}_{\mathcal{F}} f + \operatorname{ord}_{\mathcal{F}} g$.

Examples B.3.

(a) *Degree-filtration.* If we consider a graded ring $S = \bigoplus_{i\in\mathbb{Z}} S_i$ as an algebra over $R := S_0$, and set

$$\mathcal{F}_i = \bigoplus_{\rho \le i} S_\rho \qquad (i \in \mathbb{Z}),$$

then we get a separated filtration $\mathcal{F} = \{\mathcal{F}_i\}_{i\in\mathbb{Z}}$ of S/R. It is called the *degree-filtration.* The order of an element $f \in S \setminus \{0\}$ is, in this case, the largest degree of any nonzero homogeneous component of f. In particular, each polynomial algebra $S = R[\{X_\lambda\}_{\lambda\in\Lambda}]$ in any family $\{X_\lambda\}_{\lambda\in\Lambda}$ of indeterminates has a degree filtration.

(b) *I-adic filtration.* Let S/R be an algebra and $I \subset S$ an ideal. For $k \in \mathbb{N}_+$ let $\mathcal{F}_{-k} := I^k$ be the kth power of I, i.e., the ideal of S generated by all products $a_1 \cdots a_k$, where $a_i \in I$ $(i = 1, \dots, k)$. For $k \in \mathbb{N}$, let $\mathcal{F}_k := S$. Then $\mathcal{F}_k$ is an R-module for all $k \in \mathbb{Z}$, and

$$\cdots \subset I^k \subset I^{k-1} \subset \cdots \subset I^1 = I \subset I^0 = S = S = \cdots .$$

$\mathcal{F} = \{\mathcal{F}_i\}_{i\in\mathbb{Z}}$ is a filtration of S/R; it is called the *I-adic filtration.* The special case of $I = S$ is called the *trivial filtration* of S/R: then $\mathcal{F}_i = S$ for all $i \in \mathbb{Z}$.

If S is a local ring with maximal ideal $\mathfrak{m}$, one often uses the $\mathfrak{m}$-adic filtration of S (as a $\mathbb{Z}$-algebra).

In general, I-adic filtrations are not separated. However, the Krull intersection theorem (E.7) gives conditions under which they are separated. A simple but very important example for us is the following: Let $S = k[X_1, \dots, X_n]$ be a polynomial algebra and let $I = (X_1, \dots, X_n)$. In this case, I^k is the ideal generated by all monomials $X_1^{\alpha_1} \cdots X_n^{\alpha_n}$ with $\alpha_1 + \cdots + \alpha_n = k$, and the I-adic filtration is separated. The order of a polynomial $F \neq 0$ is the negative of the smallest degree of a nonzero homogeneous component of F.

Now let $(S/R, \mathcal{F})$ be an arbitrary filtered algebra. In the ring $S[T, T^{-1}] = \bigoplus_{i\in\mathbb{Z}} ST^i$ of "Laurent polynomials" in the indeterminate T over S one can consider the subring

$$\mathcal{R}_\mathcal{F} S := \bigoplus_{i\in\mathbb{Z}} \mathcal{F}_i T^i.$$

By the axioms in B.1 this is in fact a subring of $S[T, T^{-1}]$, and a graded algebra:

$$\mathcal{R}_\mathcal{F} S = \bigoplus_{i\in\mathbb{Z}} \mathcal{R}^i_\mathcal{F} S \qquad \text{with} \qquad \mathcal{R}^i_\mathcal{F} S := \mathcal{F}_i T^i.$$

Since $1 \in \mathcal{F}_0 \subset \mathcal{F}_1$, we have $T \in \mathcal{R}_\mathcal{F} S$, and $\mathcal{R}_F S$ can even be considered as an $R[T]$-algebra. $\mathcal{R}_\mathcal{F} S$ is called the *Rees algebra* of $(S/R, \mathcal{F})$.

The *associated graded algebra* $\operatorname{gr}_\mathcal{F} S$ of $(S/R, \mathcal{F})$ will be constructed as follows: For $i \in Z$, let $\operatorname{gr}^i_\mathcal{F} S := \mathcal{F}_i / \mathcal{F}_{i-1}$ and let

$$\operatorname{gr}_\mathcal{F} S := \bigoplus_{i\in\mathbb{Z}} \operatorname{gr}^i_\mathcal{F} S$$

be the direct sum of these R-modules. We define a multiplication on $\operatorname{gr}_{\mathcal{F}} S$ as follows: for $a + \mathcal{F}_{i-1} \in \operatorname{gr}^i_{\mathcal{F}} S$, $b + \mathcal{F}_{j-1} \in \operatorname{gr}^j_{\mathcal{F}} S$, set

$$(a + \mathcal{F}_{i-1}) \cdot (b + \mathcal{F}_{j-1}) := a \cdot b + \mathcal{F}_{i+j-1}.$$

The result is independent of the choice of representatives a, b for each residue class. We have defined the product of homogeneous elements of $\operatorname{gr}_{\mathcal{F}} S$, and we can extend this product to arbitrary elements by the distributive law. Then $\operatorname{gr}_{\mathcal{F}} S$ is a graded R-algebra.

Now let $\mathcal{F}$ be separated. For $f \in S \setminus \{0\}$, we call

$$f^* := f \cdot T^{\operatorname{ord}_{\mathcal{F}} f} \in \mathcal{R}_{\mathcal{F}} S$$

the *homogenization* of f, and

$$L_{\mathcal{F}} f := f + \mathcal{F}_{\operatorname{ord} f - 1} \in \operatorname{gr}_{\mathcal{F}} S$$

the *leading form* of f. For $f = 0$ we set $f^* = 0$ and $L_{\mathcal{F}} f = 0$.

Examples B.4.

(a) *Degree filtration.* In the situation of example B.3(a),

$$\operatorname{gr}^i_{\mathcal{F}} S = \mathcal{F}_i / \mathcal{F}_{i-1} = \bigoplus_{\rho \le i} S_\rho / \bigoplus_{\rho \le i-1} S_\rho \cong S_i \qquad (i \in \mathbb{Z}),$$

and we get a canonical isomorphism of graded R-algebras

$$\operatorname{gr}_{\mathcal{F}} S \cong S.$$

Under this isomorphism we identify the leading form $L_{\mathcal{F}} f = f + \mathcal{F}_{\operatorname{ord} f - 1}$ of an $f \in S \setminus \{0\}$ with the homogeneous component of f of largest degree, which we often call the "degree form" of f.
If $f = f_m + f_{m+1} + \cdots + f_d$ is the decomposition of f into homogeneous components $f_i \in S_i$ ($m \le d, f_m \neq 0, f_d \neq 0$), then the homogenization f^* of f in $\mathcal{R}_{\mathcal{F}} S$ has the form

$$(1) \qquad f^* = fT^d = (f_m T^m) T^{d-m} + \cdots + (f_{d-1} T^{d-1}) T + (f_d T^d).$$

We now consider the special case in which $S = R[X_1, \ldots, X_n]$ is a polynomial algebra. Then $\mathcal{F}_i = 0$ for $i < 0$, $\mathcal{F}_0 = R$, and

$$\mathcal{R}_{\mathcal{F}} S = R \oplus \mathcal{F}_1 T \oplus \mathcal{F}_2 T^2 \oplus \cdots \subset S[T] = R[T, X_1, \ldots, X_n].$$

The homogenization X_i^* of X_i is the element $X_i^* = X_i T \in \mathcal{F}_1 T$ ($i = 1, \ldots, n$), and one easily sees that

$$\mathcal{R}_{\mathcal{F}} S = R[T, X_1^*, \ldots, X_n^*],$$

where $\{T, X_1^*, \ldots, X_n^*\}$ are algebraically independent over R. In other words, $\mathcal{R}_{\mathcal{F}} S$ is a polynomial algebra over R in $T, X_1^*, \ldots, X_n^*$, and all these variables have degree 1. For $f \in R[X_1, \ldots, X_n]$, equation (1) becomes

$$(1') \qquad \begin{aligned} f^* = fT^{-m} = & f_m(X_1^*, \ldots, X_n^*)T^{d-m} + \cdots \\ & \cdots + f_{d-1}(X_1^*, \ldots X_n^*)T + f_d(X_1^*, \ldots X_n^*); \end{aligned}$$

i.e., f^* is what one usually understands by the homogenization of a polynomial (only one must write X_i instead of X_i^*):

$$f^*(T, X_1^*, \ldots, X_n^*) = T^{\deg(f)} f\left(\frac{X_1^*}{T}, \ldots, \frac{X_n^*}{T}\right).$$

(b) *I-adic filtration.* In the situation of B.3(b), $\operatorname{gr}_{\mathcal{F}}^k S = \{0\}$ for $k > 0$, and

$$\operatorname{gr}_{\mathcal{F}}^{-k} S = I^k / I^{k+1} \qquad \text{for } k \in \mathbb{N}.$$

Hence

$$\operatorname{gr}_{\mathcal{F}} S = \bigoplus_{k=0}^{\infty} I^k / I^{k+1}.$$

For $f \in I^k \setminus I^{k+1}$,

$$\operatorname{ord}_{\mathcal{F}} f = -k \qquad \text{and} \qquad L_{\mathcal{F}} f = f + I^{k+1} \in I^k / I^{k+1}.$$

Since it is sometimes annoying to work with negative orders, we may consider not only ascending, but also descending, filtrations, where nothing essential is changed. In order to deal with degree filtrations and I-adic filtrations simultaneously, we will stick with ascending filtrations.

In our present example, the Rees algebra has the form

$$\mathcal{R}_{\mathcal{F}} S = \bigoplus_{k \in \mathbb{N}} I^k T^{-k} \oplus \bigoplus_{k=1}^{\infty} S \cdot T^k.$$

One calls $\bigoplus_{k \in \mathbb{N}} I^k T^{-k}$ the "nonextended" Rees algebra of the I-adic filtration. It is more convenient for us to work with the "extended" Rees algebra $\mathcal{R}_{\mathcal{F}} S$.

In the special case of $S = R[X_1, \ldots, X_n]$, a polynomial algebra, and $I = (X_1, \ldots, X_n)$, we have $\operatorname{gr}_{\mathcal{F}} S \cong R[X_1, \ldots, X_n]$ with $\deg X_i = -1$ $(i = 1, \ldots, n)$. Furthermore,

$$\mathcal{R}_{\mathcal{F}} S = R[T, X_1, \ldots, X_n][X_1 T^{-1}, \ldots, X_n T^{-1}] \subset S[T, T^{-1}],$$

and therefore

$$\mathcal{R}_{\mathcal{F}} S = R[T, X_1^*, \ldots, X_n^*] \qquad \text{with} \qquad X_i^* := X_i T^{-1} \qquad (i = 1, \ldots, n).$$

Here, $\{T, X_1^*, \ldots, X_n^*\}$ is algebraically independent over R and $\deg T = 1$ as well as $\deg X_i^* = -1$ ($i = 1, \ldots, n$). If one writes $f \in S$ in the form $f = f_m + \cdots + f_d$ with homogeneous polynomials f_i of degree i, where $m \leq d$ and $f_m \neq 0$, then $\operatorname{ord}_{\mathcal{F}} f = -m$, and $L_{\mathcal{F}} f$ is identified in $\operatorname{gr}_{\mathcal{F}} S = R[X_1, \ldots, X_n]$ with f_m. Also

$$f^* = fT^{-m} = f_m(X_1^*, \ldots, X_n^*) + Tf_{m-1}(X_1^*, \ldots, X_n^*) + \cdots \\ \cdots + T^{d-m} f_d(X_1^*, \ldots, X_n^*).$$

This is a homogeneous polynomial of degree $-m$ using the given grading on $R[T, X_1^*, \ldots, X_n^*]$.

The following theorem gives simple but important relationships between the algebras explained here.

Theorem B.5. *Let $(S/R, \mathcal{F})$ be a filtered algebra. Then T is not a zerodivisor on $\mathcal{R}_{\mathcal{F}} S$. There is a canonical isomorphism of graded R-algebras*

$$\mathcal{R}_{\mathcal{F}} S / T \cdot \mathcal{R}_{\mathcal{F}} S \xrightarrow{\cong} \operatorname{gr}_{\mathcal{F}} S \qquad \left(\sum a_i T^i + T \cdot \mathcal{R}_{\mathcal{F}} S \mapsto \sum (a_i + \mathcal{F}_{i-1})\right)$$

and a canonical isomorphism of R-algebras

$$\mathcal{R}_{\mathcal{F}} S / (T-1) \cdot \mathcal{R}_{\mathcal{F}} S \xrightarrow{\cong} S \qquad \left(\sum a_i T^i + (T-1) \cdot \mathcal{R}_{\mathcal{F}} S \mapsto \sum a_i\right).$$

Proof. Since T is not a zerodivisor in the larger ring $S[T, T^{-1}]$, it is also not a zerodivisor in $\mathcal{R}_{\mathcal{F}} S$. The mapping $\alpha : \mathcal{R}_{\mathcal{F}} S \to \operatorname{gr}_{\mathcal{F}} S$ with $\alpha(\sum a_i T^i) = \sum (a_i + \mathcal{F}_{i-1})$ is well-defined (since $a_i \in \mathcal{F}_i$ for all $i \in \mathbb{Z}$) and is an epimorphism of graded R-algebras. Since $\alpha(T \cdot \sum a_i T^i) = \alpha(\sum a_i T^{i+1}) = \sum (a_i + \mathcal{F}_i) = 0$, we have $T \cdot \mathcal{R}_{\mathcal{F}} S \subset \ker \alpha$. Conversely, if $\alpha(\sum a_i T^i) = 0$, then $a_i \in \mathcal{F}_{i-1}$ for all $i \in \mathbb{Z}$. It follows that $\sum a_i T^{i-1}$ is already an element of $\mathcal{R}_{\mathcal{F}} S$ and therefore $\ker \alpha \subset T \cdot \mathcal{R}_{\mathcal{F}} S$. Hence $\ker \alpha = T \cdot \mathcal{R}_{\mathcal{F}} S$. By the homomorphism theorem, α induces an isomorphism

$$\mathcal{R}_{\mathcal{F}} S / T \cdot \mathcal{R}_{\mathcal{F}} S \xrightarrow{\cong} \operatorname{gr}_{\mathcal{F}} S.$$

The mapping $\beta : \mathcal{R}_{\mathcal{F}} S \to S$ with $\beta(\sum a_i T^i) = \sum a_i$ is an epimorphism of R-algebras, and $\beta((T-1)\sum a_i T^i) = \beta(\sum (a_i - a_{i+1}) T^{i+1}) = \sum a_i - \sum a_{i+1} = 0$; hence $(T-1) \cdot \mathcal{R}_{\mathcal{F}} S \subset \ker \beta$. Now suppose conversely that $\sum a_i T^i \in \ker \beta \setminus \{0\}$ is given and let $d := \max\{i \mid a_i \neq 0\}$. Then

$$\sum a_i T^i = \sum a_i T^i - \left(\sum a_i\right) T^d = \sum a_i T^i (1 - T^{d-i}) \in (T-1) \cdot \mathcal{R}_{\mathcal{F}} S,$$

and therefore $\ker \beta = (T-1) \cdot \mathcal{R}_{\mathcal{F}} S$. The isomorphism

$$\mathcal{R}_{\mathcal{F}} S / (T-1) \mathcal{R}_{\mathcal{F}} S \xrightarrow{\cong} S$$

then follows once again from the homomorphism theorem.

In other terminology, which will not be explained here, the theorem says that there is a "deformation" in which $\mathrm{gr}_{\mathcal{F}} S$ is the "fiber at the point $T = 0$" and S the "fiber at the point $T = 1$." One can use the theorem to infer "from $\mathrm{gr}_{\mathcal{F}} S$ to S." A first example is given by the following corollary, and another is given in B.10.

Corollary B.6. *Under the hypotheses of B.5, let $\mathcal{F}$ be separated and let $b_1, \dots, b_m$ be elements of S such that $\{L_{\mathcal{F}} b_1, \dots, L_{\mathcal{F}} b_m\}$ is a set of generators (a basis) of $\mathrm{gr}_{\mathcal{F}} S$ as an R-module. Then*

(a) *$\{b_1^*, \dots, b_m^*\}$ is a set of generators (a basis) of $\mathcal{R}_{\mathcal{F}} S$ as an $R[T]$-module.*
(b) *$\{b_1, \dots, b_m\}$ is a set of generators (a basis) of S as an R-module.*

Proof. (a) The epimorphism $\alpha : \mathcal{R}_{\mathcal{F}} S \to \mathrm{gr}_{\mathcal{F}} S$ described above maps b_i^* to $L_{\mathcal{F}} b_i$ $(i = 1, \dots, m)$. It follows therefore that

$$\mathcal{R}_{\mathcal{F}} S = R[T] \cdot b_1^* + \cdots + R[T] \cdot b_m^* + T \cdot \mathcal{R}_{\mathcal{F}} S.$$

We now consider $\mathcal{R}_{\mathcal{F}} S$ as a graded module over the graded ring $R[T]$. Since $\mathrm{gr}_{\mathcal{F}} S$ over R has a finite set of generators, the grading of $\mathrm{gr}_{\mathcal{F}} S$ is bounded below. Thus we have $\mathcal{F}_i/\mathcal{F}_{i-1} = 0$ for small i and therefore $\mathcal{F}_i = \mathcal{F}_{i-1} = \mathcal{F}_{i-2} = \cdots = 0$, since $\mathcal{F}$ is separated. Thus the gradings on $\mathcal{R}_{\mathcal{F}} S$ and on $\mathcal{R}_{\mathcal{F}} S/R[T]b_1^* + \cdots + R[T]b_m^*$ are also bounded below, and Nakayama's Lemma A.8 applies. It follows that

$$\mathcal{R}_{\mathcal{F}} S = R[T] \cdot b_1^* + \cdots + R[T] \cdot b_m^*.$$

It remains to show that $\{b_1^*, \dots, b_m^*\}$ are linearly independent over $R[T]$ when $\{L_{\mathcal{F}} b_1, \dots, L_{\mathcal{F}} b_m\}$ is a basis of $\mathrm{gr}_{\mathcal{F}} S$ over R. Suppose there were a relation

$$(2) \qquad \sum_{i=1}^{m} \rho_i b_i^* = 0 \qquad (\rho_i \in R[T], \text{ not all } \rho_i = 0).$$

Then there would also be such a relation with homogeneous $\rho_i \in R[T]$:

$$\rho_i = r_i T^{n_i} \qquad (r_i \in R, n_i + \mathrm{ord}_{\mathcal{F}} b_i \text{ independent of } i).$$

However, T is not a zerodivisor in $\mathcal{R}_{\mathcal{F}} S$. Therefore one can cancel the T's in B.3 until one of the coefficients ρ_i is no longer divisible by T. One goes now to $\mathrm{gr}_{\mathcal{F}} S$, and it follows from (2) that there is a nontrivial relation among $L_{\mathcal{F}} b_1, \dots, L_{\mathcal{F}} b_m$, a contradiction. Therefore $\{b_1^*, \dots, b_m^*\}$ is a basis for $\mathcal{R}_{\mathcal{F}} S$ over $R[T]$.

(b) The epimorphism $\beta : \mathcal{R}_{\mathcal{F}} S \to S$ maps b_i^* to b_i $(i = 1, \dots, m)$. Since $\ker \beta = (T - 1)\, \mathcal{R}_{\mathcal{F}} S$, the statement of (b) follows directly from that of (a).

Now let $(S/R, \mathcal{F})$ be a filtered algebra with a separated filtration and let $I \subset S$ be an ideal.

Definition B.7. The ideal $I^* \subset \mathcal{R}_\mathcal{F} S$ generated by all f^* with $f \in I$ is called the *homogenization* of I, and the ideal $\operatorname{gr}_\mathcal{F} I \subset \operatorname{gr}_\mathcal{F} S$ generated by all $L_\mathcal{F} f$ with $f \in I$ is called the *associated graded ideal* of I.

The residue class algebra $\bar{S} := S/I$ can be given a filtration $\bar{\mathcal{F}} = \{\bar{\mathcal{F}}_i\}_{i\in\mathbb{Z}}$ as follows:

$$\bar{\mathcal{F}}_i := (\mathcal{F}_i + I)/I \qquad (i \in \mathbb{Z})$$

is the image of $\mathcal{F}_i$ in $\bar{S}$. We call $\bar{\mathcal{F}} = \{\bar{\mathcal{F}}_i\}_{i\in\mathbb{Z}}$ the *residue class filtration* associated with the filtration $\mathcal{F}$ of S/R.

Theorem B.8. *Let $(S/R, \mathcal{F})$ be a filtered algebra with separated filtration $\mathcal{F}$ and let $\bar{\mathcal{F}}$ be the associated residue class filtration on $\bar{S} := S/I$. Then there is a canonical isomorphism of graded $R[T]$-algebras*

$$\mathcal{R}_{\bar{\mathcal{F}}} \bar{S} \cong \mathcal{R}_\mathcal{F} S/I^*$$

and a canonical isomorphism of graded R-algebras

$$\operatorname{gr}_{\bar{\mathcal{F}}} \bar{S} \cong \operatorname{gr}_\mathcal{F} S/\operatorname{gr}_\mathcal{F} I.$$

Proof. Denote the residue class of $a \in S$ in $\bar{S}$ by $\bar{a}$. The map $\alpha : \mathcal{R}_\mathcal{F} S \to \mathcal{R}_{\bar{\mathcal{F}}} \bar{S}$ with $\alpha(\sum a_i T^i) = \sum \bar{a}_i T^i$ is an epimorphism of graded $R[T]$-algebras and $I^* \subset \ker \alpha$. For $a_i T^i \in \mathcal{R}_\mathcal{F} S$ $(a_i \in \mathcal{F}_i)$, we have $\alpha(a_i T^i) = 0$ exactly when $a_i \in I$. Therefore $\operatorname{ord}_\mathcal{F} a_i =: k \le i$ and $a_i T^i = a_i^* T^{i-k}$ with $a_i^* \in I^*$. It follows that $\ker \alpha = I^*$, and by the homomorphism theorem α induces an isomorphism $\mathcal{R}_\mathcal{F} S/I^* \cong \mathcal{R}_{\bar{\mathcal{F}}} \bar{S}$.

The map $\beta : \operatorname{gr}_\mathcal{F} S \to \operatorname{gr}_{\bar{\mathcal{F}}} \bar{S}$ given by $\beta(\sum a_i + \mathcal{F}_{i-1}) = \sum \bar{a}_i + \bar{\mathcal{F}}_{i-1}$ is an epimorphism of graded R-algebras, and it is clear that $\operatorname{gr}_\mathcal{F} I \subset \ker \beta$. Conversely, if $\beta(a + \mathcal{F}_{i-1}) = 0$ for some $a \in \mathcal{F}_i \setminus \mathcal{F}_{i-1}$, then $\bar{a} \in \bar{\mathcal{F}}_{i-1}$ and therefore $a \in I + \mathcal{F}_{i-1}$. Then $a + \mathcal{F}_{i-1}$ is already represented by an element $b \in I$: $a + \mathcal{F}_{i-1} = b + \mathcal{F}_{i-1}$, $b \notin \mathcal{F}_{i-1}$. Therefore $a + \mathcal{F}_{i-1} = L_\mathcal{F} b \in \operatorname{gr}_\mathcal{F} I$ and $\ker \beta = \operatorname{gr}_\mathcal{F} I$. The second isomorphism of the theorem follows at once, again from the homomorphism theorem.

The following theorems are concerned with the generation of ideals in filtered algebras.

Theorem B.9. *Let $(S/R, \mathcal{F})$ be a filtered algebra with separated filtration $\mathcal{F}$ and let $I \subset S$ be an ideal. Furthermore, let $f_1, \dots, f_n$ be elements of I with $\operatorname{gr}_\mathcal{F} I = (L_\mathcal{F} f_1, \dots, L_\mathcal{F} f_n)$. If the residue class filtration $\bar{\mathcal{F}}$ of $\mathcal{F}$ on $S/(f_1, \dots, f_n)$ is separated, then*

$$I = (f_1, \dots, f_n).$$

Proof. Let $g \in I \setminus \{0\}$ and let $\operatorname{ord}_\mathcal{F} g =: a$. Then there is a representation

$$L_\mathcal{F} g = \sum_{i=1}^n L_\mathcal{F} h_i \cdot L_\mathcal{F} f_i$$

with $h_i \in S$, $\operatorname{ord}_{\mathcal{F}} h_i + \operatorname{ord}_{\mathcal{F}} f_i = a$ $(i = 1, \ldots, n)$. Then

$$g - \sum_{i=1}^{n} h_i f_i \in I \qquad \text{and} \qquad \operatorname{ord}_{\mathcal{F}}(g - \sum h_i f_i) < a.$$

By induction one shows that

$$g \in \bigcap_{i \leq n} (f_1, \ldots, f_n) + \mathcal{F}_i,$$

and because $\bar{\mathcal{F}}$ is separated, it follows that $g \in (f_1, \ldots, f_n)$.

Corollary B.10. *Let $(S/R, \mathcal{F})$ be a filtered algebra with separated filtration $\mathcal{F}$. For each finitely generated ideal $I \subset S$, assume that the residue class filtration of S/I is separated. If $\operatorname{gr}_{\mathcal{F}} S$ is a Noetherian ring, then S is also a Noetherian ring.*

For the next theorem we need a lemma about nonzerodivisors.

Lemma B.11. *Let $(S/R, \mathcal{F})$ be a filtered algebra with separated filtration and for $f \in S$, let $L_{\mathcal{F}} f$ be a nonzerodivisor on $\operatorname{gr}_{\mathcal{F}} S$. Then f^* is a nonzerodivisor on $\mathcal{R}_{\mathcal{F}} S$, and f is a nonzerodivisor on S. Also, for every $g \in S$:*

(a) $\operatorname{ord}_{\mathcal{F}}(g \cdot f) = \operatorname{ord}_{\mathcal{F}} g + \operatorname{ord}_{\mathcal{F}} f$.
(b) $L_{\mathcal{F}}(g \cdot f) = L_{\mathcal{F}} g \cdot L_{\mathcal{F}} f$.
(c) $(g \cdot f)^* = g^* \cdot f^*$.

Proof. We prove (a) first, and observe that we need to do only the case $g \neq 0$. If $\operatorname{ord}_{\mathcal{F}} f =: a$ and $\operatorname{ord}_{\mathcal{F}} g =: b$, then $L_{\mathcal{F}} f = f + \mathcal{F}_{a-1}$, $L_{\mathcal{F}} g = g + \mathcal{F}_{b-1}$, and

$$(3) \qquad L_{\mathcal{F}} f \cdot L_{\mathcal{F}} g = fg + \mathcal{F}_{a+b-1}, \qquad f \cdot g \in \mathcal{F}_{a+b}.$$

Since $L_{\mathcal{F}} f$ is a nonzerodivisor on $\operatorname{gr}_{\mathcal{F}} S$, we have $f \cdot g \notin \mathcal{F}_{a+b-1}$, and it follows that $\operatorname{ord}_{\mathcal{F}}(g \cdot f) = \operatorname{ord}_{\mathcal{F}} g + \operatorname{ord}_{\mathcal{F}} f$.

(b) follows immediately from (3), and (c) by definition of homogenization from (a). It is also clear that f^* cannot be a zerodivisor on $\mathcal{R}_{\mathcal{F}} S$. If f were a zerodivisor on S, then by (b), $L_{\mathcal{F}} f$ would also be a zerodivisor on $\operatorname{gr}_{\mathcal{F}} S$.

Theorem B.12. *Let $(S/R, \mathcal{F})$ be a filtered algebra, $I = (f_1, \ldots, f_n)$ an ideal of S, and suppose the residue class filtrations on $S/(f_1, \ldots, f_i)$, for $i = 1, \ldots, n-1$, are separated. Also suppose that the image of $L_{\mathcal{F}} f_{i+1}$ in $\operatorname{gr}_{\mathcal{F}} S/(L_{\mathcal{F}} f_1, \ldots, L_{\mathcal{F}} f_i)$, for $i = 0, \ldots, n-1$ is a nonzerodivisor in these rings. Then:*

(a) $\operatorname{gr}_{\mathcal{F}} I = (L_{\mathcal{F}} f_1, \ldots, L_{\mathcal{F}} f_n)$.
(b) $I^* = (f_1^*, \ldots, f_n^*)$.

Also, the images of f_{i+1}^* and f_{i+1} are nonzerodivisors in $\mathcal{R}_{\mathcal{F}} S/(f_1^*, \ldots, f_i^*)$ respectively $S/(f_1, \ldots, f_i)$ $(i = 0, \ldots, n-1)$.

Proof. Let $\bar{S} := S/(f_1)$, $\bar{I} := I/(f_1)$ and let $\bar{f}_i$ denote the image of f_i in $\bar{S}$ $(i = 2, \ldots, n)$. Also, let $\bar{\mathcal{F}}$ be the residue class filtration of $\mathcal{F}$ on $\bar{S}$. By Lemma B.11, $\operatorname{gr}_{\mathcal{F}}(f_1) = (L_{\mathcal{F}} f_1)$ and $(f_1)^* = (f_1^*)$. Because $L_{\mathcal{F}} f_1$ is not a zerodivisor of $\operatorname{gr}_{\mathcal{F}} S$, f_1^* respectively f_1 is not a zerodivisor of $\mathcal{R}_{\mathcal{F}} S$ respectively S.

By B.8 it follows that

$$\mathcal{R}_{\bar{\mathcal{F}}} \bar{S} \cong \mathcal{R}_{\mathcal{F}} S/(f_1^*) \qquad \text{and} \qquad \operatorname{gr}_{\bar{\mathcal{F}}} \bar{S} \cong \operatorname{gr}_{\mathcal{F}} S/(L_{\mathcal{F}} f_1).$$

We identify $\bar{I}^*$ with $I^*/(f_1^*)$ and $\operatorname{gr}_{\bar{\mathcal{F}}} \bar{I}$ with $\operatorname{gr}_{\mathcal{F}} I/(L_{\mathcal{F}} f_1)$ under these isomorphisms. Thus $\bar{f}_i^*$ corresponds to the residue class of f_i^* modulo (f_1^*) and $L_{\bar{\mathcal{F}}} \bar{f}_i$ to the residue class of $L_{\mathcal{F}} f_i$ modulo $(L_{\mathcal{F}} f_1)$ $(i = 2, \ldots, n)$. By induction,

$$\bar{I}^* = (\bar{f}_2^*, \ldots, \bar{f}_n^*) \qquad \text{and} \qquad \operatorname{gr}_{\bar{\mathcal{F}}} \bar{I} = (L_{\bar{\mathcal{F}}} \bar{f}_2, \ldots L_{\bar{\mathcal{F}}} \bar{f}_n),$$

where $\bar{f}_{i+1}^*$ is not a zerodivisor modulo $(\bar{f}_2^*, \ldots, \bar{f}_i^*)$ and $\bar{f}_{i+1}$ is not a zerodivisor modulo $(\bar{f}_2, \ldots, \bar{f}_i)$. The assertions of the theorem then follow immediately.

Exercises

1. Let $S := R[[X_1, \ldots, X_n]]$ be the algebra of formal power series in indeterminates $X_1, \ldots, X_n$ over a ring R and $I := (X_1, \ldots, X_n)$ the ideal of S generated by $X_1, \ldots, X_n$. Let $\mathcal{F}$ denote the I-adic filtration on S. Show that $\mathcal{F}$ is separated and that

$$\operatorname{gr}_{\mathcal{F}} \cong R[X_1, \ldots, X_n] \qquad \text{(a polynomial algebra).}$$

2. In the situation of Exercise 1, describe the Rees algebra $\mathcal{R}_{\mathcal{F}} S$ and the homogenization $f^* \in \mathcal{R}_{\mathcal{F}} S$ of a power series $f \in S$.

C

Rings of Quotients. Localization

The construction of rings of quotients corresponds to the way we extend the integers to the rational numbers, or more generally to the extension of an integral domain to its quotient field. Here we also need to look at rings of quotients of graded and filtered algebras.

A ring of quotients can be constructed for an arbitrary ring R and a multiplicatively closed subset $S \subset R$. We call S *multiplicatively closed* if $1 \in S$ and whenever $a, b \in S$, then also $ab \in S$. The most important special cases are the following:

Examples C.1. Some examples of multiplicatively closed sets:

(a) The set of all nonzero elements of an integral domain.
(b) The set of all nonzerodivisors of a ring.
(c) For $\mathfrak{p} \in \operatorname{Spec}(R)$, the set $R \setminus \mathfrak{p}$.
(d) For $f \in R$, the set $\{f^0, f^1, f^2, \ldots\}$ of powers of f.

Definition C.2. A *ring of quotients of R with respect to the denominator set S* is a pair (R_S, ϕ), where R_S is a ring, $\phi : R \to R_S$ is a ring homomorphism, and the following conditions are satisfied:

(a) For each $s \in S$, $\phi(s)$ is a unit in R_S.
(b) (Universal property of rings of quotients). If $\psi : R \to T$ is any ring homomorphism such that $\psi(s)$ is a unit in T for each $s \in S$, then there exists exactly one ring homomorphism $h : R_S \to T$ with $\psi = h \circ \phi$:

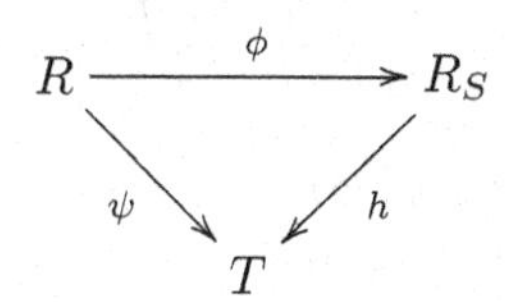

The pair (R_S, ϕ), if it exists, is unique up to isomorphism in the following sense: If (R_S^*, ϕ^*) is also a ring of quotients of R with denominator set S, then there exists an isomorphism $h : R_S \to R_S^*$ with $\phi^* = h \circ \phi$.

In fact, a homomorphism h of this kind exists by C.2(b), and similarly, there exists a homomorphism $h^* : R_S^* \to R_S$ with $\phi = h^* \circ \phi^*$. Then $\phi = h^* \circ (h \circ \phi) = (h^* \circ h) \circ \phi$, and because of the uniqueness condition in C.2(b) we get $h^* \circ h = \mathrm{id}_{R_S}$. By symmetry we also have $h \circ h^* = \mathrm{id}_{R_S^*}$, and therefore h is an isomorphism.

From now on we will call R_S the *ring of quotients of R with respect to S*, and $\phi : R \to R_S$ the *canonical mapping into the ring of quotients.*

Next we will show the *existence of the ring of quotients.* Let $\{X_s\}_{s \in S}$ be a family of indeterminates. In the polynomial ring $R[\{X_s\}]$ consider the ideal I generated by all elements of the form

$$sX_s - 1 \qquad (s \in S).$$

We set $R_S := R[\{X_s\}]/I$ and denote by ϕ the composition of the canonical injection $R \hookrightarrow R[\{X_s\}]$ followed by the canonical epimorphism $R[\{X_s\}] \to R[\{X_s\}]/I$. The residue class of X_s in R_S will be denoted by $\frac{1}{s}$. Then $\phi(s) \cdot \frac{1}{s} = 1$ in R_S; i.e., $\phi(s)$ is a unit in R_S for all $s \in S$.

Given T as in C.2(b), there exists a ring homomorphism

$$\alpha : R[\{X_s\}] \to T$$

with $\alpha(r) = \psi(r)$ for all $r \in R$ and $\alpha(X_s) = \psi(s)^{-1}$ for all $s \in S$. Now we have $\alpha(sX_s - 1) = \alpha(s)\alpha(X_s) - 1 = \psi(s)\psi(s)^{-1} - 1 = 0$ $(s \in S)$, and therefore $\alpha(I) = 0$. By the homomorphism theorem, α induces a ring homomorphism

$$h : R[\{X_s\}]/I \to T.$$

By the construction of α we have $\psi = h \circ \phi$, since for $r \in R$ we have $(h \circ \phi)(r) = \alpha(r) = \psi(r)$. Since R_S is generated over R by the elements $\frac{1}{s}$ with $s \in S$, it is clear that only one h with $h \circ \phi = \psi$ can exist: For $s \in S$ with image $\bar{s} \in R_S$ we have $h(\bar{s}) \cdot h(\frac{1}{s}) = h(1) = 1$. Therefore $h(\frac{1}{s}) = h(\bar{s})^{-1} = \psi(s)^{-1}$.

We have shown that rings of quotients exist.

For $r \in R$, $s \in S$ we write

$$\frac{r}{s} := \phi(r) \cdot \frac{1}{s}.$$

To multiply two such fractions:

$$\frac{r_1}{s_1} \cdot \frac{r_2}{s_2} = \frac{r_1 r_2}{s_1 s_2}. \tag{1}$$

In fact, $\phi(s_1 s_2) \cdot \frac{1}{s_1} \cdot \frac{1}{s_2} = \phi(s_1) \cdot \phi(s_2) \cdot \frac{1}{s_1} \cdot \frac{1}{s_2} = 1$. Conversely, $\phi(s_1 s_2) \cdot \frac{1}{s_1 s_2} = 1$ and because $\phi(s_1 s_2)$ is a unit in R_S, it follows that $\frac{1}{s_1} \cdot \frac{1}{s_2} = \frac{1}{s_1 s_2}$. Then $\frac{r_1 r_2}{s_1 s_2} = \phi(r_1) \cdot \phi(r_2) \cdot \frac{1}{s_1} \cdot \frac{1}{s_2} = \frac{r_1}{s_1} \cdot \frac{r_2}{s_2}$.

In particular, $1 = \frac{1}{1} = \frac{s}{s}$ for all $s \in S$, and this is the identity element for multiplication in R_S. The element $\frac{r}{1}$ is called the "improper" fraction associated with r.

Fractions are added according to the following rule:

$$\frac{r_1}{s_1} + \frac{r_2}{s_2} = \frac{r_1 s_2 + s_1 r_2}{s_1 s_2}. \tag{2}$$

By (1), $\frac{r_1}{s_1} = \frac{r_1 s_2}{s_1 s_2}$, $\frac{r_2}{s_2} = \frac{s_1 r_2}{s_1 s_2}$, and hence

$$\frac{r_1 s_2 + s_1 r_2}{s_1 s_2} = (\phi(r_1)\phi(s_2) + \phi(s_1)\phi(r_2)) \cdot \frac{1}{s_1 s_2} = \frac{r_1 s_2}{s_1 s_2} + \frac{s_1 r_2}{s_1 s_2} = \frac{r_1}{s_1} + \frac{r_2}{s_2}.$$

In particular, $0 = \frac{0}{s}$ (for all $s \in S$) is the identity element for addition in R_S. Since we calculate according to the usual "rules for fractions," and since $R_S = R[\{\frac{1}{s}\}]$, it is clear by formulas (1) and (2) that

$$R_S = \left\{\frac{r}{s} \mid r \in R, s \in S\right\}. \tag{3}$$

For the canonical homomorphism $\phi : R \to R_S$ $(r \mapsto \frac{r}{1})$, we have the following facts:

Theorem C.3. (a) $\ker \phi = \{r \in R \mid$ *there exists an* $s \in S$ *with* $s \cdot r = 0\}$.
(b) ϕ *is injective if and only if* S *contains no zerodivisors of* R.
(c) ϕ *is bijective if and only if* S *consists entirely of units of* R.

Proof. (a) Let $J := \{r \in R \mid \exists s \in S \ : \ s \cdot r = 0\}$. This is obviously an ideal of R, and it is clear that $J \subset \ker \phi$, because from $s \cdot r = 0$ it follows that $\phi(s) \cdot \phi(r) = 0$ and hence $\phi(r) = 0$, since $\phi(s)$ is a unit in R_S.

Conversely, let $r \in \ker \phi$; therefore $r \in R \cap (\{sX_s - 1\})$. Then there is an equation

$$r = \sum_{i=1}^{n} f_i \cdot (s_i X_{s_i} - 1) \qquad (f_i \in R[\{X_s\}]). \tag{4}$$

Let $f_i = f_i(X_{t_1}, \ldots, X_{t_m})$ $(i = 1, \ldots, n;\ t_1, \ldots, t_m \in S)$. Then there exists $(\alpha_1, \ldots, \alpha_m) \in \mathbb{N}_+^m$ such that

$$t_1^{\alpha_1} \cdots t_m^{\alpha_m} r = \sum_{i=1}^{n} g_i(t_1 X_{t_1}, \ldots, t_m X_{t_m}) \cdot (s_i X_{s_i} - 1). \tag{5}$$

Now let $\bar{R} := R/J$. We denote by $\bar{f}$ the image of $f \in R[\{X_s\}]$ in $\bar{R}[\{X_s\}]$ (under the canonical mapping). From (5) we then get in $\bar{R}[\{X_s\}]$ the equation

$$\bar{t}_1^{\alpha_1} \cdots \bar{t}_m^{\alpha_m} \bar{r} = \sum_{i=1}^{n} \bar{g}_i(\bar{t}_1 X_{t_1}, \ldots, \bar{t}_m X_{t_m}) \cdot (\bar{s}_i X_{s_i} - 1). \tag{6}$$

The $\bar{R}$-homomorphism $\beta : \bar{R}[\{Y_s\}] \to \bar{R}[\{X_s\}]$ with $Y_s \mapsto \bar{s} X_s$ is injective: For if

$$\sum r_{v_1 \ldots v_n} (\bar{s}_1 X_{s_1})^{v_1} \cdots (\bar{s}_n X_{s_n})^{v_n} = 0 \qquad (r_{v_1 \ldots v_n} \in R,\ s_i \in S),$$

then $\bar{r}_{v_1 \ldots v_n} \bar{s}_1^{v_1} \cdots \bar{s}_n^{v_n} = 0$ for all $(v_1, \ldots, v_n)$. Hence $r_{v_1 \ldots v_n} s_1^{v_1} \cdots s_n^{v_n} \in J$. Then there exists $s \in S$ with $s s_1^{v_1} \cdots s_n^{v_n} r_{v_1 \ldots v_n} = 0$. Therefore $r_{v_1 \ldots v_n} \in J$, and hence $\bar{r}_{v_1 \ldots v_n} = 0$.

By (6) and the injectivity of β we have the following equation in $\bar{R}[\{Y_s\}]$:

$$\bar{t}_1^{\alpha_1} \cdots \bar{t}_m^{\alpha_m} \bar{r} = \sum_{i=1}^{n} \bar{g}_i(Y_{t_1}, \dots, Y_{t_m}) \cdot (Y_{s_i} - 1).$$

Setting all Y_{s_i} equal to 1 gives $\bar{t}_1^{\alpha_1} \cdots \bar{t}_m^{\alpha_m} \bar{r} = 0$, and as in the above argument we get $r \in J$. This proves (a), and (b) follows immediately from (a).

(c) If ϕ is bijective, then S consists only of units, since $\phi(s)$ is a unit for $s \in S$. Conversely, if S is a set of units of R, then (R, id_R) satisfies the conditions of Definition C.2. By the uniqueness of rings of quotients (up to isomorphism) pointed out above, ϕ is bijective.

Under the conditions stated in (b), we can regard R as a subring of R_S, in which $r \in R$ is identified with the "improper" fraction $\frac{r}{1} \in R_S$. The statement (c) can also be interpreted as follows: If S already consists entirely of units, then there is no need to construct any fractions in order to "make" the elements of S units.

Corollary C.4 (Equality of Fractions). *For $\frac{r_1}{s_1}, \frac{r_2}{s_2} \in R_S$, we have $\frac{r_1}{s_1} = \frac{r_2}{s_2}$ if and only if there exists an $s \in S$ such that*

$$s \cdot (s_2 r_1 - s_1 r_2) = 0.$$

If S contains no zerodivisors, then $\frac{r_1}{s_1} = \frac{r_2}{s_2}$ if and only if $s_2 r_1 - s_1 r_2 = 0$, which is the usual equality of fractions.

Proof. First note that $\frac{r_1}{s_1} = \frac{r_2}{s_2}$ is the same as $\frac{s_2 r_1 - s_1 r_2}{s_1 s_2} = 0$. Since $\frac{1}{s_1 s_2}$ is a unit in R_S, the last equation is equivalent to $\frac{s_2 r_1 - s_1 r_2}{s_1 s_2} = \phi(s_2 r_1 - s_1 r_2) = 0$. The statement now follows from C.3(a).

Examples C.5. (a) If R is an integral domain and $S := R \setminus \{0\}$, then R_S is a field. We will write $R_S =: Q(R)$ and call $Q(R)$ the *quotient field* of R. We have

$$R \subset Q(R) = \left\{ \frac{r}{s} \mid r, s \in R,\ s \neq 0 \right\}.$$

For an arbitrary multiplicatively closed subset $S \subset R$ with $0 \notin S$, we have $R \subset R_S \subset Q(R)$.

(b) If R is an arbitrary ring and S the set of all nonzerodivisors of R, we also write $R_S =: Q(R)$ even in this case, but we call $Q(R)$ the *full ring of quotients* of R. We always have $R \subset Q(R)$.

(c) If $\mathfrak{p} \in \operatorname{Spec} R$ and $S := R \setminus \mathfrak{p}$, we will write $R_S =: R_\mathfrak{p}$ and call $R_\mathfrak{p}$ the *localization of R at the prime ideal* $\mathfrak{p}$, or the *local ring at* $\mathfrak{p}$. We will see soon that $R_\mathfrak{p}$ is in fact a local ring. In general, $\phi : R \to R_\mathfrak{p}$ is not injective.

(d) If $f \in R$ and $S = \{f^0, f^1, f^2, \dots\}$, we will write $R_S =: R_f$. We have

$$R_f = \left\{ \frac{r}{f^v} \mid r \in R,\ v \in \mathbb{N} \right\},$$

and $\phi : R \to R_f$ is injective if and only if f is not a zerodivisor on R. A special case of this kind, where X is an indeterminate, is

$$R[X]_X = \left\{ \frac{1}{X^v} \sum_{\alpha=0}^{n} r_\alpha X^\alpha \mid r_\alpha \in R,\ v, n \in \mathbb{N} \right\},$$

the ring of Laurent polynomials in X over R, which we already met in Appendix B.

The next theorems give information about the ideals in a ring of quotients.

Theorem C.6. *Let $I \subset R$ be an ideal and $S \subset R$ a multiplicatively closed set. Then*

$$I_S := \left\{ \frac{x}{s} \in R_S \mid x \in I,\ s \in S \right\}$$

is an ideal of R_S, and every ideal of R_S is of this form for some suitable ideal I of R. Moreover, $I_S \neq R_S$ if and only if $I \cap S = \emptyset$.

Proof. It is easy to check that I_S is an ideal of R_S. If $I_S = R_S$, then $1 = \frac{x}{s}$ for some $x \in I$, $s \in S$. Then there exists a $t \in S$ with $t(x - s) = 0$. Then $ts = tx \in S \cap I$ and hence $S \cap I \neq \emptyset$. Conversely, if there exists an $s \in S \cap I$, then $1 = \frac{s}{s} \in I_S$ and $I_S = R_S$.

If J is an ideal of R_S, then $I := \phi^{-1}(J)$ is an ideal of R and $I_S \subset J$, since $\phi(I) \subset J$. If $\frac{x}{s} \in J$, then $\frac{x}{1} \in J$ and therefore $x \in I$. Hence $J \subset I_S$ and therefore $J = I_S$.

For the ideal I_S we also write IR_S, since I_S is generated by $\phi(I)$.

Corollary C.7. *If R is a Noetherian ring, then so is R_S.*

Under the conditions of C.6 let $\bar{S}$ be the image of S in R/I. Of course, $\bar{S}$ is multiplicatively closed. For $r \in R$, denote the residue class of r in R/I by $\bar{r}$. According to the universal property C.2(b), there is a ring homomorphism

$$h : R_S \to (R/I)_{\bar{S}} \qquad \left(h\left(\frac{r}{s}\right) = \frac{\bar{r}}{\bar{s}} \right),$$

which is obviously surjective.

The next theorem states the permutability of rings of quotients and residue class ring constructions.

Theorem C.8. *We have* $\ker h = I_S$ *and therefore (by the homomorphism theorem)*

$$R_S/I_S \cong (R/I)_{\bar{S}} \qquad \left(\frac{r}{s} + I_S \mapsto \frac{\bar{r}}{\bar{s}} \right).$$

Proof. For $\frac{r}{s} \in R_S$ we have $h(\frac{r}{s}) = 0$ if and only if there exists $t \in S$ such that $\bar{t}\bar{r} = 0$. But this is equivalent to $\frac{r}{s} = \frac{rt}{st} \in I_S$.

Theorem C.9. $\operatorname{Spec} R_S = \{\mathfrak{p}_S \mid \mathfrak{p} \in \operatorname{Spec} R,\ \mathfrak{p} \cap S = \emptyset\}$.

Proof. For $\mathfrak{P} \in \operatorname{Spec} R_S$ we have $\mathfrak{p} := \phi^{-1}(\mathfrak{P}) \in \operatorname{Spec} R$, $\mathfrak{p} \cap S = \emptyset$, and $\mathfrak{P} = \mathfrak{p}_S$. Conversely, if $\mathfrak{p} \in \operatorname{Spec} R$ with $\mathfrak{p} \cap S = \emptyset$, we have $\mathfrak{p}_S \neq R_S$, and by C.8,

$$R_S/\mathfrak{p}_S \cong (R/\mathfrak{p})_{\bar{S}}\,.$$

Since $R/\mathfrak{p}$ is an integral domain, $(R/\mathfrak{p})_{\bar{S}}$ is also, and therefore $\mathfrak{p}_S$ is a prime ideal of R_S.

For $\mathfrak{p} \in \operatorname{Spec} R$ such that $\mathfrak{p} \cap S = \emptyset$,

$$\phi^{-1}(\mathfrak{p}_S) = \mathfrak{p},$$

so that these prime ideals of R are in one-to-one correspondence with the elements of $\operatorname{Spec}(R_S)$. In fact, if $r \in R$ and $\frac{r}{1} = \phi(r) \in \mathfrak{p}_S$, then $\frac{r}{1} = \frac{p}{s}$ for some $p \in \mathfrak{p}$, $s \in S$. Then there exists a $t \in S$ with $tsr = tp \in \mathfrak{p}$. Since $ts \in S$, $ts \notin \mathfrak{p}$. Hence $r \in \mathfrak{p}$.

Corollary C.10. *Let* $\mathfrak{p} \in \operatorname{Spec} R$. *Then* $R_\mathfrak{p}$ *is a local ring with maximal ideal* $\mathfrak{p}R_\mathfrak{p}$, *and* $R_\mathfrak{p}/\mathfrak{p}R_\mathfrak{p} \cong Q(R/\mathfrak{p})$. *The elements of* $\operatorname{Spec} R_\mathfrak{p}$ *are in one-to-one correspondence with the prime ideals of* R *that are contained in* $\mathfrak{p}$.

Proof. The last statement of the corollary was just established, and it implies in particular that $R_\mathfrak{p}$ is a local ring. The formula $R_\mathfrak{p}/\mathfrak{p}R_\mathfrak{p} \cong Q(R/\mathfrak{p})$ is a special case of C.8.

We come now to our first application of rings of quotients.

Theorem C.11. *For any ring* R, $\bigcap_{\mathfrak{p}\in\operatorname{Spec} R} \mathfrak{p}$ *is the set of all nilpotent elements of* R.

Proof. Let $f \in R$ be nilpotent, i.e, $f^n = 0$ for some $n \in \mathbb{N}$. Then $f^n \in \mathfrak{p}$ for all $\mathfrak{p} \in \operatorname{Spec} R$ and therefore $f \in \bigcap_{\mathfrak{p}\in\operatorname{Spec} R} \mathfrak{p}$. Conversely, suppose an $f \in R$ such that $f \in \bigcap_{\mathfrak{p}\in\operatorname{Spec} R} \mathfrak{p}$ is given, and let $S := \{f^0, f^1, f^2, \ldots\}$. Since $\mathfrak{p} \cap S \neq \emptyset$ for all $\mathfrak{p} \in \operatorname{Spec} R$, $R_S = R_f$ is a ring with empty spectrum according to C.9, and hence is the zero ring. In particular, $\frac{1}{1} = \frac{0}{1}$ and there exists $n \in N$ with $f^n \cdot 1 = 0$; i.e., f is nilpotent.

Corollary C.12. *If a ring has exactly one prime ideal, then this prime ideal consists of the nilpotent elements of the ring.*

If $G = \bigoplus_{n\in\mathbb{Z}} G_n$ is a graded algebra over a ring R and $S \subset G$ is a multiplicatively closed set consisting of homogeneous elements of G, then G_S is a graded R-algebra with homogeneous components

$$(G_S)_n := \left\{\frac{x}{s} \in G_S \mid x \text{ homogeneous}, \deg x - \deg s = n\right\}.$$

The condition $\deg x - \deg s = n$ is independent of any particular representation of the fraction $\frac{x}{s} \neq 0$, as the rule for equality of fractions shows immediately. Also, one can easily check that $\{(G_S)_n\}_{n\in\mathbb{Z}}$ is a grading of G_S:

$$G_S = \bigoplus_{n\in\mathbb{Z}} (G_S)_n, \qquad (G_S)_p \cdot (G_S)_q \subset (G_S)_{p+q}.$$

In particular,

$$(G_S)_0 := \left\{ \frac{x}{s} \in G_S \mid x \text{ homogeneous}, \deg x = \deg s \right\}$$

is a subring of G_S. This subring will be denoted by $G_{(S)}$.

If $\mathfrak{p} \in \operatorname{Spec} R$ is a homogeneous ideal and S the set of all homogeneous elements $s \in G$ such that $s \notin \mathfrak{p}$, then G_S is a subring of $G_\mathfrak{p}$. In particular, $G_{(S)} \subset G_\mathfrak{p}$. We write $G_{(\mathfrak{p})}$ for $G_{(S)}$ in this case. It is obvious that $G_{(\mathfrak{p})}$ is itself a local ring with maximal ideal

$$\mathfrak{m} := \left\{ \frac{x}{s} \in G_S \mid x \in \mathfrak{p} \text{ homogeneous}, \deg x = \deg s \right\}.$$

We call $G_{(\mathfrak{p})}$ the *homogeneous localization* of G at the prime $\mathfrak{p}$.

Now let S/R be an algebra, $I \subset S$ an ideal, and $N \subset S$ a multiplicatively closed subset. We view S with the I-adic and the ring of quotients S_N with the I_N-adic filtration. The corresponding Rees algebras will be denoted by $\mathcal{R}_I\, S$ and $\mathcal{R}_{I_N}\, S_N$, and the associated graded algebras by $\operatorname{gr}_I S$ and $\operatorname{gr}_{I_N} S_N$ (cf. B.4(b)). Since S is contained in $\mathcal{R}_I\, S$ as the homogeneous component of degree 0, $N \subset \mathcal{R}_I\, S$. We let $\bar{N}$ denote the image of N in $S/I = \operatorname{gr}_I^0 S \subset \operatorname{gr}_I S$.

Theorem C.13. *There are canonical isomorphisms of graded R-algebras*

$$\mathcal{R}_{I_N}\, S_N \cong (\mathcal{R}_I\, S)_N,$$
$$\operatorname{gr}_{I_N} S_N \cong (\operatorname{gr}_I S)_{\bar{N}}.$$

Proof. The canonical homomorphism $S \to S_N$ maps I^k to $(I^k)_N = (I_N)^k$ (for $k \in \mathbb{N}$). We get a homomorphism $\mathcal{R}_I\, S \to \mathcal{R}_{I_N}\, S_N$ of graded R-algebras, and by the universal property of rings of quotients also a homomorphism $\rho : (\mathcal{R}_I\, S)_N \to \mathcal{R}_{I_N}\, S_N$ of graded R-algebras. For $k \in \mathbb{N}$ we will identify $(\mathcal{R}_I^{-k}\, S)_N = (I^k T^{-k})_N = I_N^k T^{-k}$ with $\mathcal{R}_{I_N}^{-k}\, S_N$ using ρ. Similarly for $(\mathcal{R}_I^k\, S)_N$ ($k \in \mathbb{N}$). Therefore ρ is an isomorphism.

Since $\operatorname{gr}_I S \cong \mathcal{R}_I\, S / T\mathcal{R}_I\, S$ (B.5), the statement about the associated graded algebras follows from the permutability of rings of quotients and residue class rings (C.8).

Example C.14. Let $S = K[X_1, \ldots, X_n]$ be a polynomial algebra over a field K, $\mathfrak{M} := (X_1, \ldots, X_n)$, and $M := S \setminus \mathfrak{M}$. Then $\mathcal{R}_\mathfrak{M}\, S = K[T, X_1^*, \ldots, X_n^*]$ is a polynomial algebra (B.4(b)). Thus, by using $X_i = TX_i^*$, $S = K[X_1, \ldots, X_n]$ can be embedded in $K[T, X_1^*, \ldots, X_n^*]$; hence also $M \subset K[T, X_1^*, \ldots, X_n^*]$. By

C.13 the Rees algebra of the local ring $S_{\mathfrak{M}}$ with respect to the filtration given by its maximal ideal $\mathfrak{M}S_{\mathfrak{M}}$ is

$$\mathcal{R}_{\mathfrak{M}S_{\mathfrak{M}}} S_{\mathfrak{M}} \cong K[T, X_1^*, \dots, X_n^*]_M.$$

For the associated graded algebra, we have, by B.4(b),

$$\mathrm{gr}_{\mathfrak{M}} S \cong K[X_1, \dots, X_n].$$

Since the image $\bar{M}$ of M in the field $S/\mathfrak{M}$ consists entirely of units, we also have

$$\mathrm{gr}_{\mathfrak{M}S_{\mathfrak{M}}} S_{\mathfrak{M}} \cong K[X_1, \dots, X_n].$$

Exercises

Let R be a ring, $S \subset R$ a multiplicatively closed subset. Let $Id(R)$ denote the set of all ideals of R. Prove the following statements:

1. If R is a unique factorization domain, so is R_S.
2. If $I \in Id(R)$, then

$$S(I) := \{r \in R \mid \text{there exists an } s \in S \text{ with } sr \in I\}$$

is also an ideal of R. Let $Id_S(R) := \{I \in Id(R) \mid S(I) = I\}$. The mapping

$$Id_S(R) \to Id(R_S) \qquad (I \to I_S)$$

is bijective. For $\mathfrak{p} \in \operatorname{Spec} R$, we have $S(\mathfrak{p}) = \mathfrak{p}$ if and only if $\mathfrak{p} \cap S = \emptyset$.

D

The Chinese Remainder Theorem

We will derive a more ring-theoretic version of this fundamental theorem of number theory. For us it plays an essential role in the intersection theory of algebraic curves.

Two *proper*[1] ideals I_1, I_2 of a ring R are called *relatively prime* (comaximal) if $I_1 + I_2 = R$.

Theorem D.1. *Let $I_1, \dots, I_n$ $(n > 1)$ be pairwise relatively prime ideals of a ring R. Then the canonical ring homomorphism*

$$\alpha : R \to R/I_1 \times \cdots \times R/I_n,$$

$$r \mapsto (r_1 + I_1, \dots, r_n + I_n),$$

is an epimorphism with $\ker(\alpha) = \bigcap_{k=1}^n I_k$.

Proof. The statement about the kernel of α follows immediately from the definitions of α and the direct product of rings. We show the surjectivity of α by induction on n.

Let $n = 2$ and let $(r_1 + I_1, r_2 + I_2) \in R/I_1 \times R/I_2$ be given. By hypothesis we have an equation $1 = a_1 + a_2$ with $a_k \in I_k$ $(i = 1, 2)$, and it follows that

$$a_1 \equiv 1 \bmod I_2, \qquad a_2 \equiv 1 \bmod I_1.$$

Set $r := r_2 a_1 + r_1 a_2$. Then $r \equiv r_k \bmod I_k$, and this shows that α is surjective for $n = 2$.

Now suppose $n > 2$ and the theorem has already been proved for fewer than n pairwise relatively prime ideals. For each $(r_1 + I_1, \dots, r_n + I_n) \in R/I_1 \times \cdots \times R/I_n$ there is then an element $r' \in R$ with $r' \equiv r_k \bmod I_k$ for $k = 1, \dots, n-1$. We will show that $I_1 \cap \cdots \cap I_{n-1}$ is relatively prime to I_n. Since the theorem has already been shown for $n = 2$, there is an $r \in R$ with $r \equiv r' \bmod I_1 \cap \cdots \cap I_{n-1}$, $r \equiv r_n \bmod I_n$. Then we have $r \equiv r_k \bmod I_k$ for $k = 1, \dots, n$ and the theorem is proved in general.

By hypothesis we have equations

$$1 = a_1 + a_3 = a_2 + a_3' \qquad (a_k \in I_k\ (k = 1, 2, 3),\ a_3' \in I_3).$$

[1] A *proper ideal* is one that is not equal to R.

It follows that

$$1 = a_1a_2 + (a_2 + a'_3)a_3 + a_1a'_3 \in (I_1 \cap I_2) + I_3,$$

and hence $I_1 \cap I_2$ and I_3 are relatively prime. By induction it follows that $I_1 \cap \cdots \cap I_{n-1}$ and I_n are also relatively prime.

Corollary D.2 (Chinese Remainder Theorem). *If $I_1, \ldots, I_n$ are pairwise relatively prime and $\bigcap_{k=1}^n I_k = (0)$, then*

$$R \cong R/I_1 \times \cdots \times R/I_n.$$

A special case of this is of course the classical Chinese remainder theorem of elementary number theory:

$$\mathbb{Z}/(p_1^{\alpha_1} \cdots p_n^{\alpha_n}) \cong \mathbb{Z}/(p_1^{\alpha_1}) \times \cdots \times \mathbb{Z}/(p_n^{\alpha_n})$$

if $p_1, \ldots, p_n$ are distinct prime numbers and $\alpha_i \in \mathbb{N}_+$ $(i = 1, \ldots, n)$. Another variation is given in the following theorem.

Theorem D.3. *Let R be a ring for which* $\operatorname{Spec}(R) = \{\mathfrak{p}_1, \ldots, \mathfrak{p}_n\}$ *is finite and consists only of maximal ideals. Then the canonical ring homomorphism*

$$\beta : R \to R_{\mathfrak{p}_1} \times \cdots \times R_{\mathfrak{p}_n} \qquad \left(r \mapsto \left(\frac{r}{1}, \ldots, \frac{r}{1}\right)\right)$$

is an isomorphism. Here, $R_{\mathfrak{p}_i} \cong R/\mathfrak{q}_i$, where $\mathfrak{q}_i := \ker(R \to R_{\mathfrak{p}_i})$ for $i = 1, \ldots, n$.

Proof. For each $i \in \{1, \ldots, n\}$ there is a canonical injection $R/\mathfrak{q}_i \hookrightarrow R_{\mathfrak{p}_i}$, and $R_{\mathfrak{p}_i}$ has exactly one prime ideal, namely $\mathfrak{p}_i R_{\mathfrak{p}_i}$ (C.10). The ideal $\mathfrak{p}_i R_{\mathfrak{p}_i}$ consists purely of nilpotent elements (C.12), and therefore $\mathfrak{p}_i/\mathfrak{q}_i$ consists purely of nilpotent elements of $R/\mathfrak{q}_i$; i.e., for each $x \in \mathfrak{p}_i$ there exists $\rho \in \mathbb{N}_+$ with $x^\rho \in \mathfrak{q}_i$. From this it follows that $\mathfrak{p}_i$ is the only prime ideal of R that contains $\mathfrak{q}_i$: If $\mathfrak{q}_i \subset \mathfrak{p}_j$ for a $j \in \{1, \ldots, n\}$, then $x^\rho \in \mathfrak{p}_j$ for each $x \in \mathfrak{p}_i$ and certain $\rho \in \mathbb{N}_+$; but then $x \in \mathfrak{p}_j$, $\mathfrak{p}_i \subset \mathfrak{p}_j$, and $\mathfrak{p}_i = \mathfrak{p}_j$, since both ideals are maximal.

Now it follows that $\mathfrak{q}_i$ and $\mathfrak{q}_j$, for $i \neq j$, are relatively prime: None of the maximal ideals of R contains both ideals. We show furthermore that $\bigcap_{k=1}^n \mathfrak{q}_k = (0)$. For each $x \in \bigcap_{k=1}^n \mathfrak{q}_k$ and each $k \in \{1, \ldots, n\}$ there exists an $r_k \in R \setminus \mathfrak{p}_k$ with $r_k x = 0$ (C.3(a)). With $I := (r_1, \ldots, r_n)$ we have $Ix = 0$. But $I \not\subset \mathfrak{p}_k$ for $k = 1, \ldots, n$, and hence $I = R$. Since $1 \in I$, we have $x = 1x = 0$.

It now follows directly from D.2 that

$$R \cong R/\mathfrak{q}_1 \times \cdots \times R/\mathfrak{q}_n,$$

and it remains to show that the canonical injection $R/\mathfrak{q}_k \hookrightarrow R_{\mathfrak{p}_k}$ is bijective. But $R/\mathfrak{q}_k$ is a local ring with maximal ideal $\bar{\mathfrak{p}}_k := \mathfrak{p}_k/\mathfrak{q}_k$. Because of the permutability of localization and residue class ring constructions we have (see C.3(c) and C.8)

$$R/\mathfrak{q}_k \cong (R/\mathfrak{q}_k)_{\bar{\mathfrak{p}}_k} \cong R_{\mathfrak{p}_k}/\mathfrak{q}_k R_{\mathfrak{p}_k} \cong R_{\mathfrak{p}_k} \qquad (k = 1, \ldots, n).$$

Corollary D.4. *Let A be a finite-dimensional algebra over a field K. Then* $\operatorname{Spec}(A)$ *consists of only finitely many elements $\mathfrak{p}_1, \ldots, \mathfrak{p}_n$, and these are all maximal ideals of A. Furthermore, we have*

$$A \cong A_{\mathfrak{p}_1} \times \cdots \times A_{\mathfrak{p}_n} \cong A/\mathfrak{q}_1 \times \cdots \times A/\mathfrak{q}_n$$

with $\mathfrak{q}_k = \ker(A \to A_{\mathfrak{p}_k})$ *for $k = 1, \ldots, n$, and*

$$\dim_K A = \sum_{i=1}^{n} \dim_K A_{\mathfrak{p}_i} = \sum_{i=1}^{n} \dim_K A/\mathfrak{q}_i.$$

Proof. It is enough to prove that $\operatorname{Spec}(A)$ consists of only finitely many maximal ideals, for then we can apply D.3. The following lemma shows this; in fact, it shows a little more.

Lemma D.5. *If A is a finite-dimensional algebra over a field K, then* $\operatorname{Spec}(A)$ *consists of at most* $\dim_K(A)$ *elements, and these are all maximal ideals of A.*

Proof. Every $\mathfrak{p} \in \operatorname{Spec}(A)$ is maximal because $A/\mathfrak{p}$ is an integral domain that is finite-dimensional over K; this implies that $A/\mathfrak{p}$ is a field.

If $\mathfrak{p}_1, \ldots \mathfrak{p}_{k+1}$ are distinct, then $\mathfrak{p}_1 \cap \cdots \cap \mathfrak{p}_k \not\subset \mathfrak{p}_{k+1}$, for we can choose $x_i \in \mathfrak{p}_i \setminus \mathfrak{p}_{k+1}$ and then $x_1 \cdots x_k \in \mathfrak{p}_1 \cap \cdots \cap \mathfrak{p}_k$, but $x_1 \cdots x_k \notin \mathfrak{p}_{k+1}$. It follows that $\operatorname{Spec}(A)$ consists of at most $\dim_K A$ elements, for otherwise we get a chain

$$A \supsetneqq \mathfrak{p}_1 \supsetneqq (\mathfrak{p}_1 \cap \mathfrak{p}_2) \supsetneqq (\mathfrak{p}_1 \cap \mathfrak{p}_2 \cap \mathfrak{p}_3) \supsetneqq \cdots$$

of subspaces of A whose length exceeds $\dim_K A$, which is impossible.

A version of the Chinese remainder theorem for complete Noetherian rings will be given in K.11.

Exercises

1. Check that the isomorphisms

$$R \cong R/\mathfrak{q}_1 \times \cdots \times R/\mathfrak{q}_n \cong R_{\mathfrak{p}_1} \times \cdots \times R_{\mathfrak{p}_n}$$

   in the proof of D.3 actually give β.
2. How many units, zero divisors, and nilpotents are there in the rings $\mathbb{Z}/(2006)$ and $\mathbb{Z}/(2007)$?

E

Noetherian Local Rings and Discrete Valuation Rings

Certain local rings are assigned to the points of an algebraic curve and to the intersection points of two curves. In this appendix, we bring together the basic facts about such rings, and we study especially discrete valuation rings.

The following lemma is fundamental for the theory of local rings.

Nakayama's Lemma E.1. *Let R be a ring and suppose I is an ideal of R contained in the intersection of all the maximal ideals of R. Let M be an R-module and $U \subset M$ a submodule of M. If M/U is finitely generated and if $M = U + IM$, then $M = U$.*

Proof. The module $N := M/U$ is finitely generated and satisfies $N = IN$. We will show that $N = \langle 0 \rangle$. Suppose this were not the case. Let $\{n_1, \dots, n_t\}$ be a minimal set of generators for N $(t > 0)$. Then there is a relation

$$n_t = \sum_{i=1}^{t} a_i n_i \qquad (a_i \in I,\ i = 1, \dots, t),$$

and therefore

$$(1 - a_t) n_t = \sum_{i=1}^{t-1} a_i n_i .$$

Since a_t is in every maximal ideal of R, the element $1 - a_t$ is a unit of R. Hence $n_t \in \langle n_1, \dots, n_{t-1} \rangle$, in contradiction to the minimality of t.

Using Nakayama's lemma, questions about the generators of modules and ideals over local rings can be reduced to the corresponding questions for vector spaces. For a finitely generated module M over a ring R we will denote by $\mu(M)$ the number of elements in a smallest set of generators for M. A set of generators of M is called *unshortenable* if no proper subset generates the module M, and *minimal* if it consists of $\mu(M)$ elements.

Corollary E.2. *Let R be a local ring with maximal ideal $\mathfrak{m}$ and residue field $\mathfrak{k} := R/\mathfrak{m}$. Let M be a finitely generated R-module and let $m_1, \dots, m_t \in M$. The following are equivalent:*

(a) $M = \langle m_1, \dots, m_t \rangle$.

(b) *The residue classes of the* m_i *in* $M/\mathfrak{m}M$ *are a set of generators of the* $\mathfrak{k}$*-vector space* $M/\mathfrak{m}M$.

Proof. We need to show only (b) $\Rightarrow$ (a). Let $\bar{m}_i$ be the residue class of m_i $(i = 1,\dots,t)$. From $M/\mathfrak{m}M = \langle \bar{m}_1,\dots,\bar{m}_t\rangle$ it follows that $M = \langle m_1,\dots,m_t\rangle + \mathfrak{m}M$, and then from Nakayama's lemma we have $M = \langle m_1,\dots,m_t\rangle$.

From well-known theorems about vector spaces we immediately have the following facts.

Corollary E.3. *Under the assumptions of E.2,*

(a) $\mu(M) = \dim_{\mathfrak{k}} M/\mathfrak{m}M$.
(b) $m_1,\dots,m_t$ *form a minimal system of generators for* M *if and only if their residue classes* $\bar{m}_1,\dots,\bar{m}_t$ *are a basis for* $M/\mathfrak{m}M$ *as a* $\mathfrak{k}$*-vector space.*
(c) *If* $\{m_1,\dots,m_t\}$ *is a minimal set of generators for* M *and if there is a relation* $\sum_{i=1}^t r_i m_i = 0$ *for some* $r_i \in R$, *then* $r_i \in \mathfrak{m}$ $(i = 1,\dots,t)$.
(d) *Every set of generators of* M *contains a minimal set of generators. Every unshortenable set of generators is minimal.*
(e) *Elements* $m_1,\dots,m_r$ *are part of a minimal set of generators for* M *if and only if their residue classes in* $M/\mathfrak{m}M$ *are linearly independent over* $\mathfrak{k}$.

These statements can be used in the special case of ideals in Noetherian local rings, since these are of course finitely generated R-modules. In particular, they can be applied to the maximal ideal $\mathfrak{m}$ of R.

Definition E.4. For a Noetherian local ring R with maximal ideal $\mathfrak{m}$, we call

$$\operatorname{edim} R := \mu(\mathfrak{m})$$

the *embedding dimension* of R.

By E.3(a) we have $\operatorname{edim} R = \dim_{\mathfrak{k}}(\mathfrak{m}/\mathfrak{m}^2)$. It is trivial to see that $\operatorname{edim} R = 0$ if and only if R is a field.

Connected with Nakayama's lemma are the Artin–Rees lemma and the Krull intersection theorem.

Artin-Rees Lemma E.5. *Let* R *be a Noetherian ring,* $I \subset R$ *an ideal,* M *a finitely generated* R*-module, and* $U \subset M$ *a submodule. There exists a* $k \in \mathbb{N}$ *such that for all* $n \in \mathbb{N}$ *we have*

$$I^{n+k}M \cap U = I^n \cdot (I^k M \cap U).$$

Proof. Let $\mathcal{R}_I^+ R := \bigoplus_{n\in\mathbb{N}} I^n$ be the (nonextended) Rees ring of R with respect to I (cf. B.4(b)) and let $\mathcal{R}_I^+ M := \bigoplus_{n\in\mathbb{N}} I^n M$ be the corresponding graded $\mathcal{R}_I^+ R$-module.

Since R is Noetherian and therefore I is finitely generated, $\mathcal{R}_I^+ R$ is generated as an algebra by finitely many elements of degree 1, namely by the elements of a finite set of generators for the ideal I. By the Hilbert basis theorem for rings, $\mathcal{R}_I^+ R$ is then Noetherian. Since M is finitely generated as an R-module, $\mathcal{R}_I^+ M$ is finitely generated as an $\mathcal{R}_I^+ R$-module.

Now we set $U_n := I^n M \cap U$ $(n \in \mathbb{N})$ and $\bar{U} := \oplus_{n\in\mathbb{N}} U_n$. Then $\bar{U}$ is a homogeneous submodule of the $\mathcal{R}_I^+ R$-module $\mathcal{R}_I^+ M$. By the Hilbert basis theorem for modules, $\bar{U}$ has a finite set of generators $\{v_1, \dots, v_s\}$, and the v_i can be chosen to be homogeneous elements of $\mathcal{R}_I^+ M$. Let $m_i := \deg v_i$ and $k := \operatorname{Max}\{m_1, \dots, m_s\}$. We will show that $U_{n+k} = I^n U_k$ for all $n \in \mathbb{N}$, which is exactly the statement of the lemma.

Obviously, $I^n U_k \subset U_{n+k}$. Conversely, if $u \in U_{n+k}$ is given, it can be written in the form $u = \sum_{i+1}^{s} \rho_i v_i$ with homogeneous elements $\rho_i \in I^{n+k-m_i}$, and therefore $u \in I^n U_k$.

Krull Intersection Theorem E.6. *Let R be a Noetherian ring, $I \subset R$ an ideal, M a finitely generated R-module and $\widetilde{M} := \bigcap_{n\in\mathbb{N}} I^n M$. Then we have*

$$\widetilde{M} = I \cdot \widetilde{M}.$$

Proof. Use E.5 with $U := \widetilde{M}$. We have

$$\widetilde{M} = I^{k+1} M \cap \widetilde{M} = I(I^k M \cap \widetilde{M}) = I \cdot \widetilde{M}.$$

Corollary E.7. *If I is contained in the intersection of all the maximal ideals of R and if $U \subset M$ is a submodule, then $\bigcap_{n\in\mathbb{N}} (I^n M + U) = U$. In particular, $\bigcap_{n\in\mathbb{N}} I^n M = \langle 0 \rangle$.*

Proof. Set $N := M/U$ and $\widetilde{N} := \bigcap_{n\in\mathbb{N}} I^n N$. Then $\widetilde{N} = I\widetilde{N}$ by E.6 and $\widetilde{N} = \langle 0 \rangle$ by Nakayama. Hence $\bigcap_{n\in\mathbb{N}} (I^n M + U) = U$.

Corollary E.8. *Let R be a Noetherian local ring with maximal ideal $\mathfrak{m}$ and let $I \subset \mathfrak{m}$. Then $\bigcap_{n\in\mathbb{N}} I^n + J = J$ for every ideal J of R. In particular, $\bigcap_{n\in\mathbb{N}} \mathfrak{m}^n + J = J$.*

If R is a ring, we call a system $\mathfrak{p}_0 \subset \mathfrak{p}_1 \subset \cdots \subset \mathfrak{p}_n$ of elements $\mathfrak{p}_i \in \operatorname{Spec} R$ a *chain of prime ideals* if $\mathfrak{p}_{i-1} \neq \mathfrak{p}_i$ for $i = 1, \dots, n$. The chain of prime ideals is said to have length n.

Definition E.9. The *Krull dimension* $\dim R$ of a ring R is the supremum of the lengths of all chains of prime ideals.

Examples E.10.

(a) We have $\dim R = 0$ if and only if $\operatorname{Spec} R = \operatorname{Max} R$. Consequently, $\dim A = 0$ for every finite-dimensional algebra over a field K (D.5). For a local ring we have $\dim R = 0$ if and only if $\operatorname{Spec} R$ consists of exactly one element.

(b) We have $\dim R = 1$ if and only if $\operatorname{Spec} R$ consists of only maximal and minimal prime ideals, and at least one minimal prime ideal is not maximal. Examples of rings of Krull dimension 1 are $\mathbb{Z}$, $K[X]$, and $K[[X]]$ for a field K, and also the coordinate rings of affine algebraic curves (1.15). For a local ring R, $\dim R = 1$ if and only if $\operatorname{Spec} R$ consists of only the maximal ideal $\mathfrak{m}$ and minimal prime ideals $\neq \mathfrak{m}$. If one localizes $\mathbb{Z}$, $K[X]$, or the coordinate ring of an affine algebraic curve at a maximal ideal, then one gets a local ring of dimension 1.

(c) We have $\dim K[X, Y] = 2$ by 1.14. An example of a ring of infinite Krull dimension is the polynomial ring in infinitely many variables over a field K. There are even Noetherian rings of infinite Krull dimension.

Next we examine a special class of Noetherian local rings.

Definition E.11. A Noetherian local ring R with $\operatorname{edim} R = \dim R = 1$ is called a *discrete valuation ring.*

Examples of this are $\mathbb{Z}_{(p)}$ (p a prime number), $K[X]_{(f)}$ (f irreducible), and $K[[X]]$ (K a field). The next theorem gives specific information on the structure of discrete valuation rings.

Theorem E.12. *If R is a discrete valuation ring with maximal ideal $\mathfrak{m} = (\pi)$, then*

(a) *R is an integral domain and every $r \in R \setminus \{0\}$ has a unique representation*

$$r = \epsilon \cdot \pi^n \qquad (\epsilon \in R \text{ a unit}, n \in \mathbb{N}).$$

In particular, R is a unique factorization domain and π is (up to associates) the only prime element of R.

(b) *Every ideal $I \neq (0)$ of R is of the form $I = (\pi^n)$ for some $n \in \mathbb{N}$ uniquely determined by I. In particular, R is a principal ideal domain.*

(c) $\operatorname{Spec} R$ *consists of* $\mathfrak{m}$ *and* (0).

Proof. (a) By the Krull intersection theorem E.8 we have

$$\bigcap_{n \in \mathbb{N}} \mathfrak{m}^n = \bigcap_{n \in \mathbb{N}} (\pi^n) = (0).$$

There is therefore an $n \in \mathbb{N}$ such that $r \in (\pi^n)$, $r \notin (\pi^{n+1})$, and hence r can be written in the form $r = \epsilon \cdot \pi^n$ for some unit $\epsilon \in R$.

The element π is not nilpotent, for if it were, then every element of the maximal ideal $\mathfrak{m}$ would be nilpotent and $\mathfrak{m}$ would be the only prime ideal of R, in contradiction to the assumption that $\dim R = 1$. Suppose $s = \eta \cdot \pi^m$ is another element of $R \setminus \{0\}$ (η a unit, $m \in \mathbb{N}$); then $rs = \epsilon\eta \cdot \pi^{n+m} \neq 0$, and therefore R is an integral domain. Since π generates a prime ideal, π is a prime element of R.

If we also write $r = \epsilon_0 \cdot \pi^{n_0}$ with a unit $\epsilon_0 \in R$ and $n_0 \in \mathbb{N}$, then we must have $n_0 \leq n$. From $\epsilon \cdot \pi^{n-n_0} = \epsilon_0$, it then follows that $n = n_0$ and $\epsilon = \epsilon_0$. Hence (a) is proved.

(b) Choose an element $\epsilon \cdot \pi^n$ in I with n minimal ($\epsilon \in R$ a unit). Then $\pi^n \in I$. Every other element in $I \setminus \{0\}$ is of the form $\eta \cdot \pi^m$ for some unit $\eta \in R$ and some $m \geq n$. It follows that $I = (\pi^n)$.

(c) Since R is an integral domain by (a), (0) is a prime ideal of R. The ideals (π^n) with $n > 1$ are not prime ideals. Therefore $\operatorname{Spec} R$ consists only of (0) and $\mathfrak{m}$.

If K is the quotient field of a discrete valuation ring R with maximal ideal $\mathfrak{m} = (\pi)$, then every element $x \in K \setminus \{0\}$ can be written uniquely in the form

$$x = \epsilon \cdot \pi^n \qquad (\epsilon \in R \text{ a unit, } n \in \mathbb{Z}).$$

We set $v_R(x) := n$ and call n the *value* of x with respect to R. Furthermore, we set $v_R(0) := \infty$. Then $v_R : K \to \mathbb{Z} \cup \{\infty\}$ is a surjective mapping with the following properties:

(a) $v_R(x) = \infty$ if and only if $x = 0$.
(b) $v_R(x \cdot y) = v_R(x) + v_R(y)$ for all $x, y \in K$.
(c) $v_R(x + y) \geq \min\{v_R(x), v_R(y)\}$ for all $x, y \in K$.

A mapping v from a field K to $\mathbb{Z} \cup \{\infty\}$ that satisfies these conditions is called a (nontrivial) *discrete valuation* on K, and $R := \{x \in K \mid v(x) \geq 0\}$ is called the discrete valuation ring associated with v. (The trivial valuation maps all of K^* to 0 and 0 to ∞.) If $\pi \in R$ is an element with $v(\pi) = 1$, it follows easily from the valuation axioms that every ideal $I \neq (0)$ of R is of the form $I = (\pi^n)$ for some $n \in \mathbb{N}$. In particular, R is a Noetherian local ring with maximal ideal $\mathfrak{m} = (\pi)$; i.e., a discrete valuation ring according to Definition E.11.

Given a discrete valuation ring R, it is naturally the discrete valuation ring associated with v_R. In addition to (c), we have for every discrete valuation ring the following fact:
(c′) If $v_R(x) \neq v_R(y)$ for $x, y \in R$, then $v_R(x + y) = \min\{v_R(x), v_R(y)\}$.

The following dimension formula will frequently be used.

Theorem E.13. *Let R be a discrete valuation ring with maximal ideal $\mathfrak{m}$. Assume R contains a field k and the composite mapping $k \hookrightarrow R \to R/\mathfrak{m}$ is surjective. Then for every $x \in R$,*

$$\dim_k R/(x) = v_R(x).$$

Proof. Let π be a prime element of R and suppose at first $x \neq 0$. Then $x = \epsilon \cdot \pi^{v_R(x)}$ for some unit $\epsilon \in R$. There is a chain of ideals

$$(x) = (\pi^{v_R(x)}) \subset (\pi^{v_R(x)-1}) \subset (\pi^2) \subset (\pi) \subset R = (\pi^0),$$

and therefore

$$(1) \qquad \dim_k R/(x) = \sum_{i=0}^{v_R(x)-1} \dim_k (\pi^i)/(\pi^{i+1}).$$

The k-linear mapping

$$R \to (\pi^i)/(\pi^{i+1}), \qquad r \mapsto r\pi^i + (\pi^{i+1}),$$

is surjective and has (π) as its kernel. Therefore, $(\pi^i)/(\pi^{i+1}) \cong R/(\pi) \cong k$, and hence $\dim_k(\pi^i)/(\pi^{i+1}) = 1$ for all $i \in \mathbb{N}$. The proposition then follows from (1).

From this we also get that $\dim_k R = \infty$; i.e., the theorem is true for $x = 0$.

Theorem E.14. *Every discrete valuation ring is a maximal subring of its quotient field.*

Proof. Let R be a discrete valuation ring with quotient field K and let v be the associated valuation. If S is a ring with $R \subset S \subset K$, $R \neq S$, then S contains an element x with $v(x) < 0$. Noting that we can multiply by an element $r \in R$ of value $-v(x) - 1$, we get an element in S of value -1. By taking powers we get elements of every value in $\mathbb{Z}$. Hence for every $x \in K \setminus \{0\}$ we can find an $s \in S$ with $v(x) = v(s)$. Then $v(\frac{x}{s}) = 0$; hence $\frac{x}{s} \in R$ and $x \in S$. Therefore $S = K$.

Exercises

1. Show that
   (a) If K is a field, then the rings

   $$K[X]_{(f)} \quad (f \in K[X] \text{ irreducible}) \quad \text{and} \quad K[X^{-1}]_{(X^{-1})}$$

   are all discrete valuation rings containing K with quotient field $K(X)$.
   (b) The rings $\mathbb{Z}_{(p)}$ (p a prime number) are all discrete valuation rings with quotient field $\mathbb{Q}$.
2. Let R be a Noetherian local domain with quotient field K. For every $x \in K \setminus \{0\}$, let $x \in R$ or $x^{-1} \in R$. Then R is a discrete valuation ring.

F

Integral Ring Extensions

Integral ring extensions are analogous to finite field extensions in field theory. Both theories can be developed simultaneously, and for economical reasons perhaps one should do so in basic algebra courses. As a reward, one gets a simple proof of Hilbert's Nullstellensatz (see F.15 and F.16), a fundamental result of algebraic geometry.

Let S be a ring and $R \subset S$ a subring.

Theorem F.1. *For an element $x \in S$, the following statements are equivalent:*

(a) *There is a monic polynomial $f \in R[X]$ with $\deg f > 0$ such that $f(x) = 0$.*
(b) *$R[x]$ is a finitely generated R-module.*
(c) *There is a subring $S' \subset S$ with $R[x] \subset S'$ such that S' is finitely generated as an R-module.*

Proof. (a) $\Rightarrow$ (b). Let $f = X^n + r_1X^{n-1} + \cdots + r_n$ ($r_i \in R$, $n > 0$). Every $g \in R[X]$ has a remainder of degree $\leq n-1$ when g is divided by f, i.e., $g = q \cdot f + r$ ($q, r \in R[X]$, $\deg r \leq n-1$). By substituting x for X we get $g(x) \in R + Rx + \cdots + Rx^{n-1}$ and therefore

$$R[x] = R + Rx + \cdots + Rx^{n-1}.$$

(b) $\Rightarrow$ (c) is trivial. We now show (c) $\Rightarrow$ (a). Let $\{w_1, \ldots, w_n\}$ be a system of generators for S' as an R-module. Write

$$xw_i = \sum_{j=1}^{n} \rho_{ij} w_j \qquad (i = 1, \ldots, n;\ \rho_{ij} \in R).$$

then $\sum_{j=1}^{n}(x\delta_{ij} - \rho_{ij})w_j = 0$, and we have $\det(x\delta_{ij} - \rho_{ij})w_k = 0$ for $k = 1, \ldots, n$ by Cramer's rule. Since $1 \in S'$ can be written as a linear combination of the w_k, it follows that $\det(x\delta_{ij} - \rho_{ij}) = 0$. Then $f(X) := \det(X\delta_{ij} - \rho_{ij}) \in R[X]$ is a monic polynomial f of degree n with $f(x) = 0$ (the characteristic polynomial of the multiplication map $\mu_x : S' \to S'$).

Definition F.2.

(a) For $x \in S$, if the equivalent conditions of F.1 are satisfied, we say that x is *integral over* R. An equation $f(x) = 0$ as in F.1a) is called an *equation of integral dependence* or an *integral dependence relation* for x over R.
(b) The set $\overline{R}$ of all elements of S that are integral over R is called the *integral closure* of R in S.
(c) S is called an *integral extension* of R if $\overline{R} = S$.
(d) R is called *integrally closed* in S if $\overline{R} = R$.

Examples F.3.

(a) If S is finitely generated as an R-module, then S is integral over R by F.1.
(b) Every unique factorization domain R is integrally closed in its field of fractions $Q(R)$, and in particular, this is true for $\mathbb{Z}$, for polynomial rings $K[X_1, \ldots, X_n]$ over a field K, and for every discrete valuation ring (E.12). In fact:
Let $x \in Q(R)$ be integral over R and let

$$x^n + r_1 x^{n-1} + \cdots + r_n = 0$$

be an equation of integral dependence for x. Write $x = \frac{r}{s}$ in lowest terms with $r, s \in R$. Then we have

$$r^n + r_1 s r^{n-1} + \cdots + r_n s^n = 0,$$

and it follows that s is divisor of r. This is possible only if s is a unit of R. Therefore $x \in R$.

Corollary F.4. *If $x_1, \ldots, x_n \in S$ are integral over R, then $R[x_1, \ldots, x_n]$ is finitely generated as an R-module and is therefore integral over R.*

This follows from F.1 by induction on n.

Corollary F.5 (Transitivity of Integral Extensions). *Let $R \subset S \subset T$ be rings. If T is integral over S and S is integral over R, then T is integral over R.*

Proof. For $x \in T$, let $x^n + s_1 x^{n-1} + \cdots + s_n = 0$ be an equation of integral dependence for x over S. Then $R[s_1, \ldots, s_n]$ is finite over R by F.4, and $R[s_1, \ldots, s_n, x]$ is finite over $R[s_1, \ldots, s_n]$ and therefore also finite over R. Then using F.1 with $S' = R[s_1, \ldots, s_n, x]$, it follows that x is integral over R.

Corollary F.6. *The integral closure $\overline{R}$ of R in S is a subring of S that is integrally closed in S.*

Proof. If $x, y \in \overline{R}$, then $R[x, y]$ is a finitely generated R-module by F.4. Therefore $x \pm y$ and $x \cdot y$ are integral over R, and it follows that $\overline{R}$ is a subring of S. If $z \in S$ is integral over $\overline{R}$, then $z \in \overline{R}$ by F.5.

Whereas the previous results are analogous to the concepts of "algebraic" and "algebraic closure" in field theory, we come now to some facts that are specific to ring theory.

Theorem F.7. *Let R be a Noetherian integral domain that is integrally closed in its field of fractions K. Let L be a finite separable field extension of K, and let S be the integral closure of R in L. Then S is finitely generated as an R-module, and in particular is a Noetherian ring.*

Proof. Choose a primitive element x of L/K. Let $f \in K[X]$ be its minimal polynomial over K. Because $K = Q(R)$, it can be written in the form

$$f = X^n + \frac{r_1}{r} X^{n-1} + \cdots + \frac{r_n}{r} \qquad (r, r_i \in R).$$

Then rx has minimal polynomial

$$X^n + r_1 X^{n-1} + \cdots + r_n r^{n-1} \in R[X],$$

and is likewise a primitive element of L/K. We can therefore assume that $x \in S$ and $r = 1$.

Now let $x_1, \ldots, x_n$ be the conjugates of x over K, that is, the set of all zeros of f in an algebraic closure of K. For $y \in S$ there is a representation

$$y = a_1 + a_2 x + \cdots + a_n x^{n-1} \qquad (a_i \in K),$$

and the conjugates y_i of y over K are given by the equations

$$(1) \qquad y_i = a_1 + a_2 x_i + \cdots + a_n x_i^{n-1} \qquad (i = 1, \ldots, n).$$

Along with x and y the x_i and y_i are integral over R, as one sees by using a K-automorphism of $K(x_1, \ldots, x_n)$ sending x to x_i (y to y_i) and an equation of integral dependence for x (for y).

Let D be the (Vandermonde) determinant of the system (1),

$$D := \begin{vmatrix} 1 & x_1 & \cdots & x_1^{n-1} \\ 1 & x_2 & \cdots & x_2^{n-1} \\ \vdots & \vdots & & \vdots \\ 1 & x_n & \cdots & x_n^{n-1} \end{vmatrix},$$

and let D_i be the determinant that one gets by replacing the ith column of D by the column $(y_i)_{i=1,\ldots,n}$. By Cramer's rule, $a_i D = D_i$ $(i = 1, \ldots, n)$. Hence $a_i D^2 = D_i D$.

Now, D and D_i are integral over R, and D^2 as well as $D_i D$ is invariant under permutations of $x_1, \ldots, x_n$. It follows that $D^2 \in K$ and $D \cdot D_i \in K$. Since R is integrally closed in K, we must in fact have $D^2 \in R$ and $D_i D \in R$ $(i = 1, \ldots, n)$. From $a_i \in \frac{1}{D^2} R$ $(i = 1, \ldots, n)$ it follows that

$$y \in \frac{1}{D^2}(R + Rx + \cdots + Rx^{n-1})$$

and hence

$$S \subset \frac{1}{D^2}(R + Rx + \cdots + Rx^{n-1}).$$

Since S is a submodule of a finitely generated module over the Noetherian ring R, the Hilbert basis theorem for modules tells us that S is itself a finitely generated module over R.

Theorem F.8. *Every Noetherian local integral domain of Krull dimension* 1 *that is integrally closed in its field of fractions is a discrete valuation ring.*

Proof. Let R be such a ring, $\mathfrak{m}$ its maximal ideal, and $K := Q(R)$ its field of fractions. We must show that $\mathfrak{m}$ is a principal ideal (E.11).

Let $x \in \mathfrak{m} \setminus \{0\}$ be an arbitrary element. Since $\dim R = 1$, the residue class ring $R/(x)$ has only one prime ideal, namely $\mathfrak{m}/(x)$. By C.12 this is nilpotent. Hence there is a $\rho \in \mathbb{N}$ with $\mathfrak{m}^{\rho+1} \subset (x)$, and $\mathfrak{m}^\rho \not\subset (x)$. If $\rho = 0$, then we are done. So let $\rho > 0$.

For $y \in \mathfrak{m}^\rho \setminus (x)$, we have $\mathfrak{m} \cdot y \subset (x)$, and it follows that

$$\mathfrak{m} \cdot \frac{y}{x} \subset R, \quad \frac{y}{x} \notin R.$$

Therefore the R-module $\mathfrak{m}^{-1} := \{a \in K \mid \mathfrak{m} \cdot a \subset R\}$ is strictly larger than R. It is clear that $\mathfrak{m} \subset \mathfrak{m} \cdot \mathfrak{m}^{-1} \subset R$, and that $\mathfrak{m} \cdot \mathfrak{m}^{-1}$ is an ideal of R. Therefore there are only two possibilities:

$$(a) \quad \mathfrak{m} \cdot \mathfrak{m}^{-1} = \mathfrak{m};$$

$$(b) \quad \mathfrak{m} \cdot \mathfrak{m}^{-1} = R.$$

We show that (a) cannot happen: If (a) were true, then $\mathfrak{m}R[x] \subset \mathfrak{m}$ for each $x \in \mathfrak{m}^{-1}$; hence $zR[x] \subset \mathfrak{m}$ for $z \in \mathfrak{m} \setminus \{0\}$. By the Hilbert basis theorem $R[x]$ is a finitely generated R-module; i.e., x is integral over R and hence $x \in R$, since R is supposed to be integrally closed. It would then follow that $\mathfrak{m}^{-1} = R$, contrary to what was shown above. Therefore only (b) can occur.

In this case we will prove that $\mathfrak{m}$ is a principal ideal. Because $\mathfrak{m} \cdot \mathfrak{m}^{-1} = R$, there is an equation $\sum_{i=1}^m x_i y_i = 1$ with $x_i \in \mathfrak{m}$, $y_i \in \mathfrak{m}^{-1}$. Here $x_i y_i \in R$ $(i = 1, \dots, m)$. Since R is local, $x_i y_i$ must be a unit of R for at least one $i \in \{1, \dots, m\}$. For each $z \in \mathfrak{m}$ we then have

$$z = z(x_i y_i)(x_i y_i)^{-1} = x_i(z y_i)(x_i y_i)^{-1}.$$

Here $z y_i \in R$ and $(x_i y_i)^{-1} \in R$, and we have shown that $\mathfrak{m} = (x_i)$.

Lemma F.9. *Let $R \subset S$ be an integral ring extension and let $\mathfrak{P} \in \operatorname{Spec} S$. Then $\mathfrak{P} \in \operatorname{Max}(S)$ if and only if $\mathfrak{P} \cap R \in \operatorname{Max}(R)$.*

Proof. Set $\mathfrak{p} := \mathfrak{P} \cap R$. Then $R/\mathfrak{p} \subset S/\mathfrak{P}$ and $S/\mathfrak{P}$ is integral over $R/\mathfrak{p}$.

If $\mathfrak{p} \in \operatorname{Max}(R)$, then $R/\mathfrak{p}$ is a field, and $S/\mathfrak{P}$ is also a field. For if $y \in S/\mathfrak{P} \setminus \{0\}$ and if

$$y^n + \rho_1 y^{n-1} + \cdots + \rho_n = 0 \qquad (\rho_i \in R/\mathfrak{p})$$

is an equation of integral dependence for y over $R/\mathfrak{p}$, then we may assume that $\rho_n \neq 0$, since $S/\mathfrak{P}$ is a domain. From $y(y^{n-1} + \rho_1 y^{n-2} + \cdots + \rho_{n-1}) = -\rho_n$, it follows that y has an inverse in $S/\mathfrak{P}$.

Now to show the converse, we assume that $S/\mathfrak{P}$ is a field. For $x \in R/\mathfrak{p}\setminus\{0\}$ there is a $y \in S/\mathfrak{P}$ with $xy = 1$. By multiplying its equation of integral dependence by x^n we get

$$0 = (xy)^n + x\rho_1(xy)^{n-1} + \cdots + x^n\rho_n = 1 + x(\rho_1 + \rho_2 x + \cdots + \rho_n x^{n-1}),$$

and therefore x has an inverse in $R/\mathfrak{p}$.

Theorem F.10. *Let $R \subset S$ be two integral domains. Let R be Noetherian of Krull dimension* 1*, and let S as an R-module be generated by n elements. Then:*

(a) *For each $\mathfrak{p} \in \operatorname{Max}(R)$ there is at least one and there are at most n different $\mathfrak{P} \in \operatorname{Max}(S)$ such that $\mathfrak{P} \cap R = \mathfrak{p}$.*
(b) *The integral domain S also has Krull dimension* 1.

Proof. (a) For $\mathfrak{p} \in \operatorname{Max}(R)$, we have that $S/\mathfrak{p}S$ is an algebra of dimension $\leq n$ over the field $R/\mathfrak{p}$. By D.5 it has at most n distinct prime ideals, and therefore there are at most n distinct maximal ideals of S lying over $\mathfrak{p}$. Let $N := R \setminus \mathfrak{p}$. If $S/\mathfrak{p}S$ were the zero ring, then $(S/\mathfrak{p}S)_N = S_N/\mathfrak{p}S_N$ would also be the zero ring. However, S_N is a finite module over $R_N = R_\mathfrak{p}$. By Nakayama (E.1) it would follow that $S_N = (0)$, in contradiction to the fact that S and also S_N are integral domains.

Since $S/\mathfrak{p}S \neq (0)$, this ring contains at least one maximal ideal, and therefore there is also at least one $\mathfrak{P} \in \operatorname{Max}(S)$ such that $\mathfrak{P} \cap R = \mathfrak{p}$.

(b) If $\mathfrak{P} \in \operatorname{Spec}(S)$ is not maximal, then by the lemma $\mathfrak{P} \cap R$ is not a maximal ideal of R; hence $\mathfrak{P} \cap R = (0)$, since $\dim R = 1$. We must then have $\mathfrak{P} = (0)$. For if we had $x \in \mathfrak{P} \setminus \{0\}$, then x would have an equation of integral dependence

$$x^n + \rho_1 x^{n-1} + \cdots + \rho_n = 0 \qquad (\rho_i \in R,\ \rho_n \neq 0),$$

and we would have $\rho_n = -x(x^{n-1} + \rho_1 x^{n-2} + \cdots + \rho_{n-1}) \in \mathfrak{P} \cap R$. Therefore besides the zero ideal, S contains only maximal prime ideals; i.e., $\dim S = 1$.

Theorem F.11. *Let S be an integral domain with field of fractions L and let $N \subset S$ be a multiplicatively closed subset. Then:*

(a) *If S is integrally closed in L, so is S_N.*
(b) *S is integrally closed in L if and only if $S_{\mathfrak{P}}$ is integrally closed in L for all $\mathfrak{P} \in \operatorname{Max}(S)$.*

Proof. (a) If $x \in L$ is integral over S_N and

$$x^n + \rho_1 x^{n-1} + \cdots + \rho_n = 0 \qquad (\rho_i \in S_N) \tag{2}$$

is an equation of integral dependence for x, we can write $\rho_i = \frac{s_i}{s}$ with $s_i \in S$ $(i = 1, \ldots, n)$, $s \in N$. Multiplying (2) through by s^n, we see that sx is integral over S. Therefore $sx \in S$ and $x \in S_N$.

(b) Suppose $S_{\mathfrak{P}}$ is integrally closed in L for each $\mathfrak{P} \in \operatorname{Max}(S)$. An element x of L integral over S is also integral over each $S_{\mathfrak{P}}$; therefore

$$x \in \bigcap_{\mathfrak{P} \in \operatorname{Max}(S)} S_{\mathfrak{P}}.$$

The result then follows from the next lemma.

Lemma F.12. *For every integral domain S,*

$$S = \bigcap_{\mathfrak{P} \in \operatorname{Max}(S)} S_{\mathfrak{P}},$$

and for each ideal $I \subset S$,

$$I = \bigcap_{\mathfrak{P} \in \operatorname{Max}(S)} I_{\mathfrak{P}}.$$

Proof. Let $x \in \bigcap_{\mathfrak{P} \in \operatorname{Max}(S)} S_{\mathfrak{P}}$ and let $J := \{s \in S \mid sx \in S\}$. Obviously, J is an ideal of S (sometimes called the "denominator ideal" of x). For each $\mathfrak{P} \in \operatorname{Max}(S)$ there is an $s_{\mathfrak{P}} \in S_{\mathfrak{P}}$ with $s_{\mathfrak{P}} \cdot s \in S$. Hence J is not contained in any maximal ideal of S, and therefore $J = S$. From $1 \in J$ it follows that $x \in S$. The proof for ideals is similar.

Theorem F.13. *Let R be a discrete valuation ring with maximal ideal $\mathfrak{m}$ and field of fractions K, let L/K be a finite separable field extension of degree n, and let S be the integral closure of R in L. Then*

(a) *$S_{\mathfrak{P}}$ is a discrete valuation ring (with field of fractions L) for each $\mathfrak{P} \in \operatorname{Max}(S)$. The set $\operatorname{Max}(S)$ contains only finitely many elements.*
(b) *Let $\operatorname{Max}(S) = \{\mathfrak{P}_1, \ldots, \mathfrak{P}_h\}$, let $\mathfrak{m}S_{\mathfrak{P}_i} = \mathfrak{P}_i^{e_i} S_{\mathfrak{P}_i}$, and for $i = 1, \ldots, h$ let $f_i := [S_{\mathfrak{P}_i}/\mathfrak{P}_i S_{\mathfrak{P}_i} : R/\mathfrak{m}]$. Then*

$$n = \sum_{i=1}^{h} e_i f_i \qquad \textit{(Degree formula)}.$$

Proof. (a) By F.11(a) the ring $S_{\mathfrak{P}}$ is integrally closed in L; by F.7 the ring S is a finitely generated R-module, and therefore $S_{\mathfrak{P}}$ is Noetherian, and $S_{\mathfrak{P}}$ has Krull dimension 1 by F.10. Hence $S_{\mathfrak{P}}$ is a discrete valuation ring according to F.8. The maximal spectrum $\operatorname{Max}(S)$ is finite by F.9 and F.10(a).

(b) Since R is a principal ideal domain and S is a finitely generated torsion-free R-module, by the fundamental theorem for modules over a principal ideal domain, S has a basis over R, necessarily of length n. By the Chinese remainder theorem we furthermore have

$$n = \dim_{R/\mathfrak{m}} S/\mathfrak{m}S = \sum_{i=1}^{h} \dim_{R/\mathfrak{m}} S_{\mathfrak{P}_i}/\mathfrak{P}_i^{e_i} S_{\mathfrak{P}_i}.$$

In the chain of ideals

$$\mathfrak{P}_i^{e_i} S_{\mathfrak{P}_i} \subset \mathfrak{P}_i^{e_i-1} S_{\mathfrak{P}_i} \subset \cdots \subset \mathfrak{P}_i S_{\mathfrak{P}_i} \subset S_{\mathfrak{P}_i}$$

all the quotients $\mathfrak{P}_i^j S_{\mathfrak{P}_i}/\mathfrak{P}_i^{j+1} S_{\mathfrak{P}_i}$ are isomorphic to $S_{\mathfrak{P}_i}/\mathfrak{P}_i S_{\mathfrak{P}_i}$, and it follows that

$$\dim_{R/\mathfrak{m}} S_{\mathfrak{P}_i}/\mathfrak{P}_i^{e_i} S_{\mathfrak{P}_i} = e_i \cdot [S_{\mathfrak{P}_i}/\mathfrak{P}_i S_{\mathfrak{P}_i} \ : \ R/\mathfrak{m}] = e_i f_i.$$

Theorem F.14. *Let R be an integral domain integrally closed in its field of fractions K, let L be an extension field of K, and let $a \in L$ be integral over R. Then for the minimal polynomial f of a over K we have*

$$f \in R[X].$$

Proof. Decompose f into linear factors in the algebraic closure $\overline{K}$ of K:

$$f = (X - a_1) \cdots (X - a_n), \quad a_1 = a.$$

There is a K-isomorphism $K(a) \xrightarrow{\sim} K(a_i)$ sending a to a_i $(i = 1, \ldots, n)$. Then a_i is integral over R, along with a. Therefore the coefficients of f are integral over R. But they lie in K, and R is integrally closed in K. Hence $f \in R[X]$.

We shall now give a proof of the Hilbert Nullstellensatz. As ingredients to the proof we need only the following facts:
(a) $K[X]$ is a unique factorization domain for any field K.
(b) $K[X]$ has infinitely many prime polynomials.
(c) Corollary F.4.

Field-Theoretic Form of Hilbert's Nullstellensatz F.15. *Let L/K be a field extension and suppose there are elements $x_1, \ldots, x_n \in L$ with $L = K[x_1, \ldots, x_n]$ (ring adjunction!). Then L/K is algebraic.*

Proof. The proof is by induction on n. We may assume that $n > 0$, and that the theorem has already been proved for $n - 1$ generators. Suppose x_1 were transcendental over K. Since $K[x_1, \ldots, x_n] \cong K(x_1)[x_2, \ldots, x_n]$, it follows by the induction hypothesis that $L/K(x_1)$ is algebraic. The minimal polynomial f_i for x_i over $K(x_1)$ $(i = 2, \dots, n)$ can be written in the form

$$f_i = X^{n_i} + \frac{a_1^{(i)}}{u} X^{n_i - 1} + \dots + \frac{a_{n_i}^{(i)}}{u} \qquad (a_j^{(i)} \in K[x_1])$$

with a common denominator $u \in K[x_1]$. Using F.4 it follows that $L = K[x_1, \dots, x_n]$ is integral over $K[x_1, \frac{1}{u}]$. Now let $p \in K[x_1]$ be a prime polynomial that does not divide u, and let

$$\left(\frac{1}{p}\right)^m + \frac{s_1}{u^t}\left(\frac{1}{p}\right)^{m-1} + \dots + \frac{s_m}{u^t} \qquad (s_i \in K[x_1])$$

be an equation of integral dependence for $\frac{1}{p}$ over $K[x_1, \frac{1}{u}]$ in which the coefficients have been adjusted so that they all have the same denominator u^t. Multiplying through by $p^m u^t$ we get

$$u^t + s_1 p + \dots + s_m p^m = 0,$$

in contradiction to the assumption that p does not divide u. This proves Theorem F.15.

Corollary F.16 (Hilbert's Nullstellensatz). *Let K be a field and let $\overline{K}$ be its algebraic closure. Then every ideal $I \subset K[X_1, \dots, X_n]$ with $I \neq K[X_1, \dots, X_n]$ has a zero in $\overline{K}^n$; i.e., there exists $(\xi_1, \dots, \xi_n) \in \overline{K}^n$ such that $f(\xi_1, \dots, \xi_n) = 0$ for all $f \in I$.*

Proof. We can assume without loss of generality that $I = \mathfrak{M}$ is a maximal ideal. Then the field $L := K[X_1, \dots, X_n]/\mathfrak{M}$ is generated over K by the residue classes x_i of the X_i. By F.15 we know that L/K is algebraic. We therefore have an injective K-homomorphism $L \to \overline{K}$ and hence a K-homomorphism $\phi : K[X_1, \dots, X_n] \to \overline{K}$ with kernel $\mathfrak{M}$. If $\xi_i := \phi(X_i)$ $(i = 1, \dots, n)$, then $(\xi_1, \dots, \xi_n)$ is the desired zero of $\mathfrak{M}$.

Exercises

1. Show that the polynomial ring $K[X]$ in one variable over a field K is integral over every subring $R \subset K[X]$ with $K \subset R$, $K \neq R$. Also show that R is a finitely generated algebra over K and R has Krull dimension 1.
2. Deduce Theorem F.15 from F.16.
3. Show that the maximal ideals of a polynomial ring $K[X_1, \dots, X_n]$ over an algebraically closed field K are in one-to-one correspondence with the points of K^n.

G

Tensor Products of Algebras

The tensor product of two algebras S_1/R and S_2/R is an R-algebra that contains images of S_1 and S_2 so that these images are true to the original as much as possible. More precisely,

Definition G.1. A *tensor product* of algebras S_1/R and S_2/R is a triple $(T/R, \alpha_1, \alpha_2)$, where T/R is an algebra, $\alpha_i\colon S_i \to T$ are R-algebra homomorphisms ($i = 1, 2$), and where the following universal property is satisfied: If $(U/R, \beta_1, \beta_2)$ is an arbitrary triple consisting of an algebra U/R and homomorphisms $\beta_i\colon S_i \to U$ ($i = 1, 2$), then there is a unique R-algebra homomorphism $h\colon T \to U$ with $\beta_i = h \circ \alpha_i$ ($i = 1, 2$)

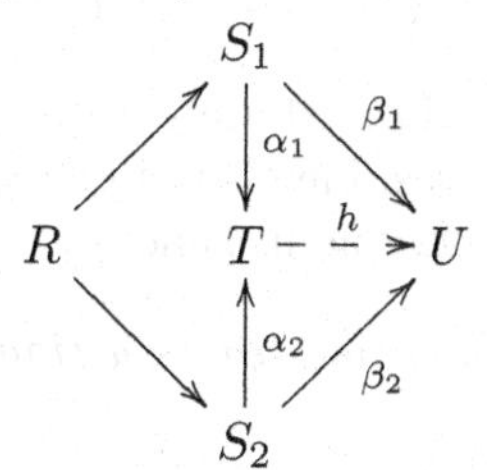

As usual, the tensor product—if it exists—is unique up to canonical isomorphism. We then write $T = S_1 \otimes_R S_2$ and call $\alpha_i : S_i \to T$ ($i = 1, 2$) the canonical homomorphisms into the tensor product.

Examples G.2. (a) Let $S_1 = R[\{X_\lambda\}_{\lambda\in\Lambda}]$ and $S_2 = R[\{Y_\mu\}_{\mu\in M}]$ be two polynomial algebras and let $T = R[\{X_\lambda\}_{\lambda\in\Lambda} \cup \{Y_\mu\}_{\mu\in M}]$ be the polynomial algebra in the variables X_λ, Y_μ ($\lambda \in \Lambda$, $\mu \in M$). Let $\alpha_i : S_i \to T$ be the obvious mappings ($i = 1, 2$). Then (T, α_1, α_2) is a tensor product of S_1/R and S_2/R.

In fact, if $\beta_i : S_i \to U$ are R-homomorphisms to an R-algebra U, then let $x_\lambda := \beta_1(X_\lambda)$, $y_\mu := \beta_2(Y_\mu)$. By the universal property of polynomial algebras there is a unique R-homomorphism $h : R[\{X_\lambda\}\cup\{Y_\mu\}] \to U$ with $h(X_\lambda) = x_\lambda$, $h(Y_\mu) = y_\mu$ ($\lambda \in \Lambda$, $\mu \in M$), which is all that is needed. We write

(1) $$R[\{X_\lambda\}] \otimes_R R[\{Y_\mu\}] = R[\{X_\lambda\} \cup \{Y_\mu\}].$$

In particular,

$$(1') \qquad R[X_1,\ldots,X_m]\otimes_R R[Y_1,\ldots,Y_n] = R[X_1,\ldots,X_m,Y_1,\ldots,Y_n].$$

(b) Let S_1/R and S_2/R be algebras such that $S_1 \otimes_R S_2$ exists, and let $I_k \subset S_k$ $(k=1,2)$ be ideals. Let J be the ideal of $S_1 \otimes_R S_2$ generated by $\alpha_k(I_k)$ $(k=1,2)$. Then there exists $S_1/I_1 \otimes_R S_2/I_2$, and we have $S_1/I_1 \otimes_R S_2/I_2 = (S_1 \otimes_R S_2)/J$, where the canonical homomorphisms $\overline{\alpha}_k : S_k/I_k \to S_1/I_1 \otimes_R S_2/I_2$ are induced on the residue class rings by $\alpha_k : S_k \to S_1 \otimes_R S_2$.

In fact, if $\overline{\beta}_k : S_k/I_k \to U$ are two R-homomorphisms to an R-algebra U, let $\beta_k : S_k \to U$ be its composition with the canonical epimorphism $S_k \to S_k/I_k$ $(k=1,2)$. There is then a unique homomorphism $h : S_1 \otimes_R S_2 \to U$ with $\beta_k = h \circ \alpha_k$ $(k=1,2)$.

Since $\beta_k(I_k) = 0$ $(k=1,2)$, we have $h(J)=0$, and therefore h induces a homomorphism $\overline{h} : (S_1 \otimes_R S_2)/J \to U$ with $\overline{\beta}_k = \overline{h} \circ \overline{\alpha}_k$ $(k=1,2)$. There can be only one such homomorphism: If h' were another and we denote by $\varepsilon : S_1 \otimes_R S_2 \to (S_1 \otimes_R S_2)/J$ the canonical epimorphism, then we would have $\overline{h} \circ \varepsilon = h' \circ \varepsilon$, according to the uniqueness condition in the universal property of $S_1 \otimes_R S_2$. Since ε is an epimorphism, it follows that $h' = \overline{h}$.

We write $J =: I_1 \otimes_R S_2 + S_1 \otimes_R I_2$. The assertion just proved can then be briefly noted by the formula

$$(2) \qquad S_1/I_1 \otimes_R S_2/I_2 = (S_1 \otimes_R S_2)/(I_1 \otimes_R S_2 + S_1 \otimes_R I_2).$$

From G.2(a) and (b) we immediately get the *existence of tensor products*, since every algebra is a residue class algebra of a polynomial algebra. In particular, equation (2) is available in general.

Theorem G.3. *$S_1 \otimes_R S_2$ is generated as a ring by $\alpha_1(S_1)$ and $\alpha_2(S_2)$:*

$$S_1 \otimes_R S_2 = \alpha_1(S_1) \cdot \alpha_2(S_2).$$

Proof. Obviously, $\alpha_1(S_1) \cdot \alpha_2(S_2)$ satisfies the universal property of $S_1 \otimes_R S_2$. The inclusion mapping $\alpha_1(S_1) \cdot \alpha_2(S_2) \hookrightarrow S_1 \otimes_R S_2$ is bijective because of the uniqueness condition in G.1, and then the result follows immediately.

We set

$$a \otimes 1 := \alpha_1(a) \quad \text{for} \quad a \in S_1,$$
$$1 \otimes b := \alpha_2(b) \quad \text{for} \quad b \in S_2,$$

and

$$a \otimes b := (a \otimes 1)(1 \otimes b).$$

By G.3 an arbitrary element of $S_1 \otimes_R S_2$ is of the form

$$\sum_{k=1}^{n} a_k \otimes b_k \quad (a_k \in S_1,\ b_k \in S_2,\ n \in \mathbb{N}),$$

but this presentation is in general not unique. The elements of $S_1 \otimes_R S_2$ are called *tensors.* Tensors of the form $a \otimes b$ are called *decomposable.* In general, not every tensor is decomposable.

Since α_1 and α_2 are R-homomorphisms, the diagram

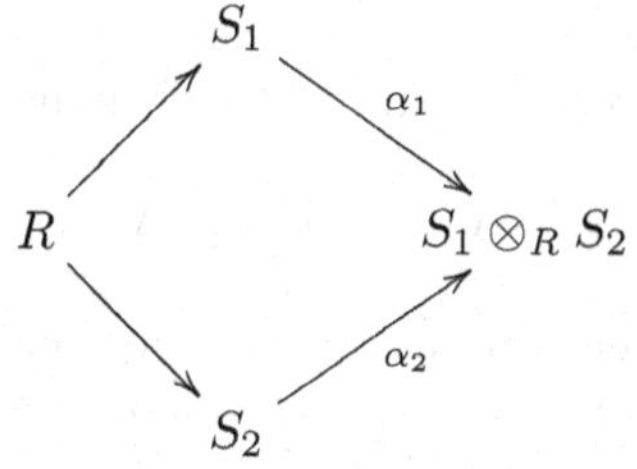

is commutative. Therefore we have

(3) $r \otimes 1 = 1 \otimes r$ for all $r \in R$, $\quad ra \otimes b = a \otimes rb$ for $r \in R$, $a \in S_1$, $b \in S_2$.

Further calculation rules for tensors follow from the fact that α_1 and α_2 are ring homomorphisms:

$$\begin{aligned} (a+b)\otimes c &= a\otimes c + b\otimes c, & a\otimes(c+d) &= a\otimes c + a\otimes d,\\ (a\otimes c)(b\otimes d) &= ab\otimes cd, & a\otimes 0 &= 0 = 0\otimes c, \end{aligned}$$

for $a, b \in S_1$ and $c, d \in S_2$. Furthermore, $1 \otimes 1 = 1$ is the identity element for multiplication.

With α_2 as structure homomorphism, $S_1 \otimes S_2$ is an S_2-algebra (similarly also an S_1-algebra). We say that the algebra $(S_1 \otimes_R S_2)/S_2$ comes from S_1/R using the *base change* $R \to S_2$. Frequently, it is an important problem to investigate how properties of algebras behave under base change.

Theorem G.4. *Suppose S_1/R has a basis $\{b_\lambda\}_{\lambda \in \Lambda}$. Then:*

(a) *$\{b_\lambda \otimes 1\}_{\lambda \in \Lambda}$ is a basis of $(S_1 \otimes_R S_2)/S_2$.*
(b) *If $c \in S_2$ is not a zero divisor of S_2, then $1 \otimes c$ is not a zero divisor of $S_1 \otimes_R S_2$.*

Proof. (a) In S_1 there are relations

$$\begin{aligned} b_\lambda b_{\lambda'} &= \sum r_{\lambda\lambda'}^{\lambda''} b_{\lambda''} \qquad && (r_{\lambda\lambda'}^{\lambda''} \in R),\\ 1 &= \sum \rho_\lambda b_\lambda && (\rho_\lambda \in R). \end{aligned}$$

To say that S_1 is commutative is equivalent to saying that $r_{\lambda\lambda'}^{\lambda''} = r_{\lambda'\lambda}^{\lambda''}$. If one writes out $(b_\lambda b_{\lambda'})b_{\lambda''}$ and $b_\lambda(b_{\lambda'} b_{\lambda''})$ in terms of the basis elements and equates coefficients, then one gets formulas in the $r_{\lambda\lambda'}^{\lambda''}$ that are equivalent to the validity of the associative property. That $\sum \rho_\lambda b_\lambda$ is the identity of S_1 is likewise equivalent to a family of formulas in the ρ_λ and $r_{\lambda\lambda'}^{\lambda''}$.

We now construct an algebra T/S_2 in which the indeterminates X_λ ($\lambda \in \Lambda$) generate the free S_2-module

$$T := \bigoplus_{\lambda \in \Lambda} S_2 X_\lambda.$$

We give T a multiplication by means of the formula

$$X_\lambda X_{\lambda'} := \sum r_{\lambda\lambda'}^{\lambda''} X_{\lambda''}.$$

(It is enough to define the product for the basis elements.) The formulas in R, which are equivalent to the associativity and commutativity of S_1, are also valid for the images in S_2; hence T is an associative and commutative S_2-algebra. Also, $\sum_{\lambda \in \Lambda} \rho_\lambda X_\lambda$ is the unit element 1_T of T.

There are two obvious R-homomorphisms:

$$\beta_1 \colon S_1 \to T \quad (\sum r_\lambda b_\lambda \mapsto \sum r_\lambda X_\lambda \text{ for } r_\lambda \in R),$$

$$\beta_2 \colon S_2 \to T \quad (s \mapsto s \cdot 1_T \text{ for } s \in S_2),$$

and therefore an R-homomorphism

$$h \colon S_1 \otimes_R S_2 \to T$$

with $h(b_\lambda \otimes 1) = X_\lambda$ ($\lambda \in \Lambda$). Using G.3 it is clear that $\{b_\lambda \otimes 1\}$ is a system of generators for $S_1 \otimes_R S_2$ as an S_2-module. Since the images of the $b_\lambda \otimes 1$ in T are linearly independent over S_2, the $b_\lambda \otimes 1$ are themselves linearly independent over S_2, and therefore $\{b_\lambda \otimes 1\}$ is a basis for $S_1 \otimes_R S_2/S_2$.

(b) Let $x(1 \otimes c) = 0$ for an $x \in S_1 \otimes_R S_2$. Write $x = \sum_\lambda b_\lambda \otimes s_\lambda$ ($s_\lambda \in S_2$). Then $x(1 \otimes c) = \sum_\lambda (b_\lambda \otimes 1)(1 \otimes cs_\lambda) = 0$, and it follows that $1 \otimes cs_\lambda = 0$ for all $\lambda \in \Lambda$. Since $S_2 \to S_1 \otimes_R S_2$ is injective by (a), we have $cs_\lambda = 0$ and therefore $s_\lambda = 0$ for all $\lambda \in \Lambda$.

Corollary G.5. *If $\{b_\lambda\}_{\lambda \in \Lambda}$ is a basis for S_1/R and $\{c_\mu\}_{\mu \in M}$ is a basis for S_2/R, then $\{b_\lambda \otimes c_\mu\}_{\lambda \in \Lambda, \mu \in M}$ is a basis for $(S_1 \otimes_R S_2)/R$:*

$$\left(\bigoplus_\lambda Rb_\lambda\right) \otimes \left(\bigoplus_\mu Rc_\mu\right) = \bigoplus_{\lambda,\mu} R(b_\lambda \otimes c_\mu).$$

Proof. Consider the ring homomorphisms

$$R \to S_2 \to S_1 \otimes_R S_2.$$

Since $\{c_\mu\}_{\mu \in M}$ is a basis for S_2/R, and $\{b_\lambda \otimes 1\}_{\lambda \in \Lambda}$ is a basis for $(S_1 \otimes_R S_2)/S_2$, it follows that $\{(b_\lambda \otimes 1)(1 \otimes c_\mu)\}_{\lambda \in \Lambda, \mu \in M}$ is a basis for $(S_1 \otimes_R S_2)/R$.

Of the many possible formulas for the tensor product, we choose to give only the following.

Formulas G.6.

(a) For an algebra S/R and a polynomial algebra $R[\{X_\lambda\}]$ we have, in a canonical way,

$$S \otimes_R R[\{X_\lambda\}] = S[\{X_\lambda\}],$$

where $s \otimes \sum r_{\nu_1 \ldots \nu_n} X_{\lambda_1}^{\nu_1} \cdots X_{\lambda_n}^{\nu_n}$ is identified with $\sum (s r_{\nu_1 \ldots \nu_n}) X_{\lambda_1}^{\nu_1} \cdots X_{\lambda_n}^{\nu_n}$. In fact, $S[\{X_\lambda\}]$ has the universal property for $S \otimes_R R[\{X_\lambda\}]$. As a special case we have

$$S \otimes_R R = S = R \otimes_R S \quad (s \otimes r = sr = r \otimes s),$$

and in particular,

$$R \otimes_R R = R \quad (a \otimes b = ab).$$

(b) For an algebra S/R and an ideal $I \subset R$ we have

$$S \otimes_R (R/I) = S/IS.$$

In fact, by G.2(b) we have $S \otimes_R (R/I) = S \otimes_R R / S \otimes_R I = S/IS$.

(c) Permutability of tensor products and localization. Let S_k/R be two algebras, and let $N_k \subset S_k$ be two multiplicatively closed subsets ($k = 1, 2$). Then we have, in a canonical way,

$$S_{1N_1} \otimes_R S_{2N_2} = (S_1 \otimes_R S_2)_{N_1 \otimes N_2} \quad \left(\frac{x}{a} \otimes \frac{y}{b} = \frac{x \otimes y}{a \otimes b} \right).$$

Here $N_1 \otimes N_2 := \{a \otimes b \in S_1 \otimes_R S_2 \mid a \in N_1, b \in N_2\}$.

Proof. There is an R-homomorphism

$$\alpha\colon (S_1)_{N_1} \otimes_R (S_2)_{N_2} \to (S_1 \otimes_R S_2)_{N_1 \otimes N_2} \quad \left(\frac{x}{a} \otimes \frac{y}{b} \mapsto \frac{x \otimes y}{a \otimes b} \right)$$

induced by $(S_k)_{N_k} \to (S_1 \otimes_R S_2)_{N_1 \otimes N_2}$. Accordingly, from $S_k \to (S_k)_{N_k} \to (S_1)_{N_1} \otimes_R (S_2)_{N_2}$ there is an induced R-homomorphism $S_1 \otimes_R S_2 \to (S_1)_{N_1} \otimes_R (S_2)_{N_2}$ $(x \otimes y \mapsto \frac{x}{1} \otimes \frac{y}{1})$. Here the elements of $N_1 \otimes N_2$ are mapped to units, and so there is an induced R-homomorphism

$$\beta\colon (S_1 \otimes_R S_2)_{N_1 \otimes N_2} \to (S_1)_{N_1} \otimes_R (S_2)_{N_2} \quad \left(\frac{x \otimes y}{1 \otimes 1} \mapsto \frac{x}{1} \otimes \frac{y}{1} \right).$$

It is clear that α and β are inverses of each other.

(d) If $N \subset R$ is multiplicatively closed, then canonically

$$(S_1)_N \otimes_{R_N} (S_2)_N = (S_1 \otimes_R S_2)_N,$$

where the N on the right side of the formula is to be understood as the set $\{a \otimes 1\}_{a \in N} = \{1 \otimes a\}_{a \in N}$.

The proof is similar to that of (c).

(e) In addition to S_i/R, suppose two other algebras S'_i/R are given and suppose $\gamma_i\colon S_i \to S'_i$ are R-homomorphisms ($i = 1, 2$). Then there is a canonical R-homomorphism

$$S_1 \otimes_R S_2 \to S'_1 \otimes_R S'_2 \quad (a \otimes b \mapsto \gamma_1(a) \otimes \gamma_2(b)).$$

This will be denoted by $\gamma_1 \otimes \gamma_2$.

Proof. The homomorphism $S_1 \xrightarrow{\gamma_1} S'_1 \to S'_1 \otimes_R S'_2$ maps $a \in S_1$ to $\gamma_1(a) \otimes 1$, and the corresponding homomorphism $S_2 \xrightarrow{\gamma_2} S'_2 \to S'_1 \otimes_R S'_2$ maps $b \in S_2$ to $1 \otimes \gamma_2(b)$. By the universal property of $S_1 \otimes_R S_2$ we get the desired homomorphism $S_1 \otimes_R S_2 \to S'_1 \otimes_R S'_2$ immediately.

(f) Permutability of tensor products and direct products. Let S'/R be an R-algebra. Then there is a canonical isomorphism of R-algebras

$$(S_1 \times S_2) \otimes_R S' \xrightarrow{\sim} (S_1 \otimes_R S') \times (S_2 \otimes_R S') \quad ((s_1, s_2) \otimes s' \mapsto (s_1 \otimes s', s_2 \otimes s')).$$

Proof. The canonical projections $p_k : S_1 \times S_2 \to S_k$ ($k = 1, 2$) furnish by (e) an R-epimorphism

$$p_k \otimes \mathrm{id}_{S'}\colon (S_1 \times S_2) \otimes_R S' \to S_k \otimes_R S'$$

and therefore an R-homomorphism

$$\alpha\colon (S_1 \times S_2) \otimes_R S' \to (S_1 \otimes_R S') \times (S_2 \otimes_R S') \quad ((s_1, s_2) \otimes s' \mapsto (s_1 \otimes s', s_2 \otimes s')).$$

The image of α contains the elements of the form $(s_1 \otimes s', 0)$ and $(0, s_2 \otimes s')$ with $s_k \in S_k$ ($k = 1, 2$), $s' \in S'$. From this we see that α is surjective. The kernel of p_1 is the principal ideal in $S_1 \times S_2$ generated by $(0, 1)$, so the kernel I_1 of $p_1 \otimes \mathrm{id}_{S'}$ is generated by $(0, 1) \otimes 1$. Similarly, we have $I_2 = ((1, 0) \otimes 1)(S_1 \times S_2) \otimes_R S'$. Furthermore, $\ker \alpha = I_1 \cap I_2$. But this intersection is 0, for if $((0, 1) \otimes 1) \cdot x = ((1, 0) \otimes 1) \cdot y$ with $x, y \in (S_1 \times S_2) \otimes_R S'$, multiplication of this equation by $(1, 0) \otimes 1$ immediately gives us that $((1, 0) \otimes 1) \cdot y = 0$. It follows that α is an isomorphism.

For an algebra S/R we call $S^e := S \otimes_R S$ the *enveloping algebra* of S/R. This is the entry point for the construction of several invariants of an algebra. Consider the diagram

$$\begin{array}{ccc} S & & \\ {\scriptstyle \alpha_1}\downarrow & \searrow^{\mathrm{id}} & \\ S^e & \overset{\mu}{\dashrightarrow} & S \\ {\scriptstyle \alpha_2}\uparrow & \nearrow_{\mathrm{id}} & \\ S & & \end{array}$$

with $\alpha_1(a) = a \otimes 1$, $\alpha_2(b) = 1 \otimes b$. By the universal property of $S \otimes_R S$ there is an induced surjective R-homomorphism called the *canonical multiplication map*

$$\mu\colon S \otimes_R S \to S \quad \text{with} \quad \mu(a \otimes b) = a \cdot b \quad (a, b \in S).$$

The kernel I of this map is called the *diagonal* of S^e.

Theorem G.7. *We have*

$$I = (\{a \otimes 1 - 1 \otimes a\}_{a \in S}).$$

If S is generated as an R-algebra by $x_1 \dots, x_n \in S$, then

$$I = (\{x_i \otimes 1 - 1 \otimes x_i\}_{i=1,\dots,n}).$$

Proof. Let $I' := (\{a \otimes 1 - 1 \otimes a\}_{a \in S})$. Clearly, $I' \subset I$ and there is an epimorphism

$$S \otimes_R S \twoheadrightarrow S \otimes_R S/I' \xrightarrow{\mu'} S \otimes_R S/I \cong S.$$

For $a, b \in S$ we have $a \otimes b = (a \otimes 1)(1 \otimes b) = -(a \otimes 1)(b \otimes 1 - 1 \otimes b) + (ab \otimes 1)$. It follows that the mapping $S \xrightarrow{\alpha_1} S \otimes_R S \to S \otimes_R S/I'$ is surjective. The composition of this map with μ' is the identity. Therefore μ' must be bijective, and hence $I' = I$.

The second assertion of the theorem follows easily from the formula

$$ab \otimes 1 - 1 \otimes ab = (b \otimes 1)(a \otimes 1 - 1 \otimes a) + (1 \otimes a)(b \otimes 1 - 1 \otimes b).$$

Example G.8. If $S = R[X_1, \dots, X_n]$ is a polynomial algebra, then $S^e = R[X_1, \dots, X_n, X_1', \dots, X_n']$ is a polynomial algebra by (1′) "with double variables." Here we identify $X_i \otimes 1$ with X_i and $1 \otimes X_i$ with X_i'. Using $\mu : S^e \to S$, every polynomial in S^e will have each X_i' replaced by X_i. The diagonal $I := \ker \mu$ is the ideal generated by $X_1 - X_1', \dots, X_n - X_n'$.

Now (in general) let

$$\operatorname{Ann}_{S^e}(I) := \{x \in S^e \mid x \cdot I = 0\}$$

be the annihilator of the ideal I. The ring S^e can be considered as an S-module in two ways, namely by

$$S \to S^e \quad (a \to a \otimes 1) \quad \text{and} \quad S \to S^e \quad (a \to 1 \otimes a).$$

Similarly, I and $\operatorname{Ann}_{S^e}(I)$ are S-modules in two ways. However, on $\operatorname{Ann}_{S^e}(I)$ these two S-module structures coincide, since

$$(a \otimes 1 - 1 \otimes a) \cdot \operatorname{Ann}_{S^e}(I) = 0$$

by definition of the annihilator. We can therefore consider $\operatorname{Ann}_{S^e}(I)$ as an S-module in a unique way.

Let $N \subset S$ be a multiplicatively closed subset. By G.6(c),

$$(S_N)^e = S^e_{N \otimes N},$$

and by G.7 the kernel of the mapping

$$\mu : S_N \otimes_R S_N \to S_N$$

is $I_{N \otimes N}$.

Rule G.9. If the diagonal I is a finitely generated ideal of S^e, then

$$\mathrm{Ann}_{S^e_N}(I_{N\otimes N}) = \mathrm{Ann}_{S^e}(I)_{N\otimes N}.$$

Considering $\mathrm{Ann}_{S^e_N}(I_{N\otimes N})$ as an S_N-module, this module is generated by the image of the canonical homomorphism $\mathrm{Ann}_{S^e}(I) \to \mathrm{Ann}_{S^e}(I)_{N\otimes N}$.

In fact, the construction of the annihilator of a finitely generated ideal commutes with the formation of fractions, as one can easily show.

Remarks G.10.

(a) $\vartheta(S/R) := \mu(\mathrm{Ann}_{S^e}(I))$ is an ideal of S. It is called the (*Noether*) *different* of the algebra S/R.

(b) $\Omega^1_{S/R} := I/I^2$ can, like I, be considered as an S-module in two ways. Since, however,

$$(a \otimes 1 - 1 \otimes a) \cdot I \subset I^2 \quad \text{for all } a \in S,$$

the two structures on $\Omega^1_{S/R}$ coincide. The S-module $\Omega^1_{S/R}$ is called the *module of* (*Kähler*) *differentials* of the algebra S/R.

In the exercises we will learn properties of the different and the module of differentials. A systematic treatment of these invariants of an algebra can be found in [Ku2]. The construction of $\Omega^1_{S/R}$ is the basis of "algebraic differential calculus."

Exercises

1. Let S_i/R be algebras $(i = 1, 2, 3)$. Show that there are R-algebra isomorphisms

$$S_1 \otimes_R S_2 \xrightarrow{\sim} S_2 \otimes_R S_1 \quad (a \otimes b \mapsto b \otimes a)$$

and

$$(S_1 \otimes_R S_2) \otimes_R S_3 \xrightarrow{\sim} S_1 \otimes_R (S_2 \otimes_R S_3) \quad ((a \otimes b) \otimes c \mapsto a \otimes (b \otimes c)).$$

2. Let S/R be an algebra of the form $S = R[X]/(f)$ with a monic polynomial $f \in R[X]$. Let x denote the residue class of X in S and f' the (formal) derivative of f. Show that the Noether different of the algebra satisfies

$$\vartheta(S/R) = (f'(x)).$$

3. Let $\Omega^1_{S/R} = I/I^2$ be the module of differentials of an algebra S/R, where I is the kernel of $S^e \to S$ ($a \otimes b \mapsto ab$). Show that
   (a) The mapping $d : S \to \Omega^1_{S/R}$ defined by $dx = x \otimes 1 = 1 \otimes x + I^2$ for $x \in S$ is a derivation of S/R; i.e., d is R-linear and satisfies the product rule $d(xy) = xdy + ydx$ $(x, y \in S)$.
   (b) $\Omega^1_{S/R}$ is generated as an S-module by the differentials dx $(x \in S)$.
4. Let $S = R[X_1, \ldots, X_n]$ be a polynomial algebra. Show that
   (a) The differentials $dX_1, \ldots, dX_n$ form a basis for $\Omega^1_{S/R}$ as an S-module.
   (b) For the mapping d defined in Exercise 3 and for every $f \in S$ we have

$$df = \sum_{i=1}^{n} \frac{\partial f}{\partial X_i} dX_i.$$

   (In particular, if $S = R[X]$, then $df = f'(X)\, dX$.)

H

Traces

In field theory the concept of the trace map for finite field extensions is well known. Here we generalize this concept to algebras that have a finite basis. This generalization is of central importance for "higher-dimensional residue theory" (Chapters 11–12) and plays a role in the proof of the Riemann–Roch theorem (Chapter 13).

Let S/R be an algebra. The R-module

$$\omega_{S/R} := \mathrm{Hom}_R(S, R)$$

of all R-linear forms $\ell\colon S \to R$ is an S-module in the following way: For $s \in S$ and $\ell \in \omega_{S/R}$ set

$$(s\ell)(x) = \ell(sx) \qquad \text{for all } x \in S.$$

Then $s\ell \in \mathrm{Hom}_R(S, R)$ and $\omega_{S/R}$ is an S-module by $S \times \omega_{S/R} \to \omega_{S/R}$ $(s, \ell) \mapsto s\ell$. It is called the *canonical module* (or *dualizing module*) of the algebra S/R.

If, for example, S/R is a finite field extension, then $\mathrm{Hom}_R(S, R)$ is an S-vector space. As an R-vector space $\mathrm{Hom}_R(S, R)$ has the same dimension as S; hence necessarily $\mathrm{Hom}_R(S, R)$ is an S-vector space of dimension 1:

$$\omega_{S/R} \cong S. \tag{1}$$

In the following let S be a free R-module with basis $B = \{s_1, \dots, s_m\}$. The linear forms $s_i^* \in \omega_{S/R}$ with

$$s_i^*(s_j) = \delta_{ij} \qquad (i, j = 1, \dots, m).$$

form a basis of $\omega_{S/R}$ as an R-module, the *dual basis* B^* of B.

One special element of $\omega_{S/R}$ is the *canonical trace* (or *standard trace*) $\sigma_{S/R}\colon S \to R$, which is defined as follows: For $x \in S$, let $\sigma_{S/R}(x)$ be the trace of the homothety

$$\mu_x\colon S \longrightarrow S \qquad (s \mapsto xs).$$

In other words, if one uses the basis B to describe μ_x by an $m \times m$ matrix A with coefficients in R, then $\sigma_{S/R}(x)$ is the sum of the elements on the main diagonal of A. It is well known that this sum does not depend on the choice of the basis.

Rule H.1.

$$\sigma_{S/R} = \sum_{i=1}^{m} s_i \cdot s_i^*.$$

Proof. Let $s_j s_k = \sum_{\ell=1}^{m} \rho_{jk}^{\ell} s_\ell$ ($j, k = 1, \ldots, m$; $\rho_{jk}^{\ell} \in R$). Then $\sigma_{S/R}(s_k) = \sum_{i=1}^{m} \rho_{ik}^{i}$ by the definition of trace. But on the other hand we also have

$$\left(\sum_{i=1}^{m} s_i s_i^*\right)(s_k) = \sum_{i=1}^{m} s_i^*(s_i s_k) = \sum_{i=1}^{m} s_i^* \left(\sum_{\ell=1}^{m} \rho_{ik}^{\ell} s_\ell\right) = \sum_{i=1}^{m} \rho_{ik}^{i}.$$

Now let

$$S = S_1 \times \cdots \times S_h$$

be a direct product of algebras S_i/R ($i = 1, \ldots, h$) each of which has a finite basis. Then

$$\omega_{S_1/R} \times \cdots \times \omega_{S_h/R}$$

becomes an S-module when one defines the scalar multiplication of $x = (x_1, \ldots, x_h) \in S_1 \times \cdots \times S_h$ and $\ell = (\ell_1, \ldots, \ell_h) \in \omega_{S_1/R} \times \cdots \times \omega_{S_h/R}$ by

$$x \cdot \ell = (x_1\ell_1, \ldots, x_h\ell_h).$$

Conversely, given $\ell \in \omega_{S/R} = \mathrm{Hom}_R(S, R)$, let $\ell_i \colon S_i \to R$ be the composition of the inclusion $S_i \hookrightarrow S$ with ℓ.

Rule H.2. $\psi \colon \omega_{S/R} \longrightarrow \omega_{S_1/R} \times \cdots \times \omega_{S_h/R}$ ($\ell \mapsto (\ell_1, \ldots, \ell_h)$) is an isomorphism of S-modules.

Proof. It is clear that ψ is S-linear. For $(\ell_1, \ldots, \ell_h) \in \omega_{S_1/R} \times \cdots \times \omega_{S_h/R}$ consider the R-linear mapping $\ell \colon S \to R$ defined by $\ell(x_1, \ldots, x_h) = \sum_{i=1}^{h} \ell_i(x_i)$; then $(\ell_1, \ldots, \ell_h) \mapsto \ell$ is the inverse map of ψ.

Rule H.3. With ψ as in H.2, we have $\psi(\sigma_{S/R}) = (\sigma_{S_1/R}, \ldots, \sigma_{S_h/R})$. In other words, for $x = (x_1, \ldots, x_h) \in S_1 \times \cdots \times S_h = S$ we have

$$\sigma_{S/R}(x) = \sum_{i=1}^{h} \sigma_{S_i/R}(x_i).$$

Proof. Choose for each R-module S_i a basis B_i and the corresponding dual basis B_i^* of $\omega_{S_i/R}$. Then $B := \cup_{i=1}^{h} B_i$ is a basis of S and $B^* := \cup_{i=1}^{h} B_i^*$ can be identified by ψ^{-1} with the dual basis of B in $\omega_{S/R}$. The result then follows easily from H.1.

Example H.4. Let $S = R \times \cdots \times R$ be a finite direct product of copies of the ring R. Then for each $x = (x_1, \ldots, x_h) \in S$

$$\sigma_{S/R}(x) = \sum_{i=1}^{h} x_i.$$

Let $\mathfrak{a}$ be an ideal of R. We set

$$\overline{R} := R/\mathfrak{a}, \qquad \overline{S} := S/\mathfrak{a}S,$$

and denote the residue class of elements of R and S with a bar. Then $\overline{B} := \{\overline{s}_1, \dots, \overline{s}_m\}$ is a basis of $\overline{S}$ as an $\overline{R}$-module. Each linear form $\ell \in \operatorname{Hom}_R(S, R)$ maps $\mathfrak{a}S$ to $\mathfrak{a}$ and therefore induces a linear form $\overline{\ell} \in \operatorname{Hom}_{\overline{R}}(\overline{S}, \overline{R})$:

$$\overline{\ell}(\overline{x}) = \overline{\ell(x)} \quad \text{for all} \quad x \in S.$$

The dual basis B^* of B thereby is mapped to the dual basis $\{\overline{s_1^*}, \dots, \overline{s_m^*}\}$ of $\overline{B}$.

Theorem H.5. *The S-linear mapping*

$$\alpha\colon \omega_{S/R} \longrightarrow \omega_{\overline{S}/\overline{R}} \qquad (\alpha(\ell) = \overline{\ell})$$

induces an isomorphism of $\overline{S}$-modules

$$\omega_{\overline{S}/\overline{R}} \cong \omega_{S/R}/\mathfrak{a}\omega_{S/R}.$$

Proof. It is clear that α is S-linear. The above statement on dual bases shows that α is surjective. Also, $\mathfrak{a}\omega_{S/R}$ is contained in the kernel of α. If $\sum_{i=1}^m r_i s_i^* \in \omega_{S/R}$ and the image $\sum_{i=1}^m \overline{r_i} \overline{s_i^*}$ is 0, then $\overline{r_i} = 0$ $(i = 1, \dots, m)$, and therefore $\sum_{i=1}^m r_i s_i^* \in \mathfrak{a}\omega_{S/R}$. Since $\ker \alpha = \mathfrak{a}\omega_{S/R}$, the result follows from the first isomorphism theorem.

Rule H.6. For the standard traces we obtain $\sigma_{\overline{S}/\overline{R}} = \alpha(\sigma_{S/R})$. In other words, $\sigma_{\overline{S}/\overline{R}}(\overline{x}) = \overline{\sigma_{S/R}(x)}$ for all $x \in S$.

Proof. This follows immediately from H.1.

We sometimes use the word "trace" in another sense:

Definition H.7. The algebra S/R has a *trace* σ if there exists an element $\sigma \in \omega_{S/R}$ with

$$\omega_{S/R} = S \cdot \sigma.$$

Observe that $\sigma_{S/R}$ is in general not a trace in this sense: If S/R is a finite field extension, it is well known that $\sigma_{S/R} \neq 0$ if and only if S/R is separable. Also, we have $\omega_{S/R} \cong S$ by (1), and hence $\sigma_{S/R}$ is a trace exactly in the separable case. Specific traces will be constructed in Chapter 11. See also [KK] for a related theory.

Rules H.8. Let σ be a trace of S/R.

(a) If $s \cdot \sigma = 0$ for an $s \in S$, then $s = 0$. Therefore $\omega_{S/R} = S \cdot \sigma \cong S$; i.e., $\omega_{S/R}$ is a free S-module with basis $\{\sigma\}$.
(b) $\sigma' \in \omega_{S/R}$ is a trace of S/R if and only if there exists a unit $\varepsilon \in S$ with $\sigma' = \varepsilon \cdot \sigma$.

Proof. We need to prove only (a). For each $x \in S$ we have $0 = (s\sigma)(x) = \sigma(sx) = (x\sigma)(s)$. Therefore $\ell(s) = 0$ for each $\ell \in \omega_{S/R}$. If we write $s = \sum_{i=1}^m r_i s_i$ $(r_i \in R)$, then we get $r_j = s_j^*(\sum r_i s_i) = s_j^*(s) = 0$ $(j = 1, \dots, m)$ and therefore $s = 0$.

Rule H.9. If σ is a trace of S/R, there exists a dual basis $\{s'_1, \dots, s'_m\}$ of S/R to the basis B with respect to σ; i.e., there are elements $s'_1 \dots, s'_m \in S$ with

$$\sigma(s_i s'_j) = \delta_{ij} \qquad (i, j = 1, \dots, m).$$

Also,

$$\sigma_{S/R} = \left(\sum_{i=1}^m s_i s'_i\right) \cdot \sigma.$$

Proof. Write $s_j^* = s'_j \cdot \sigma$ $(s'_j \in S,\ j = 1, \dots, m)$ for the elements of the dual basis B^*. Then

$$\sigma(s_i s'_j) = s_j^*(s_i) = \delta_{ij} \qquad (i, j = 1, \dots, m).$$

Being the images of the s_i^* under the isomorphism $\omega_{S/R} \cong S$, the s'_i form a basis of S/R. By H.1,

$$\sigma_{S/R} = \sum_{i=1}^m s_i s_i^* = \left(\sum_{i=1}^m s_i s'_i\right) \cdot \sigma.$$

We now turn to the question of the existence of a trace.

Rule H.10. Under the assumptions of H.2 let

$$\sigma = (\sigma_1, \dots, \sigma_h) \in \omega_{S_1/R} \times \cdots \times \omega_{S_h/R} = \omega_{S/R}$$

be given. Then σ is a trace of S/R if and only if σ_i is a trace of S_i/R for $i = 1, \dots, h$. In particular, S/R has a trace if and only if each S_i/R has a trace $(i = 1, \dots, h)$.

This can be seen immediately from the description of the isomorphism ψ in H.2.

Rule H.11. Under the assumptions of H.5 suppose $\mathfrak{a}$ is contained in the intersection of all the maximal ideals of R. Then $\sigma \in \omega_{S/R}$ is a trace of S/R if and only if $\overline{\sigma} := \alpha(\sigma)$ is a trace of $\overline{S}/\overline{R}$. In particular, S/R has a trace if and only if $\overline{S}/\overline{R}$ has a trace.

This follows from H.5 and Nakayama's Lemma E.1.

We now assume that $R = \oplus_{k\in\mathbb{Z}} R_k$ and $S = \oplus_{k\in\mathbb{Z}} S_k$ are graded rings and that the structure homomorphism $\rho\colon R \to S$ is homogeneous (i.e., $\rho(R_k) \subset S_k$ for all $k \in \mathbb{Z}$). Also suppose S is an R-module with a basis $B = \{s_1, \dots, s_m\}$, where s_i is homogeneous of degree d_i $(i = 1, \dots, m)$. Then $S_k = \oplus_{i=1}^m R_{k-d_i} s_i$ for all $k \in \mathbb{Z}$ (a direct sum of R_0-modules).

A linear form $\ell \in \omega_{S/R}$ is called *homogeneous of degree d* if $\ell(S_k) \subset R_{k+d}$ for all $k \in \mathbb{Z}$. This condition is equivalent to saying that $\deg \ell(s_i) = d_i + d$ $(i = 1, \dots, m)$. For instance, the dual basis $B^* = \{s_1^*, \dots, s_m^*\}$ of the basis B consists of homogeneous linear forms with

$$\deg s_i^* = -\deg s_i \qquad (i = 1, \dots, m). \tag{2}$$

If we write $\ell = \sum_{i=1}^m r_i s_i^*$ $(r_i \in R)$, then ℓ is homogeneous of degree d if and only if r_i is homogeneous of degree $d_i + d$ $(i = 1, \dots, m)$. Denoting by $(\omega_{S/R})_d$ the R_0-module of all homogeneous linear forms $\ell\colon S \to R$ of degree d, it is clear that

$$\omega_{S/R} = \bigoplus_{d\in\mathbb{Z}} (\omega_{S/R})_d$$

and that

$$S_k \cdot (\omega_{S/R})_d \subset (\omega_{S/R})_{k+d} \qquad (k, d \in \mathbb{Z}).$$

Hence $\omega_{S/R}$ is a graded module over the graded ring S.

Rule H.12. $\sigma_{S/R}$ is homogeneous of degree 0.

In fact, $\sigma_{S/R} = \sum_{i=1}^m s_i s_i^*$ by H.1, and $\deg(s_i s_i*) = 0$ $(i = 1, \dots, m)$ by (2).

If S/R has a trace σ, it is called *homogeneous* if it is a homogeneous element of $\omega_{S/R}$.

Now let R be positively graded. Then the grading of $\omega_{S/R}$ is bounded below. For a homogeneous ideal $\mathfrak{a} \subset R$,

$$\overline{R} = R/\mathfrak{a} \qquad \text{and} \qquad \overline{S} = S/\mathfrak{a}S$$

are graded rings. Nakayama's lemma for graded modules (A.8) then gives the following result analogous to H.11:

Rule H.13. Let $\sigma \in \omega_{S/R}$ be a homogeneous linear form. Then σ is a trace of S/R if and only if the induced linear form $\overline{\sigma} \in \omega_{\overline{S}/\overline{R}}$ is a trace of $\overline{S}/\overline{R}$.

In the following let $R = K$ be a field, which is considered as a graded ring with the trivial grading $R = R_0$. Let $S = G$ be a finite-dimensional, positively graded K-algebra with $G_0 = K$:

$$G = \bigoplus_{k=0}^{p} G_k, \qquad G_p \neq \{0\}.$$

We assume that G is generated as a K-algebra by G_1. Then the ideal G_+ is also generated by G_1. Further, it is clear that G/K has a homogeneous basis. Therefore the above can be used in this situation.

Lemma H.14. *If G/K has a trace, then G/K has a homogeneous trace.*

Proof. Let σ be a trace of G/K and let $\sigma = \sum \sigma_d$ be the decomposition of σ into homogeneous linear forms σ_d of degree d. Write $\sigma_d = a_d \cdot \sigma$ with $a_d \in G$ ($d \in \mathbb{Z}$). It follows from $\sigma = \sum \sigma_d = (\sum a_d) \cdot \sigma$ that $\sum a_d = 1$. It cannot be the case that all $a_d \in G_+$, so some $a_\delta \notin G_+$. Write $a_\delta = \kappa \cdot (1 - u)$ with $\kappa \in K^*$, $u \in G_+$. Since $u^{p+1} = 0$, we have that a_δ is a unit of G:

$$\kappa(1-u) \cdot \kappa^{-1} \cdot \sum_{i=0}^{\infty} u^i = 1.$$

It follows that $\sigma_\delta = a_\delta \cdot \sigma$ is a homogeneous trace of G/K (H.8b).

Lemma H.15. *Assume that the socle of G,*

$$\mathfrak{S}(G) := \{x \in G \mid G_+ \cdot x = \{0\}\},$$

is a 1-dimensional K-vector space. Then $\mathfrak{S}(G) = G_p$, and for $i = 0, \ldots, p$ the multiplication

$$G_i \times G_{p-i} \longrightarrow G_p,$$
$$(a, b) \mapsto ab,$$

is a nondegenerate bilinear form.

Proof. Because $G_p \subset \mathfrak{S}(G)$, the first assertion is clear. For each $a \in G_i \setminus \{0\}$ we must find $b \in G_{p-i}$ such that $a \cdot b \neq 0$. In case $i = p$ we can take $b = 1$. Now let $k < p$ and suppose the statement has already been proved for $i = k + 1$.

We then have $a \notin \mathfrak{S}(G)$; hence $a \cdot G_1 \neq \{0\}$. Therefore there is an $a' \in G_1$ with $aa' \in G_{k+1} \setminus \{0\}$. By the induction hypothesis there exists $b' \in G_{p-k-1}$ such that $aa'b' \neq 0$. Now set $b := a'b'$.

The existence of traces can be shown in many cases by means of the following theorem:

Theorem H.16.

(a) *G/K has a (homogeneous) trace if and only if* $\dim_K \mathfrak{S}(G) = 1$.
(b) *In this case, a homogeneous element $\sigma \in \omega_{G/K}$ is a trace if and only if* $\sigma(\mathfrak{S}(G)) \neq \{0\}$. *We then have* $\deg \sigma = -p$.

Proof. A homogeneous linear form $\ell\colon G \to K$ of degree $-i$ maps G_k to $\{0\}$ for $k \neq i$; therefore $(\omega_{G/K})_{-i}$ can be identified with $\operatorname{Hom}_K(G_i, K)$, and hence

$$\dim_K(\omega_{G/K})_{-i} = \dim_K G_i \qquad (i = 0, \dots, p).$$

Now let σ be a homogeneous trace of G/K, $\deg \sigma = -d$. Then

$$(\omega_{G/K})_{-i} = G_{d-i} \cdot \sigma \qquad (i = 0, \dots, p).$$

Then we must necessarily have $d = p$ and $\dim_K G_p = \dim_K(\omega_{G/K})_{-p} = \dim_K G_0 = 1$. Furthermore, $\sigma(G_p) \neq \{0\}$ and hence also $\sigma(\mathfrak{S}(G)) \neq \{0\}$. On the other hand, $\sigma(G_k) = 0$ for $k < p$.

Now let $g \in \mathfrak{S}(G)$ be homogeneous of degree $< p$. Then $(g\sigma)(G_+) = \sigma(g \cdot G_+) = \{\sigma(0)\} = \{0\}$. Also, $(g\sigma)(G_0) = \sigma(gG_0) = 0$, since $\deg g < p$. Now from $g\sigma = 0$ it follows that $g = 0$.

We have proved that $\mathfrak{S}(G) = G_p$ is 1-dimensional, and we have also shown that $\sigma(\mathfrak{S}(G)) \neq \{0\}$ for each homogeneous trace σ of G/K.

Now let $\sigma \in \omega_{G/K}$ be an arbitrary homogeneous element with $\sigma(\mathfrak{S}(G)) \neq \{0\}$, where $\dim_K \mathfrak{S}(G) = 1$ by assumption. In order to show that σ is a homogeneous trace, it is sufficient to show that

$$\dim_K G_k\sigma = \dim_K G_k \quad \text{for} \quad k = 0, \dots, p.$$

To do this it is enough to show that if $g \in G_k \setminus \{0\}$, then $g\sigma \neq 0$. By Lemma H.15 choose an element $h \in G_{p-k}$ with $gh \neq 0$. Then $gh \in \mathfrak{S}(G)$ and it follows that $(g\sigma)(h) = \sigma(gh) \neq 0$. Therefore $g\sigma \neq 0$.

Now let $(A/K, \mathcal{F})$ be a filtered algebra, where K is a field. The corresponding Rees algebra will be denoted by A^*, and we let $G = \operatorname{gr}_{\mathcal{F}} A$ be the associated graded algebra. We assume that

$$G_0 = K \qquad \text{and} \qquad \dim_K G < \infty.$$

Also suppose that G is generated as a K-algebra by G_1 and that $\mathcal{F}$ is separated.

According to B.5 and H.5 we have

Remark H.17.

$$\omega_{G/K} \cong \omega_{A^*/K[T]}/T\omega_{A^*/K[T]},$$
$$\omega_{A/K} \cong \omega_{A^*/K[T]}/(T-1)\omega_{A^*/K[T]}.$$

Corollary H.18. *If* $\dim_K \mathfrak{S}(G) = 1$, *then* $A^*/K[T]$ *has a homogeneous trace and* A/K *has a trace.*

Proof. By H.16, G/K has a homogeneous trace σ^0. If $\sigma^* \in \omega_{A^*/K[T]}$ is a homogeneous preimage of σ^0, then it is a trace of $A^*/K[T]$ by H.13. Because $A = A^*/(T-1)$, the image σ of σ^* in $\omega_{A/K}$ is a trace of A/K.

We now return to the case where S/R is an arbitrary algebra with a basis $B = \{s_1, \dots, s_m\}$, and let $S^e = S \otimes_R S$ be the enveloping algebra of S/R. As in Appendix G, let $\mu\colon S^e \to S$ $(a \otimes b \mapsto a \cdot b)$ be the canonical multiplication map and $I = \ker \mu$ the diagonal of S^e. We will see that the S-module $\operatorname{Ann}_{S^e}(I)$ is very closely related to $\omega_{S/R}$, and we will get a bijection between the traces of S/R and the generators of $\operatorname{Ann}_{S^e}(I)$.

There is a homomorphism of R-modules

$$\phi\colon S \otimes_R S \longrightarrow \operatorname{Hom}_R(\omega_{S/R}, S),$$

where $s_i \otimes s_j$ $(i, j = 1, \dots, m)$ is assigned to the R-linear mapping $\omega_{S/R} \to S$, which sends each $\ell \in \omega_{S/R}$ to $\ell(s_i)s_j$. One can easily check that for an arbitrary element $\sum a_i \otimes b_i \in S \otimes_R S$ and each $\ell \in \omega_{S/R}$ we have

$$\phi\left(\sum a_i \otimes b_i\right)(\ell) = \sum \ell(a_i)b_i. \tag{3}$$

Therefore ϕ is independent of the choice of the basis B. If $B^* = \{s_1^*, \dots, s_m^*\}$ is the dual basis of B, then $\phi(s_i \otimes s_j)(s_k^*) = \delta_{ik}s_j$ $(i, j, k = 1, \dots, m)$. From this one sees that ϕ is bijective.

Observe that $\operatorname{Hom}_S(\omega_{S/R}, S) \subset \operatorname{Hom}_R(\omega_{S/R}, S)$ in a natural way.

Theorem H.19. ϕ *induces a canonical isomorphism of* S*-modules*

$$\phi\colon \operatorname{Ann}_{S^e}(I) \xrightarrow{\sim} \operatorname{Hom}_S(\omega_{S/R}, S)$$

described by formula (3). If S/R *has a trace, then* $\operatorname{Ann}_{S^e}(I)$ *is free of rank* 1, *and by dualizing* ϕ *we get a canonical isomorphism*

$$\psi\colon \omega_{S/R} \xrightarrow{\sim} \operatorname{Hom}_S(\operatorname{Ann}_{S^e}(I), S).$$

Proof. For $x = \sum a_i \otimes b_i \in \operatorname{Ann}_{S^e}(I)$ and $s \in S$ we have $\sum sa_i \otimes b_i = \sum a_i \otimes sb_i$ and therefore

$$\phi(x)(s\ell) = \sum \ell(sa_i)b_i = \sum \ell(a_i)sb_i = s\sum \ell(a_i)b_i = s\phi(x)(\ell)$$

for each $\ell \in \omega_{S/R}$. Hence $\phi(x)\colon \omega_{S/R} \to S$ is an S-linear mapping. Furthermore, $\phi(sx) = s\phi(x)$, and so ϕ is S-linear.

Conversely, if for $x = \sum a_i \otimes b_i \in S \otimes_R S$ the mapping $\phi(x)$ is S-linear and we set $x_1 := \sum sa_i \otimes b_i$, $x_2 := \sum a_i \otimes sb_i$, then for each $\ell \in \omega_{S/R}$

$$\phi(x_1)(\ell) = \phi(x)(s\ell) = s\phi(x)(\ell) = \phi(x_2)(\ell),$$

from which it follows that $x_1 = x_2$, and so $x \in \mathrm{Ann}_{S^e}(I)$. If S/R has a trace, then along with $\omega_{S/R}$, $\mathrm{Ann}_{S^e}(I)$ is a free S-module of rank 1. Passing to the dual modules, we get from ϕ a canonical isomorphism $\psi\colon \omega_{S/R} \xrightarrow{\sim} \mathrm{Hom}_S(\mathrm{Ann}_{S^e}(I), S)$.

Corollary H.20. *Suppose S/R has a trace. Then ϕ induces a bijection between the set of all traces of S/R and the set of all generators of the S-module $\mathrm{Ann}_{S^e}(I)$: Each trace $\sigma \in \omega_{S/R}$ is mapped to the unique element $\Delta_\sigma = \sum_{i=1}^m s'_i \otimes s_i \in \mathrm{Ann}_{S^e}(I)$ such that $\sum_{i=1}^m \sigma(s'_i)s_i = 1$. Furthermore:*

(a) *Δ_σ generates the S-module $\mathrm{Ann}_{S^e}(I)$, and $\{s'_1, \ldots, s'_m\}$ is the dual basis of B with respect to σ; i.e.,*

$$\sigma(s'_i s_j) = \delta_{ij} \qquad (i, j = 1, \ldots, m).$$

(b) *If $\sum_{i=1}^m a_i \otimes s_i$ generates the S-module $\mathrm{Ann}_{S^e}(I)$ and if $\sigma \in \omega_{S/R}$ is a linear form with*
$sum_{i=1}^m \sigma(a_i)s_i = 1$, then σ is a trace of S/R and $\Delta_\sigma = \sum_{i=1}^m a_i \otimes s_i$; hence $\{a_1, \ldots, a_m\}$ is the dual basis of B with respect to σ.

(c) *For each trace σ of S/R,*

$$\sigma_{S/R} = \mu(\Delta_\sigma) \cdot \sigma.$$

Proof. By H.19 we know that $\mathrm{Ann}_{S^e}(I)$, as well as $\omega_{S/R}$, is a free S-module of rank 1. The isomorphism ψ maps basis elements of $\omega_{S/R}$ to basis elements of $\mathrm{Hom}_S(\mathrm{Ann}_{S^e}(I), S)$. These are in one-to-one correspondence with the basis elements of $\mathrm{Ann}_{S^e}(I)$. To each trace σ we associate the preimage Δ_σ under ϕ^{-1} of the linear form $\omega_{S/R} \to S$ given by $\sigma \mapsto 1$; i.e., if $\Delta_\sigma = \sum_{i=1}^m s'_i \otimes s_i$ ($s'_i \in S$), then $\phi(\Delta_\sigma)(\sigma) = \sum_{i=1}^m \sigma(s'_i)s_i = 1$.

(a) We have $s_j = s_j \cdot \phi(\Delta_\sigma)(\sigma) = \phi(\Delta_\sigma)(s_j\sigma) = \sum_{i=1}^m \sigma(s'_i s_j)s_i$, and it follows that $\sigma(s'_i s_j) = \delta_{ij}$.

(b) Set $x := \sum_{i=1}^m a_i \otimes s_i$. Then $\{\phi(x)\}$ is a basis of $\mathrm{Hom}_S(\omega_{S/R}, S)$. Because $\phi(x)(\sigma) = \sum_{i=1}^m \sigma(a_i)s_i = 1$ and $\omega_{S/R} \cong S$, it must also be the case that $\{\sigma\}$ is a basis of $\omega_{S/R}$ and $x = \Delta_\sigma$.

(c) Let $\{s'_1, \ldots, s'_m\}$ be the dual basis to B with respect to σ. By H.9,

$$\sigma_{S/R} = \left(\sum s_i s'_i\right) \cdot \sigma = \mu\left(\sum s'_i \otimes s_i\right)\sigma = \mu(\Delta_\sigma) \cdot \sigma.$$

Now let R and S be graded rings, as in the discussion after H.11, and let B be a homogeneous basis. An element $x \in S^e$ is called *homogeneous* of degree d if it can be written in the form $x = \sum a_i \otimes b_i$, where $a_i, b_i \in S$ are homogeneous and $\deg a_i + \deg b_i = d$ for all i. It is clear that this gives a grading on S^e and that $\mathrm{Ann}_{S^e}(I)$ is a homogeneous ideal of S^e, hence also a graded S-module. Since $\omega_{S/R}$ is a graded S-module, $\mathrm{Hom}_S(\omega_{S/R}, S)$ is also a graded S-module, and one sees immediately that the canonical isomorphism $\phi : \mathrm{Ann}_{S^e}(I) \xrightarrow{\sim} \mathrm{Hom}_S(\omega_{S/R}, S)$ is homogeneous of degree 0. In the situation of H.20 there is a one-to-one correspondence between the homogeneous traces of $\omega_{S/R}$ and the homogeneous generators of $\operatorname{Ann}_{S^e}(I)$.

We now study the behavior of Ann_{S^e} under base change. Let R'/R be an algebra, $S' := R' \otimes_R S$, and $S'^e := S' \otimes_{R'} S'$. Then $1 \otimes B := \{1 \otimes s_1, \ldots, 1 \otimes s_m\}$ is a basis of S'/R' (G.4). Set $I^S := \ker(S^e \to S)$ and $I^{S'} := \ker(S'^e \to S')$. We have canonical homomorphisms

$$A : S \otimes_R S \to S' \otimes_{R'} S' \qquad (a \otimes b \mapsto (1 \otimes a) \otimes (1 \otimes b))$$

and

$$\alpha : \omega_{S/R} \to \omega_{S'/R'},$$

where $\alpha(l)(r' \otimes x) = l(x) \cdot r'$ for $l \in \omega_{S/R}$, $r' \in R'$ and $x \in S$. If $B^* := \{s_1^*, \ldots, s_m^*\}$ is the basis of $\omega_{S/R}$ that is dual to B, then $1 \otimes B^*$ is dual to $1 \otimes B$. Hence α is injective and $\omega_{S'/R'}$ is generated as an S'-module by $\operatorname{im} \alpha$. Therefore, if σ is a trace of S/R, then $\alpha(\sigma)$ is a trace of S'/R'.

By G.7 we have

$$\begin{aligned} I^S &= (\{s_i \otimes 1 - 1 \otimes s_i\}_{i=1,\ldots,m}), \\ I^{S'} &= (\{(1 \otimes s_i) \otimes (1 \otimes 1) - (1 \otimes 1) \otimes (1 \otimes s_i)\}_{i=1,\ldots,m}), \end{aligned}$$

and hence $I^{S'} = I^S \cdot S'^e$. From this we see that A induces an S-linear mapping

$$\gamma\colon \operatorname{Ann}_{S^e}(I^S) \to \operatorname{Ann}_{S'^e}(I^{S'}).$$

We denote by β the composition of the canonical homomorphisms

$$\operatorname{Hom}_S(\omega_{S/R}, S) \xrightarrow{\phi^{-1}} \operatorname{Ann}_{S^e}(I^S) \xrightarrow{\gamma} \operatorname{Ann}_{S'^e}(I^{S'}) \xrightarrow{\phi'} \operatorname{Hom}'_S(\omega_{S'/R'}, S'),$$

where ϕ and ϕ' are the bijections from H.19.

We first consider the case in which S/R has a trace.

Lemma H.21. *Let $\Delta \in \operatorname{Ann}_{S^e}(I^S)$ be the element corresponding to a trace σ of S/R and let $\Delta' := \gamma(\Delta)$. Then Δ' generates the S'-module $\operatorname{Ann}_{S'^e}(I^{S'})$. If σ' is the trace corresponding to Δ', then $\sigma' = \alpha(\sigma)$.*

Proof. With the dual basis $\{s'_1, \ldots, s'_m\}$ of B with respect to σ we have $\Delta = \sum_{i=1}^{m} s'_i \otimes s_i$ by H.20, hence $\Delta' = \sum_{i=1}^m (1 \otimes s'_i) \otimes (1 \otimes s_i)$. Since $\{1 \otimes s'_1, \ldots, 1 \otimes s'_m\}$ is obviously the dual basis of $1 \otimes B$ with respect to the trace $\alpha(\sigma)$, it follows again from H.20 that Δ' is the element corresponding to $\alpha(\sigma)$. Hence Δ' generates $\operatorname{Ann}_{S'}(I^{S'})$ and $\sigma' = \alpha(\sigma)$.

We now treat the special case where $R' = R/\mathfrak{a}$ is a residue class algebra of R, with $S' = S/\mathfrak{a}S$ and $\omega_{S'/R'} \cong \omega_{S/R}/\mathfrak{a}\omega_{S/R}$ via α (H.5).

Lemma H.22. *In this case the above mapping*

$$\beta\colon \operatorname{Hom}_S(\omega_{S/R}, S) \to \operatorname{Hom}_{S'}(\omega_{S'/R'}, S')$$

is given by reduction modulo $\mathfrak{a}$.

Proof. We denote residue classes modulo $\mathfrak{a}$ with a bar. For $\sum a_i \otimes b_i \in \operatorname{Ann}_{S^e}(I^S)$, the corresponding element under α is $\sum \overline{a}_i \otimes \overline{b}_i \in \operatorname{Ann}_{S'^e}(I^{S'})$. This element corresponds to the linear form in $\operatorname{Hom}_{S'}(\omega_{S'/R'}, S')$ given by $\ell' \mapsto \sum \ell'(\overline{a}_i)\overline{b}_i$. Choose $\ell \in \omega_{S/R}$ with $\overline{\ell} = \ell'$. Then

$$\sum \ell'(\overline{a}_i)\overline{b}_i = \sum \overline{\ell}(\overline{a}_i)\overline{b}_i = \overline{\sum \ell(a_i)b_i}.$$

Since $\phi(\sum a_i \otimes b_i)$ is given by $\ell \mapsto \sum \ell(a_i)b_i$, the result follows.

Lemma H.23. *Under the assumptions of H.22 let σ' be a trace of S'/R' and $\Delta' \in \operatorname{Ann}_{S'^e}(I^{S'})$ the element corresponding to σ'. Assume that either $\mathfrak{a}$ is in the intersection of all the maximal ideals of R, or that R and S are positively graded rings, $\mathfrak{a}$ is a homogeneous ideal, and Δ' is a homogeneous element. Then*

(a) $\alpha\colon \operatorname{Ann}_{S^e}(I^S) \to \operatorname{Ann}_{S'^e}(I^{S'})$ *is surjective.*
(b) *If $\Delta \in \operatorname{Ann}_{S^e}(I^S)$ is a (homogeneous) element with $\alpha(\Delta) = \Delta'$, then Δ generates the S-module $\operatorname{Ann}_{S^e}(I^S)$ and the trace σ of S/R corresponding to Δ is a (homogeneous) preimage of σ' under the epimorphism $\alpha\colon \omega_{S/R} \to \omega_{S'/R'}$.*

Proof. In H.22 it was shown that in the commutative diagram

$$\begin{array}{ccc} \operatorname{Ann}_{S^e}(I^S) & \xrightarrow{\alpha} & \operatorname{Ann}_{S'^e}(I^{S'}) \\ \downarrow{\scriptstyle\phi} & & \downarrow{\scriptstyle\phi'} \\ \operatorname{Hom}_S(\omega_{S/R}, S) & \xrightarrow{\beta} & \operatorname{Hom}_{S'}(\omega_{S'/R'}, S') \end{array}$$

the mapping β is given by reduction mod $\mathfrak{a}$. Since $\omega_{S'/R'} \cong S'$, it follows that $\omega_{S/R} \cong S$ by H.11 respectively H.13. In particular, β is surjective and therefore so is α. From Nakayama's Lemma it follows that $\operatorname{Ann}_{S^e}(I^S)$ is generated as an S-module by Δ.

Write $\Delta = \sum_{i=1}^m s_i' \otimes s_i$. Then $\Delta' = \sum_{i=1}^m \overline{s}_i' \otimes \overline{s}_i$, and $\sigma(s_i' s_j) = \delta_{ij}$, $\sigma'(\overline{s}_i' \overline{s}_j) = \delta_{ij}$ $(i, j = 1, \dots, m)$. By these equations σ and σ' are uniquely determined, for if $1 = \sum_{i=1}^m r_i s_i'$ $(r_i \in R)$, then $\sigma(s_j) = \sigma(\sum r_i s_i' s_j) = r_j$ and $\sigma'(\overline{s_j}) = \overline{r_j}$ $(j = 1, \dots, m)$. It follows that σ' is in fact the reduction of σ mod $\mathfrak{a}$.

Exercises

1. For S'/R' and α as in H.21 show that

$$\sigma_{S'/R'} = \alpha(\sigma_{S/R}).$$

2. Let L/K be a finite separable field extension of degree n and let $\overline{K}$ be the algebraic closure of K. Show that
   (a) $\overline{K} \otimes_K L$ is a direct product of n copies of the K-algebra $\overline{K}$. The canonical homomorphism $L \to \overline{K} \otimes_K L$ sends each $x \in L$ to the n-tuple $(x_1, \dots, x_n)$ of conjugates of x.
   (b) For each $x \in L$ we have $\sigma_{L/K}(x) = \sum_{i=1}^n x_i$.
3. Let A be a finite-dimensional algebra over a field K and $x \in A$ a nilpotent element. Show that $\sigma_{A/K}(x) = 0$.
4. Let A/K be as in Exercise 3. Additionally let A be a local ring with maximal ideal $\mathfrak{m}$. The socle of A is defined as

$$\mathfrak{S}(A) := \{x \in A \mid \mathfrak{m} \cdot x = \{0\}\}.$$

   Show that
   (a) $\mathfrak{S}(A)$ is an ideal of A, $\mathfrak{S}(A) \neq \{0\}$.
   (b) If A/K has a trace σ, then $\sigma(\mathfrak{S}(A)) \neq \{0\}$.

I

Ideal Quotients

At the end of Appendix H it was shown that in order for a trace to exist, certain annihilator ideals must be principal ideals. Here we give conditions under which this is the case (see I.5).

Let I and J be two ideals of a ring R.

Definition I.1. The *ideal quotient* $I : J$ is defined as

$$I : J := \{x \in R \mid xJ \subset I\}.$$

It is clear that $I : J$ is an ideal of R with $I \subset I : J$.

Lemma I.2. *Let $\mathfrak{a}$ be an ideal of R with $\mathfrak{a} \subset I \cap J$. The images of ideals of R in $\overline{R} := R/\mathfrak{a}$ will be denoted with a bar. Then in $\overline{R}$,*

$$\overline{I} : \overline{J} = \overline{I : J}.$$

Proof. Let $\overline{x}$ be the residue class of $x \in R$. From $\overline{x} \cdot \overline{J} \subset \overline{I}$ it follows that $xJ \subset I$. Therefore $\overline{x} \in \overline{I : J}$. Conversely, if $x \in I : J$, then of course $\overline{x} \in \overline{I} : \overline{J}$.

Corollary I.3. *Let $I \subset J$ and $\mathfrak{a} = I$. Then $\overline{I : J} = (0) : \overline{J} = \mathrm{Ann}_{\overline{R}}(\overline{J})$.*

This can be used as follows: Instead of calculating an annihilator in a residue class ring, it is sometimes advisable to determine an ideal quotient in the original ring.

Now let

$$J = (a_1, a_2), \quad I = (b_1, b_2), \qquad (a_i, b_i \in R),$$

and $I \subset J$. We write

$$\begin{aligned} b_1 &= r_{11}a_1 + r_{12}a_2, \\ b_2 &= r_{21}a_1 + r_{22}a_2, \end{aligned} \tag{1}$$

with $r_{ij} \in R$ and set $\Delta := \det(r_{ij})$. By Cramer's rule, $a_i\Delta \in (b_1, b_2)$ and therefore

$$\Delta \in I : J.$$

Lemma I.4. *Suppose b_1 is a nonzerodivisor on $R/(b_2)$ and b_2 is a nonzerodivisor on $R/(b_1)$. Then the image of Δ in $R/(b_1, b_2)$ is independent of the choice of the coefficients r_{ij} in equation (1).*

Proof. Write $b_2 = r'_{21}a_1 + r'_{22}a_2$ ($r'_{2i} \in R$) and apply Cramer's rule to the system of equations

$$\begin{aligned} b_1 &= r_{11}a_1 + r_{12}a_2, \\ 0 &= (r_{21} - r'_{21})a_1 + (r_{22} - r'_{22})a_2. \end{aligned}$$

With $\Delta' := \det \begin{pmatrix} r_{11} & r_{12} \\ r'_{21} & r'_{22} \end{pmatrix}$ we get

$$(a_1, a_2) \cdot (\Delta - \Delta') \subset (b_1),$$

and so in particular,

$$b_2(\Delta - \Delta') \subset (b_1) \subset (b_1, b_2).$$

Therefore, because b_2 is not a zerodivisor mod (b_1),

$$\Delta - \Delta' \in (b_1) \subset (b_1, b_2).$$

By symmetry it follows that the choice of any other representation of b_1 does not change the image of Δ in $R/(b_1, b_2)$.

Theorem I.5. *Suppose the following conditions are satisfied for the elements $a_1, a_2, b_1, b_2 \in R$ given above:*

(a) *a_1 and b_1 are nonzerodivisors on $R/(b_2)$.*
(b) *a_2 is a nonzerodivisor on $R/(a_1)$.*

Then

$$(b_1, b_2) : (a_1, a_2) = (\Delta, b_1, b_2).$$

If $\widetilde{R} := R/(b_1, b_2)$ and $\widetilde{J} := J/(b_1, b_2)$, then

$$\operatorname{Ann}_{\widetilde{R}}(\widetilde{J}) = (\widetilde{\Delta}),$$

where $\widetilde{\Delta}$ is the image of Δ in $\widetilde{R}$.

Proof. In the following, all calculations will be done in $\overline{R} := R/(b_2)$. We denote the residue class in $\overline{R}$ of an element of R with a bar. By (1) we have an equation

$$\overline{r}_{21}\overline{a}_1 + \overline{r}_{22}\overline{a}_2 = 0, \tag{2}$$

and using Cramer's rule it also follows from (1) that

$$\overline{a}_1\overline{\Delta} = \overline{b}_1\overline{r}_{22}. \tag{3}$$

By I.2 it suffices to prove the formula

$$(\overline{\Delta}, \overline{b}_1) = (\overline{b}_1) : (\overline{a}_1, \overline{a}_2).$$

Multiplication by $\overline{a}_1$ is injective in $\overline{R}$; therefore it is enough to prove the equation

(4) $$\overline{a}_1 \cdot ((\overline{b}_1) : (\overline{a}_1, \overline{a}_2)) = (\overline{a}_1 \overline{\Delta}, \overline{a}_1 \overline{b}_1) \overset{(3)}{=} (\overline{b}_1 \overline{r}_{22}, \overline{a}_1 \overline{b}_1).$$

We first show that

(5) $$(\overline{r}_{22}, \overline{a}_1) = (\overline{a}_1) : (\overline{a}_1, \overline{a}_2).$$

In fact, by (2) it is clear that $(\overline{r}_{22}, \overline{a}_1) \subset (\overline{a}_1) : (\overline{a}_1, \overline{a}_2)$. Suppose conversely that for an $x \in R$ the condition

$$\overline{x}\overline{a}_2 \in (\overline{a}_1)$$

is satisfied. Then one has an equation in R

$$xa_2 = c_1 a_1 + c_2 b_2 = c_1 a_1 + c_2 r_{21} a_1 + c_2 r_{22} a_2.$$

Since a_2 is a nonzerodivisor on $R/(a_1)$, we have

$$\overline{x} \in (\overline{r}_{22}, \overline{a}_1).$$

Therefore (5) has been proved, and by multiplication by $\overline{b}_1$ we get

$$(\overline{b}_1 \overline{r}_{22}, \overline{a}_1 \overline{b}_1) = \overline{b}_1 \cdot ((\overline{a}_1) : (\overline{a}_1, \overline{a}_2)).$$

Instead of (4) we now see that we have to prove the equation

(6) $$\overline{a}_1 \cdot ((\overline{b}_1) : (\overline{a}_1, \overline{a}_2)) = \overline{b}_1 \cdot ((\overline{a}_1) : (\overline{a}_1, \overline{a}_2)).$$

For $x \in R$ with $\overline{x}\overline{a}_i \in (\overline{b}_1)$ $(i = 1, 2)$ there is a $y \in R$ with $\overline{a}_1 \overline{x} = \overline{y}\overline{b}_1$. From $\overline{y}\overline{b}_1 \overline{a}_i = \overline{a}_1 \overline{x}\overline{a}_i \in (\overline{a}_1 \overline{b}_1)$ $(i = 1, 2)$ it follows that $\overline{y} \in (\overline{a}_1) : (\overline{a}_1, \overline{a}_2)$, since $\overline{b}_1$ is a nonzerodivisor on $\overline{R}$, and therefore

$$\overline{a}_1 \overline{x} \in \overline{b}_1 \cdot ((\overline{a}_1) : (\overline{a}_1, \overline{a}_2)).$$

That is, the left side of equation (6) is contained in the right side.

Now let $y \in R$ with $\overline{y}\overline{a}_i \in (\overline{a}_1)$ $(i = 1, 2)$ be given. Then $\overline{b}_1 \overline{y}\overline{a}_i \in (\overline{a}_1 \overline{b}_1)$ $(i = 1, 2)$, so in particular, $\overline{b}_1^2 \overline{y} \in (\overline{a}_1 \overline{b}_1)$ and therefore $\overline{b}_1 \overline{y} \in (\overline{a}_1)$. If one writes $\overline{b}_1 \overline{y} = \overline{a}_1 \overline{x}$ and uses $\overline{a}_1 \overline{x}\overline{a}_i \in (\overline{a}_1 \overline{b}_1)$, then one sees that $\overline{x} \in (\overline{b}_1) : (\overline{a}_1, \overline{a}_2)$, since $\overline{a}_1$ is a nonzerodivisor on $\overline{R}$, and hence

$$\overline{b}_1 \overline{y} \in \overline{a}_1 \cdot ((\overline{b}_1) : (\overline{a}_1, \overline{a}_2)).$$

The observations we have gone through here generalize to ideals with n generators, (see, e.g., [Ku$_2$], Appendix E).

K

Complete Rings. Completion

We restrict ourselves to I-adic filtered rings. The theory is developed in more generality in, e.g., Bourbaki [B], Greco-Salmon [GS], and Matsumura [M]. The completion of a filtered ring corresponds to the passage from the rationals to the real numbers or—in number theory—from the integers to the p-adic numbers. Completion of local rings is an important tool for studying singularities of algebraic curves.

Let R be a ring, $I \subset R$ an ideal, and $(a_n)_{n\in\mathbb{N}}$ a sequence of elements $a_n \in R$.

Definition K.1.

(a) The sequence $(a_n)_{n\in\mathbb{N}}$ *converges to* $a \in R$ (or has the limit a) if for each $\varepsilon \in \mathbb{N}$ there exists an $n_0 \in \mathbb{N}$ such that $a_n - a \in I^\varepsilon$ for all $n \geq n_0$. One then writes $a = \lim_{n\to\infty} a_n$.
(b) A sequence that converges to 0 will be called a *zero sequence.*
(c) The infinite series $\sum_{n\in\mathbb{N}} a_n$ converges to $a \in R$ if the sequence of partial sums $(\sum_{n=0}^k a_n)_{k\in\mathbb{N}}$ converges to a. One then writes $a = \sum a_n$.
(d) The sequence $(a_n)_{n\in\mathbb{N}}$ is called a *Cauchy sequence* if for each $\varepsilon \in \mathbb{N}$ there exists $n_0 \in \mathbb{N}$ such that $a_m - a_n \in I^\varepsilon$ for all $m, n \geq n_0$.

Remarks K.2.

(a) If the I-adic filtration on R is separated, then the limit of a convergent sequence is unique: If $a = \lim_{n\to\infty} a_n = a'$, then $a - a' \in \bigcap_{\varepsilon\in\mathbb{N}} I^\varepsilon = (0)$, therefore $a' = a$.
(b) Convergent sequences are Cauchy sequences.
(c) Every subsequence of a Cauchy sequence (of a sequence converging to $a \in R$) is a Cauchy sequence (a sequence converging to a).
(d) A sequence (a_n) is a Cauchy sequence if and only if $(a_{n+1} - a_n)_{n\in\mathbb{N}}$ is a zero sequence.
(e) If (a_n) is a Cauchy sequence, then we can assume by passing to a subsequence that

$$a_{n+1} - a_n \in I^n \qquad \text{for all } n \in \mathbb{N}.$$

(f) If $\sum_{n\in\mathbb{N}} a_n$ is convergent, then (a_n) is a zero sequence.
(g) If $a = \sum a_n$ and $b = \sum b_n$ are convergent series in R, then $\sum(a_n + b_n)$ converges to $a + b$ and the Cauchy product series $\sum_n(\sum_{\rho+\sigma=n} a_\rho \cdot b_\sigma)$ converges to ab.

Many other rules from analysis can be transferred over to our situation here. A few are even simpler here than there.

Definition K.3. A ring R is called *I-complete (or I-adically complete)* if every Cauchy sequence (with respect to I) converges to a limit in R.

Rules K.4.

(a) In a complete ring an infinite series $\sum a_n$ converges if and only if (a_n) is a zero sequence. (This is of course not the case in analysis.)

For if (a_n) is a zero sequence, then to each $\varepsilon \in \mathbb{N}$ there is an $n_0 \in \mathbb{N}$ such that $a_n \in I^\varepsilon$ for all $n \geq n_0$. Then also $\sum_{i=n}^m a_i \in I^\varepsilon$ for $m \geq n \geq n_0$; i.e., $(\sum_{i=0}^n a_i)_{n\in\mathbb{N}}$ is a Cauchy sequence. Since R is complete, $\sum_{n\in\mathbb{N}} a_n$ exists.

(b) R is I-complete if and only if for every zero sequence (a_n) in R, the infinite series $\sum a_n$ converges.

If this is the case and if (b_n) is an arbitrary Cauchy sequence, then $(b_{n+1} - b_n)$ is a zero sequence. The kth partial sum of the series $a = \sum_{n\in\mathbb{N}}(b_{n+1} - b_n)$ is $\sum_{n=0}^k (b_{n+1} - b_n) = b_{k+1} - b_0$. Therefore (b_n) converges to $a + b_0$.

(c) Let $k \in \mathbb{N}$. If R is I-complete and (a_n) is a Cauchy sequence with $a_n \in I^k$ for almost all $n \in \mathbb{N}$, then also $\lim_{n\to\infty} a_n \in I^k$ (I^k is closed with respect to taking limits).

In fact, if $a := \lim_{n\to\infty} a_n$, then $a - a_n \in I^k$ for large n. Since also $a_n \in I^k$ for large n, it follows that $a \in I^k$.

(d) Let $J \subset R$ be another ideal. Suppose there are numbers $\rho, \sigma \in \mathbb{N}$ with $J^\rho \subset I$, $I^\sigma \subset J$. Then R is I-complete if and only if R is J-complete.

One sees easily that (a_n) is a Cauchy sequence with respect to I if and only if it is a Cauchy sequence with respect to J. A similar statement holds for limits.

(e) Let R be I-complete and $\mathfrak{a} \subset R$ an ideal. Then $\overline{R} := R/\mathfrak{a}$ is complete with respect to $\overline{I} := (I + \mathfrak{a})/\mathfrak{a}$.

One can show easily that every zero sequence $(\overline{a}_n)$ in $\overline{R}$ arises from a zero sequence (a_n) in R. If $a = \sum a_n$, then the residue class $\overline{a}$ of a in R is the limit of $\sum_{n\in\mathbb{N}} \overline{a}_n$. Now use (b).

One of the most important properties of complete rings is the following version of Nakayama's lemma.

Theorem K.5. *Let R be I-complete and let M be an R-module such that $\bigcap_{k\in\mathbb{N}} I^k M = (0)$. Let $x_1, \dots, x_n \in M$ be elements with the property that $M = Rx_1 + \cdots + Rx_n + IM$. Then $M = Rx_1 + \cdots + Rx_n$. In other words, if M is an I-adically separated R-module and the residue class module M/IM is finitely generated, then M is finitely generated.*

Proof. Every $m \in M$ has a representation

$$(1) \qquad m = \sum_{i=1}^{n} r_i^{(0)} x_i + m' \qquad (r_i^{(0)} \in R,\ m' \in IM).$$

Write $m' = \sum s_k m_k$ ($s_k \in I,\ m_k \in M$) and choose for each m_k a representation (1). Then there is a new representation for m given by

$$m = \sum_{i=1}^{n} (r_i^{(0)} + r_i^{(1)}) x_i + m'' \qquad (r_i^{(1)} \in I,\ m'' \in I^2 M).$$

By induction, for every $k \in \mathbb{N}$ there is a representation

$$m = \sum_{i=1}^{n} \left(\sum_{j=0}^{k} r_i^{(j)} \right) x_i + m^{(k+1)} \qquad (r_i^{(j)} \in I^j,\ m^{(k+1)} \in I^{k+1} M).$$

Set $r_i := \sum_{j \in \mathbb{N}} r_i^{(j)}$ $(i = 1, \dots, n)$. This series converges by K.4(a), because its terms form a zero sequence and R is complete. Furthermore, $\sum_{j=k+1}^{\infty} r_i^{(j)} \in I^{k+1}$ by K.4(c). Hence for all $k \in \mathbb{N}$,

$$m - \sum_{i=1}^{n} r_i x_i = m^{(k+1)} - \sum_{i=1}^{n} \left(\sum_{j=k+1}^{\infty} r_i^{(j)} \right) x_i \in I^{k+1} M.$$

and from $\bigcap I^{k+1} M = (0)$ it follows that $m = \sum_{i=1}^{n} r_i x_i$.

Example K.6. The ring $R = P[[X_1, \dots, X_n]]$ of all formal power series in the indeterminates $X_1, \dots, X_n$ over a ring P is complete and separated with respect to $I = (X_1, \dots, X_n)$. In the following we shall write $X^\alpha := X_1^{\alpha_1} \cdots X_n^{\alpha_n}$ for $\alpha = (\alpha_1, \dots, \alpha_n) \in \mathbb{N}^n$.

If (f_k) is a zero sequence in R, then $\lim_{k \to \infty} (\operatorname{ord}_I f_k) = -\infty$, and if $f_k = \sum_\alpha a_\alpha^{(k)} X^\alpha$ ($a_\alpha^{(k)} \in P$), then

$$f := \sum_\alpha \left(\sum_k a_\alpha^{(k)} \right) X^\alpha$$

is well-defined, since in $\sum_k a_\alpha^{(k)}$ for each α only a finite number of nonzero summands appear. It is then clear that $f = \sum_{k \in \mathbb{N}} f_k$. From K.4(b) it follows that R is I-complete. That the I-adic filtration is separated is in any case clear.

Hilbert Basis Theorem for Power Series Rings K.7.
If P is a Noetherian ring, then $P[[X_1,\dots,X_n]]$ is also Noetherian.

Proof. Set $R := P[[X_1,\dots,X_n]]$ and $I := (X_1,\dots,X_n)$. It is clear that $\operatorname{gr}_I R \cong P[X_1,\dots,X_n]$ with $\deg X_i = -1$ ($i = 1,\dots,n$). By the Hilbert basis theorem for polynomial rings, every ideal in $\operatorname{gr}_I R$ is finitely generated. In particular, this holds for $\operatorname{gr}_I \mathfrak{a}$ if $\mathfrak{a}$ is an ideal of R.

For any such ideal suppose

$$\operatorname{gr}_I \mathfrak{a} = (L_I f_1,\dots,L_I f_m) \qquad \text{with } f_1,\dots,f_m \in \mathfrak{a}.$$

We will show that

$$\mathfrak{a} = (f_1,\dots,f_m).$$

If $f \in \mathfrak{a}$ is an arbitrary element, then

$$L_I f = \sum_{j=1}^{m} g_j^{(0)} L_I f_j,$$

where $g_j^{(0)} \in P[X_1,\dots,X_n]$ is homogeneous and $\deg g_j^{(0)} = \operatorname{ord}_I f - \operatorname{ord}_I f_j$. We have $f^{(1)} := f - \sum g_j^{(0)} f_j \in \mathfrak{a}$ and $\operatorname{ord}_I f^{(1)} < \operatorname{ord}_I f$. Now write

$$L_I f^{(1)} = \sum_{j=1}^{m} g_j^{(1)} L_I f_j,$$

where $g_j^{(1)} \in P[X_1,\dots,X_n]$ is homogeneous, $\deg g_j^{(1)} = \operatorname{ord}_I f^{(1)} - \operatorname{ord}_I f_j$, and we get

$$f^{(2)} := f^{(1)} - \sum_{j=1}^{m} g_j^{(1)} f_j = f - \sum_{j=1}^{m} \left(g_j^{(0)} + g_j^{(1)}\right) f_j \in \mathfrak{a},$$

with $\operatorname{ord}_I f^{(2)} < \operatorname{ord}_I f^{(1)} < \operatorname{ord}_I f$. By induction we construct a zero sequence $(f^{(k)})_{k\in\mathbb{N}}$ with

$$f^{(k)} = f - \sum_{j=1}^{m} \left(\sum_{i=0}^{k-1} g_j^{(i)}\right) f_j \in \mathfrak{a} \quad \text{and} \quad \operatorname{ord}_I f^{(k)} \le (\operatorname{ord}_I f) - k,$$

where also $(g_j^{(i)})_{i\in\mathbb{N}}$ is a zero sequence in $P[[X_1,\dots,X_n]]$. It follows that

$$f = \sum_{j=1}^{m} g_j f_j \quad \text{with} \quad g_j := \sum_{i=0}^{\infty} g_j^{(i)} \quad (j = 1,\dots,m).$$

We show next that complete rings are frequently homomorphic images of power series rings.

Theorem K.8. *Let R be I-complete and $\rho : P \to R$ a ring homomorphism. Furthermore, let $x_1, \dots, x_n \in I$ and $\bigcap_{k\in\mathbb{N}} I^k = (0)$.*

(a) *There exists a unique P-algebra homomorphism (called a substitution homomorphism)*

$$\varepsilon : P[[X_1, \dots, X_n]] \to R \quad \text{with} \quad \varepsilon(X_i) = x_i \quad (k = 1, \dots, n).$$

(b) *If $I = (x_1, \dots, x_n)$ and if the composition of ρ with the canonical epimorphism $R \to R/I$ is bijective, then ε is surjective. In this case if P is Noetherian, then R is also Noetherian.*

Proof. (a) For an arbitrary power series $\sum a_\alpha X^\alpha \in P[[X_1, \dots, X_n]]$, the series $\sum_n \sum_{|\alpha|=n} \rho(a_\alpha)x^\alpha$ converges in R, since its terms form a zero sequence. Using K.2(g) it is easy to show that the assignment $\sum a_\alpha X^\alpha \mapsto \sum \rho(a_\alpha)x^\alpha$ gives a P-homomorphism ε with $\varepsilon(X_k) = x_k$. On the basis of continuity there can be only one such ε.

(b) Consider R as a module over $P[[X_1, \dots, X_n]]$ and set $J := (X_1, \dots, X_n)$. By assumption, ε induces a bijection of $P = P[[X_1, \dots, X_n]]/J$ onto R/I; i.e., R/JR is generated as a $P[[X_1, \dots, X_n]]$-module by the image of the unit element of R. The hypotheses of K.5 are fulfilled (K.6). Hence R as a $P[[X_1, \dots, X_n]]$-module is generated by 1; i.e., ε is surjective. The last statement of (b) follows from K.7.

Corollary K.9. *Let R be a Noetherian local ring with maximal ideal $\mathfrak{m} = (x_1, \dots, x_n)$. Suppose R contains a field K that is mapped bijectively under $R \to R/\mathfrak{m}$ onto $R/\mathfrak{m}$. If R is $\mathfrak{m}$-complete, then there exists a unique K-epimorphism*

$$K[[X_1, \dots, X_n]] \to R \qquad (X_i \mapsto x_i).$$

Proof. By the Krull intersection theorem (E.8) we know that $\bigcap \mathfrak{m}^k = (0)$, and then K.8 can be applied.

Corollary K.10. *Under the assumptions of K.9 let R be a complete discrete valuation ring and $\mathfrak{m} = (t)$. Then there exists a unique K-isomorphism*

$$K[[T]] \xrightarrow{\sim} R \qquad (T \mapsto t).$$

Proof. By K.9 there is a unique K-epimorphism $\varepsilon : K[[T]] \to R$ with $\varepsilon(T) = t$. If $\ker \varepsilon \neq (0)$, then $T^n \in \ker \varepsilon$ for some $n \in \mathbb{N}$, and then $t^n = 0$, a contradiction. Therefore ε is bijective.

We identify the rings R and $K[[T]]$ in the situation of K.10 and denote by ν the discrete valuation belonging to R, so for every $r \in R \setminus \{0\}$ the value $\nu(r)$ is precisely the order of the power series in $K[[T]]$ represented by r.

In the following, when we speak of complete local rings, we will always mean that they are complete with respect to their maximal ideals, and furthermore that they are separated.

We have the following version of the Chinese remainder theorem.

Theorem K.11. *Let R be a complete Noetherian local ring with maximal ideal $\mathfrak{m}$ and let S be an R-algebra that is finitely generated as an R-module. Suppose $\mathfrak{M}_1, \dots, \mathfrak{M}_h$ are the maximal ideals of S. Then the canonical ring homomorphism*

$$\alpha : S \to S_{\mathfrak{M}_1} \times \cdots \times S_{\mathfrak{M}_h}$$

is an isomorphism. Furthermore, S is $\mathfrak{m}S$-complete and the $S_{\mathfrak{M}_i}$ are complete Noetherian local rings $(i = 1, \dots, h)$.

Proof. Consider $S_{\mathfrak{M}_1} \times \cdots \times S_{\mathfrak{M}_h}$ as an R-module. Since the $\mathfrak{M}_i$ have $\mathfrak{m}$ in R as preimage (F.9) and $S_{\mathfrak{M}_i}$ is Noetherian, $\bigcap_{k \in \mathbb{N}} \mathfrak{m}^k (S_{\mathfrak{M}_1} \times \cdots \times S_{\mathfrak{M}_h}) = (0)$ by the Krull intersection theorem. Set $\overline{S} := S/\mathfrak{m}S$, $\overline{\mathfrak{M}}_i := \mathfrak{M}_i/\mathfrak{m}S$ $(i = 1, \dots, h)$. Then $\overline{S}$ is a finitely generated $R/\mathfrak{m}$-algebra and $\overline{S}$ has only the prime ideals $\overline{\mathfrak{M}}_i$ $(i = 1, \dots, h)$. By the Chinese remainder theorem (D.3) there are canonical isomorphisms

$$\begin{aligned} \overline{S} &\cong \overline{S}_{\overline{\mathfrak{M}}_1} \times \cdots \times \overline{S}_{\overline{\mathfrak{M}}_h} \cong S_{\mathfrak{M}_1}/\mathfrak{m}S_{\mathfrak{M}_1} \times \cdots \times S_{\mathfrak{M}_h}/\mathfrak{m}S_{\mathfrak{M}_h} \\ &\cong S_{\mathfrak{M}_1} \times \cdots \times S_{\mathfrak{M}_h}/\mathfrak{m}(S_{\mathfrak{M}_1} \times \cdots \times S_{\mathfrak{M}_h}). \end{aligned}$$

Choose elements $x_1, \dots x_n \in S$ whose images in $\overline{S}_{\overline{\mathfrak{M}}_1} \times \cdots \times \overline{S}_{\overline{\mathfrak{M}}_h}$ generate this ring as an R-module. By K.5 we have

$$S_{\mathfrak{M}_1} \times \cdots \times S_{\mathfrak{M}_h} = R \cdot \alpha(x_1) + \cdots + R \cdot \alpha(x_n).$$

Hence α is surjective.

Furthermore, $\ker \alpha = \bigcap_{i=1}^h \ker \alpha_i$ where $\alpha_i : S \to S_{\mathfrak{M}_i}$ is the canonical homomorphism. Of course, $\ker \alpha_i = \{s \in S \mid \exists\, t \in S \setminus \mathfrak{M}_i \text{ such that } ts = 0\}$. For each $s \in \ker \alpha$ we therefore have $\operatorname{Ann}(s) \not\subset \mathfrak{M}_i$ for $i = 1, \dots, h$, i.e., $\operatorname{Ann}(s) = S$, and hence $s = 0$. This shows that α is a bijection.

Now let (a_n) be a zero sequence of S with respect to $I = \mathfrak{m}S$ and let $\{s_1, \dots, s_m\}$ be a system of generators of S as an R-module. We can write each a_n in the form

$$a_n = r_1^{(n)} s_1 + \cdots + r_m^{(n)} s_m,$$

where the $(r_j^{(n)})_{n \in \mathbb{N}}$ for $j = 1, \dots, m$ are zero sequences in R. Since $r_j := \sum r_j^{(n)}$ exists in R, it is clear that $\sum a_n$ converges to $\sum_{j=1}^m r_j s_j$ in S. By K.4(b), then, S is $\mathfrak{m}S$-complete.

Since $S_{\mathfrak{M}_i}$ is a homomorphic image of S, by K.4(e), each $S_{\mathfrak{M}_i}$ is an $\mathfrak{m}S_{\mathfrak{M}_i}$-complete ring. Also, there exists a $\rho_i \in \mathbb{N}$ such that $\mathfrak{M}_i^{\rho_i} S_{\mathfrak{M}_i} \subset \mathfrak{m}S_{\mathfrak{M}_i}$ (C.12), and by K.4(d), it follows that each $S_{\mathfrak{M}_i}$ is also complete with respect to its maximal ideal $\mathfrak{M}_i S_{\mathfrak{M}_i}$.

Now let R again be an arbitrary ring with an I-adic filtration and let $\mathfrak{a} \subset R$ be an ideal. The *closure* $\overline{\mathfrak{a}}$ of $\mathfrak{a}$ is the set of all limits of convergent sequences $(a_k)_{k \in \mathbb{N}}$ with $a_k \in \mathfrak{a}$ for all $k \in \mathbb{N}$.

Theorem K.12. (a) *We always have*

$$\overline{\mathfrak{a}} = \bigcap_{k \in \mathbb{N}} (\mathfrak{a} + I^k) \quad \text{and} \quad \overline{\overline{\mathfrak{a}}} = \overline{\mathfrak{a}}.$$

(b) *If R is Noetherian as well as complete and separated with respect to I, then $\overline{\mathfrak{a}} = \mathfrak{a}$ for every ideal $\mathfrak{a} \subset R$.*

Proof. (a) Let $a = \lim_{k\to\infty} a_k$, where $a_k \in \mathfrak{a}$ for all $k \in \mathbb{N}$. By passing to a subsequence we can assume $a - a_k \in I^k$ for all $k \in \mathbb{N}$ and hence $a \in \bigcap_{k\in\mathbb{N}}(\mathfrak{a} + I^k)$.

Conversely, let $a \in \bigcap_{k\in\mathbb{N}}(\mathfrak{a} + I^k)$ be given. Then for each $k \in \mathbb{N}$ there is a representation $a = a_k + b_k$ with $a_k \in \mathfrak{a}$, $b_k \in I^k$. Hence (b_k) is a zero sequence, and therefore $a = \lim_{k\to\infty} a_k$ exists. This proves the first formula of (a), and the second follows immediately.

(b) Let $\mathfrak{a} = (f_1, \ldots, f_m)$ and let $a \in \overline{\mathfrak{a}}$, so $a = \lim_{k\to\infty} a_k$ $(a_k \in \mathfrak{a})$. We can assume that $a - a_k \in I^k$ for all $k \in \mathbb{N}$ and thus

$$a_{k+1} - a_k \in \mathfrak{a} \cap I \qquad (k \in \mathbb{N}).$$

By Artin–Rees (E.5) there exists a $k_0 \in \mathbb{N}$ such that

$$\mathfrak{a} \cap I^{k+k_0} = I^k \cdot (\mathfrak{a} \cap I^{k_0}) \quad \text{for all} \quad k \in \mathbb{N}.$$

Write

$$a_{k_0} = \sum_{j=1}^{m} r_j^{(0)} f_j \qquad (r_j^{(0)} \in R),$$

and for $t > k_0$,

$$a_t - a_{t-1} = \sum_{j=1}^{m} r_j^{(t-k_0)} f_j \qquad (r_j^{(t-k_0)} \in I^{t-k_0-1}).$$

Then

$$a_t = \sum_{j=1}^{m} \left(\sum_{s=0}^{t-k_0} r_j^{(s)} \right) f_j,$$

and with $r_j := \sum_{s=0}^{\infty} r_j^{(s)}$ we see that $a = \sum_{j=1}^{m} r_j f_j$.

Now that we have learned some good properties of complete rings, we will try to embed an arbitrary ring with an I-adic filtration into a complete ring in order to take advantage of these properties. We now assume that the ideal I of R is finitely generated. Without this assumption the theory of completions leaves the category of I-adic filtered rings.

Definition K.13. A (separated) *completion* of (R, I) is a pair $(\hat{R}, i)$, where $i : R \to \hat{R}$ is a homomorphism to a ring $\hat{R}$ such that

(a) $\hat{R}$ is $\hat{I}$-complete and separated with respect to an ideal $\hat{I}$ containing $I\hat{R}$.
(b) If $j : R \to S$ is any homomorphism to a ring S that is complete and separated with respect to an ideal J containing IS, then there is exactly one ring homomorphism $h : \hat{R} \to S$ with $h(\hat{I}) \subset J$ and $j = h \circ i$:

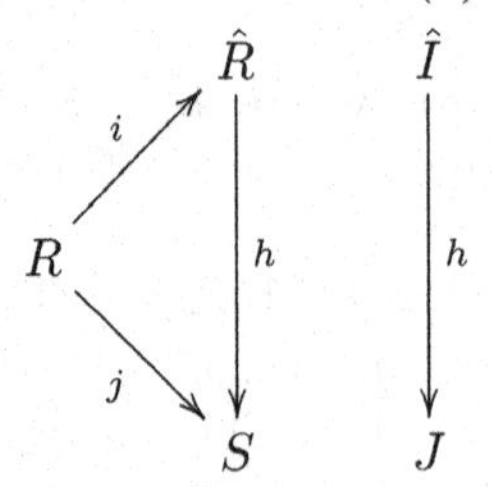

If $(\hat{R}, i)$ exists, then as with all objects that are defined by a universal property, it is unique up to isomorphism. We then also call $\hat{R}$ the completion of (R, I) and $i : R \to \hat{R}$ the canonical homomorphism into the completion.

Under the assumption that $I = (a_1, \dots, a_n)$ has a finite system of generators, the completion always exists. By K.6 the ring $R[[X_1, \dots, X_n]]$ of formal power series in $X_1, \dots, X_n$ over R is complete and separated with respect to $\mathfrak{M} := (X_1, \dots, X_n)$. Let $\mathfrak{a} = (X_1 - a_1, \dots, X_n - a_n)$ and let

$$\overline{\mathfrak{a}} = \bigcap_{k \in \mathbb{N}} (\mathfrak{a} + \mathfrak{M}^k)$$

be the closure of $\mathfrak{a}$ (K.12(a)). Then

$$\hat{R} := R[[X_1, \dots, X_n]]/\overline{\mathfrak{a}}$$

is complete with respect to

$$\hat{I} := (\mathfrak{M} + \overline{\mathfrak{a}})/\overline{\mathfrak{a}}$$

by K.4(e). Also, it is clear that the $\hat{I}$-adic filtration of $\hat{R}$ is separated, for if $z \in \bigcap_{k \in \mathbb{N}} \hat{I}^k$ and $y \in R[[X_1, \dots, X_n]]$ is a preimage of z, then $y = \overline{\mathfrak{a}} + \mathfrak{M}^k$ for all $k \in \mathbb{N}$; hence $y \in \overline{\overline{\mathfrak{a}}} = \overline{\mathfrak{a}}$ and therefore $z = 0$.

Let $i : R \to \hat{R}$ be the composition of the canonical injection $R \to R[[X_1, \dots, X_n]]$ with the canonical epimorphism

$$R[[X_1, \dots, X_n]] \to R[[X_1, \dots, X_n]]/\overline{\mathfrak{a}}.$$

Since $X_k - a_k \in \overline{\mathfrak{a}}$, we see that $i(a_k)$ is equal to the residue class x_k of X_k in $\hat{R}$ $(k = 1, \dots, n)$. Therefore

$$\hat{I} = (i(a_1), \dots, i(a_n)) = I\hat{R}.$$

Thus it has been shown that condition (a) of definition K.13 is satisfied.

Now let $j : R \to S$ be given as in K.13(b). By the universal property of power series rings (K.8(a)), j can be extended to a ring homomorphism

$$H : R[[X_1, \dots, X_n]] \to S \quad \text{with} \quad H(X_k) = j(a_k) \quad (k = 1, \dots, n).$$

Since a_k and X_k have the same image in S, $\mathfrak{a} \subset \ker H$, and since S is separated with respect to J, we even have $\overline{\mathfrak{a}} \subset \ker H$. Thus there is an induced homomorphism

$$h : R[[X_1, \dots, X_n]]/\overline{\mathfrak{a}} \to S.$$

By the construction of h it is clear that $j = h \circ i$ and $h(\hat{I}) \subset J$.

There can be only one such homomorphism h. If an arbitrary homomorphism is composed with the canonical epimorphism $R[[X_1, \dots, X_n]] \to \hat{R}$, then we get a homomorphism $H : R[[X_1, \dots, X_n]] \to S$ that agrees with j on R and maps X_k onto $j(a_k)$ $(k = 1, \dots, n)$. Under these requirements H is uniquely determined, and hence also the map h induced by H.

Remark K.14. The following formulas follow from the existence proof of the completion:

$$\hat{R} = R[[X_1, \dots, X_n]]/ \bigcap_{k \in \mathbb{N}} ((X_1 - a_1, \dots, X_n - a_n) + (X_1, \dots, X_n)^k)$$

and

$$\hat{I} = I\hat{R}.$$

Applying K.7 and K.12(b) yields the following.

Theorem K.15. *If R is Noetherian, then*

$$\hat{R} = R[[X_1, \dots, X_n]]/(X_1 - a_1, \dots, X_n - a_n)$$

and $\hat{R}$ is also Noetherian.

We are interested in the completion of a residue class ring of R. Under the assumptions of K.13 let $\mathfrak{a} \subset R$ be an ideal, let $\widehat{R/\mathfrak{a}}$ be the completion of $R/\mathfrak{a}$ with respect to $(I + \mathfrak{a})/\mathfrak{a}$, and $j : R/\mathfrak{a} \to \widehat{R/\mathfrak{a}}$ the canonical map into the completion. By K.13(b) there is a commutative diagram

$$\begin{array}{ccc} R & \longrightarrow & R/\mathfrak{a} \\ \downarrow i & & \downarrow j \\ \hat{R} & \xrightarrow[h]{} & \widehat{R/\mathfrak{a}} \end{array}$$

Theorem K.16 (Permutability of Completion and Residue Classes). *If R is Noetherian, then h induces an isomorphism*

$$\hat{R}/\mathfrak{a}\hat{R} \xrightarrow{\sim} \widehat{R/\mathfrak{a}}.$$

Proof. Let $I = (a_1, \ldots, a_n)$ and let $\overline{a}_k \in R/\mathfrak{a}$ be the residue class of a_k $(k = 1, \ldots, n)$. Using K.15 we see that

$$\begin{aligned}\widehat{R/\mathfrak{a}} &\cong (R/\mathfrak{a})[[X_1, \ldots, X_n]]/(X_1 - \overline{a}_1, \ldots, X_n - \overline{a}_n)\\ &\cong R[[X_1, \ldots, X_n]]/\mathfrak{a}R[[X_1, \ldots, X_n]] + (X_1 - a_1, \ldots, X_n - a_n)\\ &\cong \hat{R}/\mathfrak{a}\hat{R}.\end{aligned}$$

Examples K.17.

(a) If $R = P[X_1, \ldots, X_n]$ is a polynomial ring over a ring P and $I = (X_1, \ldots, X_n)$, then the I-adic completion of R is

$$\hat{R} = P[[X_1, \ldots, X_n]],$$

and $i : R \to \hat{R}$ is the canonical injection of the polynomial ring into the power series ring.

Indeed, the conditions of definition K.13 are satisfied for i by K.8(a).

(b) Now let $P = K[[u_1, \ldots, u_m]]$ itself be a power series ring over a field K, and let $\mathfrak{m} = (u_1, \ldots, u_m)$ be its maximal ideal. In $P[X_1, \ldots, X_n]$, then, $\mathfrak{M} := (\mathfrak{m}, X_1, \ldots, X_n)$ is a maximal ideal. By the universal property of localizations, there is a canonical injection

$$i : P[X_1, \ldots, X_n]_{\mathfrak{M}} \to P[[X_1, \ldots, X_n]].$$

One easily determines that $P[[X_1, \ldots, X_n]] = K[[u_1, \ldots, u_m, X_1, \ldots, X_n]]$ is the completion of the local ring $P[X_1, \ldots, X_n]_{\mathfrak{M}}$ with respect to its maximal ideal.

In general, from K.16 it follows for a local ring of the form

$$R = P[X_1, \ldots, X_n]_{\mathfrak{M}}/(f_1, \ldots, f_t) \qquad (f_i \in P[X_1, \ldots, X_n])$$

that

$$\hat{R} = P[[X_1, \ldots, X_n]]/(f_1, \ldots, f_t) \cdot P[[X_1, \ldots, X_n]].$$

We will now concern ourselves with a few properties of the ring $R = K[[X_1, \ldots, X_n]]$ of formal power series over a field K $(n > 0)$. Clearly R is a local integral domain with maximal ideal $\mathfrak{m} = (X_1, \ldots, X_n)$; i.e., the units of R are precisely those power series whose constant term does not vanish. By K.6, R is separated and complete with respect to $\mathfrak{m}$. By K.7, R is Noetherian, and hence every finitely generated R-algebra is also Noetherian; in particular, every residue class ring of R is Noetherian. We now come to the main point, that R is a unique factorization domain. This will be shown with the help of the Weierstraß preparation theorem.

Definition K.18. A power series $f \in K[[X_1, \ldots, X_n]]$ is called *X_n-general of order m* if $f(0, \ldots, 0, X_n) = \sum_{\nu=0}^{\infty} a_\nu X_n^{m+\nu}$ with $a_\nu \in K$, $a_0 \neq 0$.

Expressed differently, X_n^m occurs in f with a coefficient $\neq 0$, but no X_n^i with $i < m$ occurs in f with a nonzero coefficient.

Theorem K.19 (Weierstraß Preparation Theorem).
Let the power series $f \in K[[X_1,\dots,X_n]]$ be X_n-general of order m, set $S := K[[X_1,\dots,X_n]]/(f)$, and let x_n be the residue class of X_n in S. Then $\{1, x_n, \dots, x_n^{m-1}\}$ is a basis of S as a $K[[X_1,\dots,X_{n-1}]]$-module. In other words, for each $g \in K[[X_1,\dots,X_n]]$ there are uniquely determined series $q \in K[[X_1,\dots,X_n]]$, $r \in K[[X_1,\dots,X_{n-1}]][X_n]$ with $\deg_{X_n} r < m$ such that

$$g = q \cdot f + r.$$

Proof. If f is a unit ($m = 0$), then there is nothing to show. So let f be a nonunit. Then there is a K-homomorphism (K.8(a))

$$K[[X_1,\dots,X_{n-1},Y]] \to K[[X_1,\dots,X_n]] \quad (X_i \mapsto X_i,\ Y \mapsto f).$$

Let $\mathfrak{n} := (X_1,\dots,X_{n-1},Y)$ be the maximal ideal of $P := K[[X_1,\dots,X_{n-1},Y]]$ and set $R := K[[X_1,\dots,X_n]]$. Then we have

$$R/\mathfrak{n}R \cong K[[X_n]]/(f(0,\dots,0,X_n)) = K[[X_n]]/(X_n^m).$$

The images of the elements $1, X_n, \dots, X_n^{m-1}$ in $R/\mathfrak{n}R$ thus form a K-basis of $R/\mathfrak{n}R$. Since R is separated with respect to its maximal ideal, $\bigcap_{k\in\mathbb{N}} \mathfrak{n}^k R = 0$, and therefore K.5 is applicable. It follows that $\{1, X_n, \dots, X_n^{m-1}\}$ is a generating system of R as a P-module. Then also $\{1, x_n, \dots, x_n^{m-1}\}$ is a generating system of $S = R/(f)$ as a module over $P/(Y) = K[[X_1,\dots,X_{n-1}]]$.

We will show by induction on n that $\{1, x_n, \dots, x_n^{m-1}\}$ is even a basis of $S/K[[X_1,\dots,X_{n-1}]]$. For $n = 1$ there is nothing to show, so let $n > 1$ and suppose the claim has already been shown for $n-1$ variables.

Consider a relation

$$\sum_{i=0}^{m-1} \rho_i x_n^i = 0 \qquad (\rho_i \in K[[X_1,\dots,X_{n-1}]]).$$

Since $S/X_1S \cong K[[X_2,\dots,X_n]]/(f(0,X_2,\dots,X_n))$ and since $f(0,X_2,\dots,X_n)$ is X_n-general of order m, by the induction hypothesis applied to the above relation it must be that all the ρ_i are divisible by X_1. In R there is an equation

$$\sum_{i=0}^{m-1} \rho_i X_n^i = q \cdot f \qquad (q \in R),$$

and X_1 is a prime element of R. Since f is not divisible by X_1, it must be that q is. Write $\rho_i = X_1 \sigma_i$ ($i = 0,\dots,m-1$). Then we have

$$\sum_{i=0}^{m-1} \sigma_i x_n^i = 0.$$

Again, all of the σ_i must be divisible by X_1. By induction it follows that $\rho_i \in \bigcap_{k\in\mathbb{N}} X_1^k R = (0)$ $(i = 0, \dots, m-1)$.

Thus the first part of the preparation theorem has been proved. The existence of a representation $g = qf + r$ and the uniqueness of r follow immediately. But then q is also unique, because R is an integral domain.

Corollary K.20. *For each f as in the theorem there is a uniquely determined monic polynomial $\tilde{f} \in K[[X_1, \dots, X_{n-1}]][X_n]$ of the form*

$$\tilde{f} = X_n^m + \sum_{i=0}^{m-1} \alpha_i X_n^i \quad (\alpha_i \in (X_1, \dots, X_{n-1}))$$

and a unit $\varepsilon \in K[[X_1, \dots, X_n]]$ such that

$$f = \varepsilon \cdot \tilde{f}.$$

Proof. In $S = R/(f)$ there is an equation

$$x_n^m = -\sum_{i=0}^{m-1} \alpha_i x_n^i \quad (\alpha_i \in K[[X_1, \dots, X_{n-1}]]).$$

Set $\tilde{f} := X_n^m + \sum_{i=0}^{m-1} \alpha_i X_n^i$. Then $\tilde{f} = q \cdot f$ for some $q \in K[[X_1, \dots, X_n]]$. In this equation we set $X_1 = \dots = X_{n-1} = 0$. Then by comparing coefficients with respect to X_n, we see that all the α_i lie in $(X_1, \dots X_{n-1})$ and that q has a nonzero constant term, so q is a unit. Using $\varepsilon := q^{-1}$, we have $f = \varepsilon \cdot \tilde{f}$. The uniqueness of $\tilde{f}$ is clear, because $\{1, x_n, \dots, x_n^{m-1}\}$ is a basis of $S/K[[X_1, \dots, X_{n-1}]]$.

The polynomial $\tilde{f}$ is called the *Weierstraß polynomial* of the power series f. If f and g are X_n-general power series with Weierstraß polynomials $\tilde{f}$ respectively $\tilde{g}$, then $\tilde{f} \cdot \tilde{g}$ is the Weierstraß polynomial of $f \cdot g$. This follows from the uniqueness statement in K.20.

Lemma K.21. *Let $f_1, \dots f_r \in K[[X_1, \dots, X_n]] \setminus \{0\}$ be given. Then there is a K-automorphism α of $K[[X_1, \dots, X_n]]$ such that $\alpha(f_1), \dots, \alpha(f_r)$ are X_n-general.*

Proof. We content ourselves to prove this only in the case where K is infinite. The automorphism α can then be given by a substitution

$$X_j \mapsto X_j + \rho_j X_n \quad (j = 1, \dots, n-1), \quad X_n \mapsto X_n$$

with suitably chosen $\rho_j \in K$. Let $L(f_i)$ be the leading form of f_i with respect to the $(X_1, \dots, X_n)$-filtration, let $d_i := \deg L(f_i)$, and set $\lambda := \prod_{i=1}^r L(f_i)$. By suitable choice of ρ_j, the above substitution in λ leads to a polynomial in X_n of degree $\sum_{i=1}^r d_i$. Then all the $L(f_i)$ have degree d_i in X_n, and the f_i are X_n-general of order d_i $(i = 1, \dots, r)$.

Theorem K.22. *$K[[X_1, \dots, X_n]]$ is a unique factorization domain.*

Proof. Since $R := K[[X_1, \dots, X_n]]$ is Noetherian, it is enough to show that every irreducible element $f \in R$ generates a prime ideal. By K.21 we can assume that f is X_n-general of some order, say m. If $\tilde{f}$ is the Weierstraß polynomial of f, then

$$R/(f) \cong K[[X_1, \dots, X_{n-1}]][X_n]/(\tilde{f}),$$

and it suffices to show that $\tilde{f} \in K[[X_1, \dots, X_{n-1}]][X_n]$ generates a prime ideal.

Suppose we have already shown that $K[[X_1, \dots, X_{n-1}]]$ is a unique factorization domain. Then so is $P := K[[X_1, \dots, X_{n-1}]][X_n]$, and it is enough to show that $\tilde{f}$ is irreducible in this ring.

If $\tilde{f}$ were reducible, then there would be monic polynomials in X_n, say $g, h \in P$ with $\deg_{X_n} g < m$, $\deg_{X_n} h < m$, such that $\tilde{f} = g \cdot h$. Then $f = \varepsilon g h$ for some unit $\varepsilon \in R$. If g and h are both nonunits, then this contradicts the irreducibility of f. If, say, h is a unit in R, then we get a contradiction to the uniqueness of the Weierstraß polynomial. In every case $\tilde{f}$ must therefore be irreducible in P.

Exercises

Assume that the assumptions of K.13 are satisfied.

1. Show that $i(R)$ is "dense" in $\hat{R}$; i.e., every element $x \in \hat{R}$ is the limit of a Cauchy sequence $(i(r_k))_{k \in \mathbb{N}}$ with elements $r_k \in R$. Every $x \in \hat{R}$ can also be written as an infinite series

$$x = \sum_{k \in \mathbb{N}} i(r_k)$$

   with a zero sequence $(r_k)_{k \in \mathbb{N}}$ from R.
2. Show that the homomorphism

$$R/I^k \to \hat{R}/\hat{I}^k \qquad (k \in \mathbb{N})$$

   induced by $i : R \to \hat{R}$ is an isomorphism. Conclude that

$$\ker(i) = \bigcap_{k \in \mathbb{N}} I^k.$$

   If R is I-adically separated, then $i : R \to \hat{R}$ is injective.

L

Tools for a Proof of the Riemann–Roch Theorem

This appendix contains a few facts from linear algebra that occur in the proof of the Riemann–Roch theorem as given by F.K. Schmidt in [Sch]. The ideas of F.K. Schmidt will be formulated here in the language of Appendices B and H. The actual proof of the Riemann–Roch theorem is then rather short, and it also results instantly in a proof of the "singular case" (cf. Chapter 13).

Let K be an arbitrary field and L/K an algebraic function field of one variable. This means that there is an $x \in L$ that is transcendental over K, while L is finite algebraic over $K(x)$. In the following we assume that K is algebraically closed in L, and fix a transcendental x of L/K. Set $n := [L : K(x)]$, $R := K[x]$ and $R_\infty := K[x^{-1}]_{(x^{-1})}$, the localization of the polynomial ring $K[x^{-1}]$ with respect to its maximal ideal (x^{-1}).

The ring R_∞ is a discrete valuation ring of $K(x)$ (E.11). If ν_∞ is the corresponding discrete valuation, then for $\frac{f}{g} \in K(x)$ ($f \in K[x], g \in K[x]\setminus\{0\}$) we have the formula:

$$\nu_\infty\left(\frac{f}{g}\right) = \deg g - \deg f, \tag{1}$$

as one easily sees. We denote by S_∞ the integral closure of R_∞ in L.

Remark L.1. S_∞ is a free R_∞-module of rank n.

If $L/K(x)$ is separable, then from F.7 it follows that S_∞ is finitely generated as an R_∞-module. The remark is valid in the general case, but we will show this only when K is a perfect field, the only case that really interests us.

Let $p := \operatorname{Char} K$. Hence $K = K^p = K^{p^2} = \cdots$, and let L' be the separable closure of $K(x)$ in L, S' the integral closure of R_∞ in L'. Then S' is finitely generated as an R_∞-module (F.7). Furthermore, there is an $e \in \mathbb{N}$ such that $L^{p^e} \subset L'$. Hence $S_\infty^{p^e} \subset S'$. Then $R_\infty^{p^e} = K[x^{-p^e}]_{(x^{-p^e})} \subset S_\infty^{p^e}$. Clearly R_∞ is finitely generated over $K[x^{-p^e}]_{(x^{-p^e})}$. Then S' is also finitely generated over $K[x^{-p^e}]_{(x^{-p^e})}$, and so, of course, is $S_\infty^{p^e}$ as well. But via Frobenius, S_∞/R_∞ is isomorphic to $S_\infty^{p^e}/R_\infty^{p^e}$. It follows that S_∞ is finitely generated over R_∞.

Because S_∞ is a torsion-free R_∞-module, by the fundamental theorem for modules over principal ideal domains, it is even free. Each element of L can be written as a fraction with a numerator from S_∞ and a denominator from R_∞. Hence every R_∞-basis of S_∞ is also a $K(x)$-basis of L, and S_∞ has the same rank over R_∞ as L does over $K(x)$, namely n.

We consider now also an extension ring A of $R = K[x]$ with $Q(A) = L$, where A is a finitely generated R-module. Now, A can (but we don't need this here) be the integral closure of R in L, since this is finitely generated as an R-module, as one can show by similar arguments to those that led to L.1. Just as S_∞ is free over R_∞, it is also true that A is a free R-module of rank n. Furthermore, we have:

Remark L.2. $A \cap S_\infty = K$.

Proof. Let $y \in A \cap S_\infty$ and let $f \in K(x)[T]$ be the minimal polynomial of y over $K(x)$. Since y is integral over R and R is integrally closed in $K(x)$, all the coefficients of f are contained in R by F.14. For the same reason all the coefficients of f are contained in R_∞, hence in $R_\infty \cap R = K$. Therefore y is algebraic over K and so $y \in K$, since K is algebraically closed in L.

After these preparations we now come to the main point. For $\alpha \in \mathbb{Z}$ let $\mathcal{F}_\alpha := x^\alpha S_\infty$. Since $x^{-1} \in S_\infty$, we have $\mathcal{F}_\alpha = x^\alpha S_\infty = x^{-1}x^{\alpha+1}S_\infty = x^{-1}\mathcal{F}_{\alpha+1} \subset \mathcal{F}_{\alpha+1}$. One sees immediately that $\mathcal{F} := \{\mathcal{F}_\alpha\}_{\alpha\in\mathbb{Z}}$ is a separated filtration of the R_∞-algebra L (B.1). For the associated graded ring we have

$$(2) \qquad \operatorname{gr}_{\mathcal{F}} L = \bigoplus_{\alpha\in\mathbb{Z}} \mathcal{F}_\alpha/\mathcal{F}_{\alpha-1} = \bigoplus_{\alpha\in\mathbb{Z}} x^\alpha S_\infty/x^{\alpha-1}S_\infty = S_\infty/(x^{-1})[T, T^{-1}].$$

That is, $\operatorname{gr}_{\mathcal{F}} L$ is the ring of Laurent polynomials in T over $S_\infty/(x^{-1})$, where T corresponds to the leading form $L_{\mathcal{F}}x = x + S_\infty$ of x and T^{-1} to the leading form $L_{\mathcal{F}}x^{-1} = x^{-1} + x^{-2}S_\infty$ of x^{-1}. Recall that for $a \in L^*$,

$$(3) \qquad \operatorname{ord}_{\mathcal{F}} a = \operatorname{Min}\{\alpha \in \mathbb{Z} \mid a \in \mathcal{F}_\alpha\}$$

and

$$L_{\mathcal{F}}a = a + \mathcal{F}_{\operatorname{ord} a - 1} \in S_\infty/(x^{-1}) \cdot T^{\operatorname{ord} a}.$$

If we write $a = x^{\operatorname{ord} a} \cdot b$ with $b \in S_\infty$, then

$$(4) \qquad L_{\mathcal{F}}a = \overline{b} \cdot T^{\operatorname{ord} a},$$

where $\overline{b}$ is the residue class of b in $S_\infty/(x^{-1})$.

The restriction of $\mathcal{F}$ to R_∞ is the $\mathfrak{m}_\infty$-adic filtration of R_∞, if $\mathfrak{m}_\infty$ denotes the maximal ideal of R_∞, and we can identify $\operatorname{ord}_{\mathcal{F}}$ on R_∞ with $-\nu_\infty$, since we have

$$\mathcal{F}_\alpha \cap R_\infty = x^\alpha S_\infty \cap R_\infty = x^\alpha R_\infty \cap R_\infty = \begin{cases} R_\infty, & \alpha \geq 0, \\ x^\alpha R_\infty, & \alpha < 0. \end{cases}$$

On the other hand the restriction of $\mathcal{F}$ to R is the degree filtration $\mathcal{G}$ of the polynomial ring $R = K[x]$, for by (1),

$$\mathcal{F}_\alpha \cap K[x] = x^\alpha S_\infty \cap K[x] = x^\alpha K[x^{-1}]_{(x^{-1})} \cap K[x]$$

is the K-vector space $\mathcal{G}_\alpha$ of polynomials of degree $\leq \alpha$. In

$$\operatorname{gr}_{\mathcal{F}} L = S_\infty/(x^{-1})[T, T^{-1}]$$

we identify $\operatorname{gr}_{\mathcal{F}} R_\infty$ with $K[T^{-1}]$ and $\operatorname{gr}_{\mathcal{F}} R$ with $K[T]$.

Now we consider a finitely generated A-module $I \subset L$, $I \neq \{0\}$. Then I is also a free R-module of rank n, because I contains a submodule isomorphic to A. A fundamental idea of F.K. Schmidt is to construct an R-basis of I that can be transformed in a simple way to a basis of S_∞ over R_∞.

Set $\mathcal{F}'_\beta := \mathcal{F}_\beta \cap I$ ($\beta \in \mathbb{Z}$) and

$$\operatorname{gr}_{\mathcal{F}} I := \bigoplus_{\beta \in \mathbb{Z}} \mathcal{F}'_\beta / \mathcal{F}'_{\beta-1},$$

where $\operatorname{gr}_{\mathcal{F}} I$ is to be viewed at first as a graded K-vector space. Because

$$\mathcal{G}_\alpha \cdot \mathcal{F}'_\beta = (\mathcal{F}_\alpha \cap K[x]) \cdot (\mathcal{F}_\beta \cap I) \subset \mathcal{F}_{\alpha+\beta} \cap I = \mathcal{F}'_{\alpha+\beta},$$

it is clear that $\operatorname{gr}_{\mathcal{F}} I$ is a graded module over the graded ring $\operatorname{gr}_{\mathcal{G}} K[x] = K[T]$. The canonical mapping

$$\mathcal{F}_\beta \cap I / \mathcal{F}_{\beta-1} \cap I \to \mathcal{F}_\beta / \mathcal{F}_{\beta-1}$$

is injective. Therefore $\operatorname{gr}_{\mathcal{F}} I$ can be considered as a $K[T]$-submodule of $\operatorname{gr}_{\mathcal{F}} L$.

Lemma L.3. *The grading on* $\operatorname{gr}_{\mathcal{F}} I$ *is bounded below.*

Proof. Since I is a finitely generated A-module and $Q(A) = L$, there is an $f \in L \setminus \{0\}$ with $fI \subset A$. Because $A \cap \mathcal{F}_0 = A \cap S_\infty = K$ (L.2) and $A \cap \mathcal{F}_{-1} = (0)$, it follows that $\operatorname{ord}_{\mathcal{F}}(a) \geq 0$ for every $a \in A$. Hence for $x \in I$ (B.2b),

$$0 \leq \operatorname{ord}_{\mathcal{F}}(fz) \leq \operatorname{ord}_{\mathcal{F}}(f) + \operatorname{ord}_{\mathcal{F}}(z)$$

and therefore $\operatorname{ord}_{\mathcal{F}}(z) \geq -\operatorname{ord}_{\mathcal{F}}(f)$.

Let $\operatorname{gr}_{\mathcal{F}}^{\alpha_0} I$ be the homogeneous component of smallest degree of $\operatorname{gr}_{\mathcal{F}} I$. There are then elements $a_1, \dots, a_{\nu_1} \in I$ with $\operatorname{ord}_{\mathcal{F}} a_i = \alpha_0$ ($i = 1, \dots, \nu_1$) such that $\{L_{\mathcal{F}} a_1, \dots, L_{\mathcal{F}} a_{\nu_1}\}$ is a K-basis of $\operatorname{gr}_{\mathcal{F}}^{\alpha_0} I$. Write $a_i = x^{\alpha_0} b_i$ with $b_i \in S_\infty$, and then by formula (4) we have $L_{\mathcal{F}} a_i = \overline{b_i} \cdot T^{\alpha_0} \in (S_\infty/(x^{-1})) \cdot T^{\alpha_0}$ ($i = 1, \dots, \nu_1$), where $\overline{b_i}$ is the residue class of b_i in $S_\infty/(x^{-1})$.

Choose elements $a_{\nu_1+1}, \dots, a_{\nu_2} \in I$ such that $\{L_{\mathcal{F}} a_1 \cdot T, \dots, L_{\mathcal{F}} a_{\nu_1} \cdot T\}$ is extended by $L_{\mathcal{F}} a_{\nu_1+1}, \dots, L_{\mathcal{F}} a_{\nu_2}$ to a K-basis of $\operatorname{gr}_{\mathcal{F}}^{\alpha_0+1} I$. As above, write $L_{\mathcal{F}} a_j = \overline{b_j} \cdot T^{\alpha_0+1}$, $b_j \in S_\infty$ ($j = \nu_1 + 1, \dots, \nu_2$). It is then clear that $\{\overline{b}_1, \dots, \overline{b}_{\nu_2}\}$ are K-linearly independent elements of $S_\infty/(x^{-1})$.

By iterating this method one finds elements $a_1, \dots, a_m \in I$ such that the leading forms $L_{\mathcal{F}} a_i$ ($i = 1, \dots, m$) are a system of generators of the $K[T]$-module $\operatorname{gr}_{\mathcal{F}} I$. Here

$$L_{\mathcal{F}} a_i = \overline{b}_i \cdot T^{\operatorname{ord} a_i}$$

with $b_i := a_i x^{-\operatorname{ord} a_i} \in S_\infty$, where $\overline{b}_i$ denotes the residue class of b_i in $S_\infty/(x^{-1})$ ($i = 1, \dots, m$). Furthermore, $\{\overline{b}_1, \dots, \overline{b}_m\}$ is K-linearly independent, hence $m \leq n$.

Theorem L.4. (a) *We have* $m = n$ *and* $\{a_1, \dots, a_n\}$ *is an* R*-basis of* I.
(b) $\{a_1 x^{-\operatorname{ord} a_1}, \dots, a_n x^{-\operatorname{ord} a_n}\}$ *is an* R_∞*-basis of* S_∞.

Proof. (a) For $a \in I \setminus \{0\}$, the leading form $L_{\mathcal{F}} a$ can be written in the form $L_{\mathcal{F}} a = \sum_{i=1}^m \kappa_i T^{\mu_i} \cdot L_{\mathcal{F}} a_i$ ($\kappa_i \in K, \mu_i + \operatorname{ord}_{\mathcal{F}} a_i = \operatorname{ord}_{\mathcal{F}} a$ for $i = 1, \dots, m$). Then

$$\operatorname{ord}_{\mathcal{F}} \left(a - \sum_{i=1}^m \kappa_i x^{\mu_i} a_i \right) < \operatorname{ord}_{\mathcal{F}} a.$$

Since the orders of the elements of I are bounded below (L.3), it follows by induction that $a \in K[x]a_1 + \cdots + K[x]a_m$ (cf. B.9). Therefore $\{a_1, \dots, a_m\}$ is a generating system for the R-module I. Since I is free of rank n over R and $m \le n$, we must have $m = n$, and $\{a_1, \dots, a_n\}$ is an R-basis of I.

(b) Since $\dim_K S_\infty/(x^{-1}) = n$, it follows from (a) that $\{\bar{b}_1, \dots, \bar{b}_n\}$ is a K-basis of $S_\infty/(x^{-1})$. Because $b_i = a_i x^{-\operatorname{ord} a_i}$ ($i = 1, \dots, n$), it follows using Nakayama's lemma that $\{a_1 x^{-\operatorname{ord} a_1}, \dots, a_n x^{-\operatorname{ord} a_n}\}$ is an R_∞-basis of S_∞.

Definition L.5. An R-basis $\{a_1, \dots, a_n\}$ of I is called a *standard basis* of I if there are integers $\alpha_1, \dots, \alpha_n \in \mathbb{Z}$ such that $\{a_1 x^{-\alpha_1}, \dots, a_n x^{-\alpha_n}\}$ is an R_∞-basis of S_∞.

The existence of a standard basis was shown by L.4. If a basis as in L.5 is given, then $a_i \in x^{\alpha_i} S_\infty$, but $a_i \notin x^{\alpha_i - 1} S_\infty$, because $a_i x^{-\alpha_i} \notin (x^{-1}) S_\infty$. Hence $\alpha_i = \operatorname{ord}_{\mathcal{F}} a_i$ ($i = 1, \dots, n$).

Theorem L.6. (a) $I \cap S_\infty$ *is a finite-dimensional vector space over* K.
(b) *If* $\{a_1, \dots, a_n\}$ *is a standard basis of* I, *then*

$$\dim_K (I \cap S_\infty) = \sum_{\operatorname{ord}_{\mathcal{F}} a_i \le 0} (-\operatorname{ord}_{\mathcal{F}} a_i + 1).$$

Proof. First of all, we have

$$I \cap S_\infty = \bigoplus_{i=1}^n K[x] a_i \cap \bigoplus_{i=1}^n x^{-\operatorname{ord} a_i} K[x^{-1}]_{(x^{-1})} a_i = \bigoplus_{i=1}^n \mathcal{G}_{-\operatorname{ord} a_i} \cdot a_i.$$

Since $\mathcal{G}_\alpha = 0$ for $\alpha < 0$ and $\dim_K \mathcal{G}_\alpha = \alpha + 1$ for $\alpha \ge 0$, we get the desired dimension formula.

Now let A' be another extension ring of R in L with $Q(A') = L$ and let A' be finitely generated as an R-module. Further, let $I' \ne \{0\}$ be a finitely generated A'-module with $I' \subset L$. Then I' also has a standard basis $\{a'_1, \dots, a'_n\}$. If $I \subset I'$ and $\{a_1, \dots, a_n\}$ is a standard basis of I, then there are equations

$$(5) \qquad a_i = \sum_{j=1}^n \rho_{ij} a'_j \qquad (i = 1, \dots, n; \quad \rho_{ij} \in R).$$

The determinant of this transformation $\Delta := \det(\rho_{ij})$ is a polynomial in $R = K[x]$. Let $\deg \Delta$ be the degree of this polynomial.

Theorem L.7.

$$\dim_K I'/I = \sum_{i=1}^n \operatorname{ord}_{\mathcal{F}} a_i - \sum_{i=1}^n \operatorname{ord}_{\mathcal{F}} a'_i = \deg \Delta.$$

Proof. Let $\alpha_i := \operatorname{ord}_{\mathcal{F}} a_i$ and $\alpha'_i := \operatorname{ord}_{\mathcal{F}} a'_i$ $(i = 1, \ldots, n)$. Then

$$\{a_1 x^{-\alpha_1}, \ldots, a_n x^{-\alpha_n}\} \quad \text{and} \quad \{a'_1 x^{-\alpha'_1}, \ldots, a'_n x^{-\alpha'_n}\}$$

are two R_∞-bases of S_∞ (L.4). From (5) we obtain

$$a_i x^{-\alpha_i} = \sum_{j=1}^n (x^{\alpha'_j - \alpha_i} \rho_{ij})(a'_j x^{-\alpha'_j}).$$

Let $\delta := \sum_{i=1}^n (\alpha'_i - \alpha_i)$. Then $x^\delta \cdot \Delta$ is the determinant of this system and so is a unit in R_∞. That is, we have

$$\nu_\infty(x^\delta \Delta) = \sum_{i=1}^n (\alpha_i - \alpha'_i) + \nu_\infty(\Delta) = \sum_{i=1}^n (\alpha_i - \alpha'_i) - \deg \Delta = 0,$$

and therefore

$$\sum_{i=1}^n (\alpha_i - \alpha'_i) = \deg \Delta.$$

By the fundamental theorem for modules over a principal ideal domain, there is a basis $\{c_1, \ldots, c_n\}$ of the R-module I' and there are polynomials $e_1, \ldots, e_n \in R$ such that $\{e_1 c_1, \ldots, e_n c_n\}$ is an R-basis of I. Then $I'/I \cong R/(e_1) \oplus \cdots \oplus R/(e_n)$, and it follows that $\dim_K I'/I = \sum_{i=1}^n \deg e_i = \deg \prod_{i=1}^n e_i$. The determinant of the transformation from $\{a'_1, \ldots, a'_n\}$ to $\{c_1, \ldots, c_n\}$ is a unit of R, hence an element of K^*. The same is true for the determinant of the transformation from $\{e_1 c_1, \ldots, e_n c_n\}$ to $\{a_1, \ldots, a_n\}$. From this it follows that $\deg \Delta = \deg \prod e_i = \dim_K I'/I$.

Following F.K. Schmidt, we now dualize with respect to a trace σ of $L/K(x)$ in the sense of Appendix H. So let

$$\omega_{L/K(x)} = \operatorname{Hom}_{K(x)}(L, K(x)) = L \cdot \sigma,$$

with a fixed chosen trace σ. If $L/K(x)$ is separable, we can of course choose the canonical trace $\sigma_{L/K(x)}$, and then we have a canonical duality.

If A and I are given as above, then I and also $\operatorname{Hom}_R(I, R)$ are finitely generated A-modules that are free as R-modules. The canonical mapping $\operatorname{Hom}_R(I, R) \to \operatorname{Hom}_{K(x)}(L, K(x))$ is injective. We identify $\operatorname{Hom}_R(I, R)$ with its image in $L \cdot \sigma$. Then

$$\operatorname{Hom}_R(I, R) = I^* \cdot \sigma,$$

where $I^* \subset L$ is a finitely generated A-module with $I^* \neq \{0\}$. For example, we have

$$\mathrm{Hom}_R(A, R) = \mathfrak{C}_{A/R} \cdot \sigma$$

with a finitely generated A-module $\mathfrak{C}_{A/R}$. This is called the (Dedekind) *complementary module of A/R* (with respect to σ). Similarly, $\mathrm{Hom}_{R_\infty}(S_\infty, R_\infty) = \mathfrak{C}_{S_\infty/R_\infty} \cdot \sigma$ with a finitely generated S_∞-module $\mathfrak{C}_{S_\infty/R_\infty}$, which is called the complementary module of S_∞/R_∞. In general, we have

$$I^* = \{z \in L \mid \sigma(za) \in R \text{ for all } a \in I\}, \tag{6}$$

and in particular,

$$\mathfrak{C}_{A/R} = \{z \in L \mid \sigma(za) \in R \text{ for all } a \in A\} \tag{7}$$

as well as

$$\mathfrak{C}_{S_\infty/R_\infty} = \{z \in L \mid \sigma(zb) \in R_\infty \text{ for all } b \in S_\infty\}. \tag{8}$$

Let $B = \{a_1, \dots, a_n\}$ be a standard basis of I. The elements $a_i^\vee$ of the dual basis of B in $\mathrm{Hom}_R(I, R)$ can be written in the form $a_i^\vee = a_i^* \cdot \sigma$ ($a_i^* \in I^*$), and then $B^* := \{a_1^*, \dots, a_n^*\}$ is an R-basis of I^*, the dual basis of B with respect to σ (H.9). For these we have

$$\sigma(a_i a_j^*) = \delta_{ij} \qquad (i, j = 1, \dots, n).$$

From

$$\sigma(a_i x^{-\operatorname{ord} a_i} \cdot x^{\operatorname{ord} a_j} a_j^*) = x^{\operatorname{ord} a_j - \operatorname{ord} a_i} \delta_{ij} = \delta_{ij}$$

we see that an R_∞-basis of $\mathfrak{C}_{S_\infty/R_\infty}$ is given by $\{b_1^*, \dots, b_n^*\}$ with $b_j^* := x^{\operatorname{ord} a_j} a_j^*$ ($j = 1, \dots, n$), the dual to the basis $\{b_1, \dots, b_n\}$ of S_∞/R_∞ ($b_i := a_i x^{-\operatorname{ord} a_i}$).

Theorem L.8. $\dim_K(I^* \cap x^{-2}\mathfrak{C}_{S_\infty/R_\infty}) = \sum_{\operatorname{ord}_{\mathcal{F}} a_i \geq 1} (\operatorname{ord}_{\mathcal{F}} a_i - 1)$.

Proof. We have

$$I^* = \bigoplus_{i=1}^{n} K[x] a_i^*$$

and

$$x^{-2}\mathfrak{C}_{S_\infty/R_\infty} = \bigoplus_{i=1}^{n} x^{\operatorname{ord} a_i - 2} K[x^{-1}]_{(x^{-1})} a_i^*.$$

As in the proof of L.6, the formula in the statement of the theorem follows immediately.

Setting $\chi(I) := \dim_K(I \cap S_\infty) - \dim_K(I^* \cap x^{-2}\mathfrak{C}_{S_\infty/R_\infty})$, from L.6 and L.8 we get the following

Corollary L.9. $\chi(I) = n - \sum_{i=1}^{n} \mathrm{ord}_{\mathcal{F}}\, a_i$.

Furthermore, from this corollary and theorem L.7 we obtain

Corollary L.10. *Under the assumptions of L.7 we have*

$$\chi(I') - \chi(I) = \dim_K(I'/I).$$

This is already in essence the Riemann–Roch theorem; we need only to interpret this formula in the language of divisors and functions.

Let $\{a_1, \ldots, a_n\}$ be a standard basis of A, so that by L.9 we have

$$\chi(A) = \dim_K(A \cap S_\infty) - \dim_K(\mathfrak{C}_{A/R} \cap x^{-2}\mathfrak{C}_{S_\infty/R_\infty}) = n - \sum_{i=1}^{n} \mathrm{ord}_{\mathcal{F}}\, a_i.$$

Since $A \cap S_\infty = K$ (L.2), we have

Corollary L.11. $\dim_K(\mathfrak{C}_{A/R} \cap x^{-2}\mathfrak{C}_{S_\infty/R_\infty}) = \sum_{i=1}^{n} \mathrm{ord}_{\mathcal{F}}\, a_i - n + 1$.

Hint. This formula has an interpretation in terms of differentials, which, however, we will not use. In case $L/K(x)$ is separable and σ is the canonical trace, we can consider the intersection $\mathfrak{C}_{A/R}dx \cap x^{-2}\mathfrak{C}_{S_\infty/R_\infty}dx$ inside the module of differentials $\Omega^1_{L/K}$ (G.10). Because $x^{-2}dx = -dx^{-1}$, we get using L.11 the formula

$$\dim_K(\mathfrak{C}_{A/K[x]}dx \cap \mathfrak{C}_{S_\infty/K[x^{-1}]_{(x^{-1})}}dx^{-1}) = \sum_{i=1}^{n} \mathrm{ord}_{\mathcal{F}}\, a_i - n + 1.$$

The mysterious factor x^{-2} from the earlier formula does not appear here. The vector space $\mathfrak{C}_{A/K[x]}dx \cap \mathfrak{C}_{S_\infty/K[x^{-1}]_{(x^{-1})}}dx^{-1}$ is called the vector space of "global regular differentials" with respect to A.

There is also the following formula for the "dual module" I^* of I.

Theorem L.12. *We have* $I^* = \mathfrak{C}_{A/R} :_L I := \{f \in L \mid f \cdot I \subset \mathfrak{C}_{A/R}\}$ *and* $(I^*)^* = I$.

Proof. By definition of I^* and $\mathfrak{C}_{A/R}$ (cf. (6) and (7)),

$$I^* = \{z \in L \mid \sigma(zb) \in R \text{ for all } b \in I\}$$

and

$$\mathfrak{C}_{A/R} = \{u \in L \mid \sigma(ua) \in R \text{ for all } a \in A\}.$$

For $z \in \mathfrak{C}_{A/R} :_L I$ and an arbitrary $b \in I$ we have $\sigma(zb) \in \sigma(\mathfrak{C}_{A/R}) \subset R$. Hence $\mathfrak{C}_{A/R} :_L I \subset I^*$. Conversely, if $z \in I^*$ and $b \in I$, then $\sigma(zba) \in R$ for an arbitrary $a \in A$, since $ba \in I$. It follows that $zb \in \mathfrak{C}_{A/R}$ and $I^* \subset \mathfrak{C}_{A/R} :_L I$.

Using the above notation, B is the dual basis to the R-basis B^* of I^*, i.e., $I = (I^*)^*$.

In addition to I^*, one frequently considers also

$$I' := \{z \in L \mid zI \subset A\} =: A :_L I. \tag{9}$$

This is also a finitely generated A-module $\neq \{0\}$. For $z \in L^*$, we have the formula

$$(z \cdot I)' = z^{-1} \cdot I'. \tag{10}$$

If $\mathfrak{C}_{A/R}$ is generated as an A-module by an element z, then by L.12,

$$I^* = \mathfrak{C}_{A/R} : I = (z \cdot A) : I = zI', \tag{11}$$

and as a result we have by the second statement of L.12,

Corollary L.13. *If $\mathfrak{C}_{A/R}$ is generated as an A-module by one element, then*

$$(I')' = I.$$

List of Symbols

The numbers indicate the pages where the symbol is defined

References

[AM] Abhyankar, S.S. and Moh, T.T. Newton-Pusieux expansion and generalized Tschirnhausen transformations I, II. *J. reine angew. Math.* 260, (1973), 47–83 and 261 (1973), 29–54.

[An] Angermüller, G. Die Wertehalbgruppe einer ebenen irreduziblen algebroiden Kurve. *Math. Z.* 153 (1977), 267–282.

[Ap] Apéry, R. Sur les branches superlinéaires des courbes algébriques. *C.R. Acad. Sci. Paris 222* (1946), 1198–1200.

[Az] Azevedo, A. The Jacobian ideal of a plane algebroid curve. Thesis. Purdue Univ. 1967.

[BDF] Barucci, V., Dobbs, D.E., Fontana, M. Maximality properties in numerical semigroups and applications to one-dimensional analytically irreducible local domains. *Memoirs Am. Math. Soc.* 598 (1997).

[BDFr$_1$] Barucci, V., D'Anna, M., Fröberg, R. On plane algebroid curves. *Commutative ring theory and applications* (Fez 2001), 37–50, Lect. Notes Pure Appl. Math., 231, Dekker, New York, 2003.

[BDFr$_2$] — The Apéry algorithm for a plane singularity with two branches. *Beiträge Algebra und Geometrie* (to appear).

[B] Bourbaki, N. *Commutative Algebra.* Springer 1991.

[BC] Bertin, J. and P. Carbonne. Semi-groupe d'entiers et application aux branches. *J. Algebra* 49 (1977), 81–95.

[BK] Brieskorn, E. and H. Knörrer. *Plane Algebraic Curves.* Birkhäuser 1986.

[CDK] Campillo, A., Delgado, F., Kiyek, K. Gorenstein property and symmetry for one-dimensional local Cohen-Macaulay rings. *Manuscripta Math.* 83 (1994), 405–423.

[C] Chevalley, C. *Introduction to the Theory of Algebraic Functions of One Variable.* Math. Surveys VI. Am. Math. Soc. New York 1951.

[De] Delgado, F. The semigroup of values of a curve singularity with several branches. *Manuscripta Math.* 59 (1987), 347–374.

[E] Eisenbud, D. *Commutative Algebra with a View Towards Algebraic Geometry,* Springer 1995.

[EH] Eisenbud, D. and J. Harris. Existence, decomposition, and limits of certain Weierstrass points. *Invent. Math.* 87 (1987), 495–515.

[Fa] Faltings, G. Endlichkeitssätze für abelsche Varietäten über Zahlkörpern. *Invent. Math.* 73 (1983), 349–366.

[F] Fischer, G. *Plane Algebraic Curves*, AMS student math. library, Vol. 15 (2001).

[Fo] Forster, O. *Lectures on Riemann Surfaces*, Springer 1999.

[Fu] Fulton, W. *Algebraic Curves*, Benjamin, New York 1969.

[Ga] Garcia, A. Semigroups associated to singular points of plane curves. *J. reine angew. Math.* 336 (1982), 165–184.

[GSt] Garcia, A. and K.O. Stöhr. On semigroups of irreducible plane curves. *Comm. Algebra* 15 (1987), 2185–2192.

[Go] Gorenstein, D. An arithmetic theory of adjoint plane curves. *Trans. AMS* 72 (1952), 414–436.

[GS] Greco, S. and P. Salmon. *Topics in* $\mathfrak{m}$*-adic Topologies*, Springer 1971.

[GH] Griffiths, P. and J. Harris, *Principles of Algebraic Geometry*, Wiley, New York, 1978.

[H] Hartshorne, R. *Residues and Duality*, Springer Lecture Notes in Math. 20 (1966).

[Hü] Hübl, R. Residues of regular and meromorphic differential forms. *Math. Ann.* 300 (1994), 605–628.

[HK] Hübl, R. and E. Kunz, On the intersection of algebraic curves and hypersurfaces. *Math. Z.* 227 (1998), 263–278.

[Hu] Humbert, G. Application géométrique d'un théorème de Jacobi. *J. Math.* (4) 1 (1885), 347–356.

[Hus] Husemöller, D. *Elliptic Curves*, Springer 1986.

[J] Jacobi, K.G. Teoremata nova algebraica circa systema duarum aequationem inter duas variabilis propositarum. *J. reine angew. Math.* 14 (1835), 281–288.

[K] Koblitz, N. *Algebraic Aspects of Cryptography*, Springer 1998.

[KK] Kreuzer, M. and E. Kunz. Traces in strict Frobenius algebras and strict complete intersections. *J. reine angew. Math.* 381 (1987), 181–204.

[KR] Kreuzer, M. and L. Robbiano. *Computational Commutative Algebra I.* Springer 2000.

[Ku$_1$] Kunz, E. *Introduction to Commutative Algebra and Algebraic Geometry*, Birkhäuser 1985.

[Ku$_2$] — *Kähler Differentials*, Advanced Lectures in Math. Vieweg, Braunschweig 1986.

[Ku$_3$] — Über den n-dimensionalen Residuensatz. *Jahresber. deutsche Math.-Verein.* 94 (1992), 170–188.

[Ku$_4$] — Geometric applications of the residue theorem on algebraic curves. In: *Algebra, Arithmetic and Geometry with Applications.* Papers from Shreeram S. Abhyankar's 70th Birthday Conference. C. Christensen, G. Sundaram, A. Sathaye, C. Bajaj, editors. (2003), 565–589.

[KW] Kunz, E. and R. Waldi. Deformations of zero dimensional intersection schemes and residues. *Note di Mat.* 11 (1991), 247–259.

[L] Lang, S. *Elliptic Functions*, Springer 1987.

[Li$_1$] Lipman, J. Dualizing sheaves, differentials and residues on algebraic varieties. *Astérisque* 117 (1984).

[Li$_2$] — Residues and traces of differential forms via Hochschild homology. *Contemporary Math.* 61 (1987).

[M] Matsumura, H. *Commutative Algebra*, Benjamin, New York 1980.

[Mo] Moh, T.T. On characteristic pairs of algebroid plane curves for characteristic p. *Bull. Inst. Math. Acad. Sinica 1* (1973), 75–91.

[P] Pretzel, O. *Codes and Algebraic Curves*, Oxford Science Publishers. Oxford 1998.

[Q] Quarg, G. Über Durchmesser, Mittelpunkte and Krümmung projektiver algebraischer Varietäten. Thesis. Regensburg 2001.

[R] Roquette, P. Über den Riemann–Rochschen Satz in Funktionenkörpern vom Transzendenzgrad 1. *Math. Nachr.* 19 (1958), 375–404.

[SS$_1$] Scheja, G. and U. Storch. Über Spurfunktionen bei vollständigen Durchschnitten. *J. reine angew. Math.* 278/279 (1975), 174–190.

[SS$_2$] — Residuen bei vollständigen Durchschnitten. *Math. Nachr.* 91 (1979), 157–170.

[Sch] Schmidt, F.K. Zur arithmetischen Theorie der algebraischen Funktionen I. Beweis der Riemann–Rochschen Satzes für algebraische Funktionen mit beliebigem Konstantenkörper. *Math. Z.* 41 (1936), 415–438.

[Se] Segre, B. Sui theoremi di Bézout, Jacobi e Reiss. *Ann. di Mat.* (4) 26 (1947), 1–16.

[S$_1$] Silverman, J. *The Arithmetic of Elliptic Curves*, Springer 1994.

[S$_2$] — *Advanced Topics in the Arithmetic of Elliptic Curves*, Springer 1999.

[Si] Singh, S. *Fermat's Last Theorem. The Story of a Riddle That Confounded the World's Greatest Minds for 358 Years*, Fourth Estate, London (1997).

[St] Stichtenoth, H. *Algebraic Function Fields and Codes*, Universitext. Springer 1993.

[TW] Taylor, R. and A. Wiles, Ring theoretic properties of certain Hecke algebras. *Annals of Math.* 141 (1995), 533-572.

[W] Washington, L. *Elliptic Curves, Number Theory and Cryptography*, Chapman and Hall 2003.

[Wa$_1$] Waldi, R. Äquivariante Deformationen monomialer Kurven. *Regensburger Math. Schriften* 4 (1980).

[Wa$_2$] Waldi, R. On the equivalence of plane curve singularities. *Comm. Algebra* 28 (2000), 4389–4401.

[Wi] Wiles, A. Modular elliptic curves and Fermat's Last Theorem. *Annals of Math.* 141 (1995), 443-551.

Index